Engineering Materials

This series provides topical information on innovative, structural and functional materials and composites with applications in optical, electrical, mechanical, civil, aeronautical, medical, bio- and nano-engineering. The individual volumes are complete, comprehensive monographs covering the structure, properties, manufacturing process and applications of these materials. This multidisciplinary series is devoted to professionals, students and all those interested in the latest developments in the Materials Science field, that look for a carefully selected collection of high quality review articles on their respective field of expertise.

Indexed at Compendex (2021) and Scopus (2022)

Mohd Azli Salim · Chonlatee Photong ·
Norhisham Ismail

Nanotechnology in Conductive Inks

Physical Performance Under Dynamic Mechanical Loading

Springer

Mohd Azli Salim
Fakulti Teknologi dan Kejuruteraan Mekanikal
Universiti Teknikal Malaysia Melaka
Durian Tunggal, Melaka, Malaysia

Norhisham Ismail
Fakulti Teknologi dan Kejuruteraan Mekanikal
Universiti Teknikal Malaysia Melaka
Durian Tunggal, Melaka, Malaysia

Chonlatee Photong
Faculty of Engineering
Mahasarakham University
Maha Sarakham, Thailand

ISSN 1612-1317 ISSN 1868-1212 (electronic)
Engineering Materials
ISBN 978-981-95-5707-3 ISBN 978-981-95-5708-0 (eBook)
https://doi.org/10.1007/978-981-95-5708-0

This Springer imprint is published by the registered company Springer Nature Singapore Pte Ltd.
The registered company address is: 152 Beach Road, #21-01/04 Gateway East, Singapore 189721, Singapore

Preface

Over the past two decades, the field of printed and flexible electronics has grown rapidly. Devices that were once rigid and bulky to fracture are increasingly being replaced by lightweight, stretchable, and wearable solutions that can conform to complex surfaces and withstand repeated mechanical deformation. This paradigm shift is driven by advances in conductive materials, fabrication processes, and a deeper understanding of the physical principles that govern their behaviour. However, as printed electronics are subjected to ever more demanding environments, from foldable displays to medical biosensors and wearable antennas, there remains a pressing need to thoroughly explore the complex interplay between materials, design geometry, and mechanical durability under long-term dynamic loads.

This book addresses this challenge by systematically investigating the electrical, mechanical, and morphological properties of graphene nanoplatelet (GNPs) and carbon black (CB) conductive inks across a broad spectrum of trace geometries and loading scenarios. In particular, it explores how different patterns such as straight-line, square, curved, and zigzag shapes, influence durability under cyclic torsion and tensile loads, and how these designs, coupled with careful formulation adjustments, can dramatically improve performance in flexible and stretchable electronic devices. Throughout this work, attention is given to understanding not only the empirical trends observed but also the underlying physical and engineering principles that explain these behaviours at the micro- and nanoscale.

From a materials science perspective, GNPs and CB offer distinct advantages as fillers. The GNPs, due to their high aspect ratio, excellent electrical conductivity, and mechanical strength, provide highly percolating networks at relatively low volume fractions. The CB, though spherical and often lower in intrinsic conductivity, forms dense conductive networks that can rearrange under stress, imparting resilience under repeated deformation. By fine-tuning the ink composition and applying advanced mixing techniques, this work has successfully achieved significant improvements in sheet resistance and mechanical durability, opening new avenues for environmentally sustainable and cost-effective conductive inks.

One of the central contributions of this work lies in the careful examination of the structure–property relationships governing the response of these inks to cyclic mechanical stresses. Scanning Electron Microscopy (SEM) and Energy-Dispersive X-ray (EDX) analyses have provided critical insights into microstructural evolution under torsional and tensile loading, revealing phenomena such as micro-crack formation, filler agglomeration, and binder debonding. Importantly, these findings underscore the role of pattern design in mitigating localised stress concentrations and highlight the superior performance of curved and zigzag traces, which distribute stresses more evenly and prolong the electrical and mechanical functionality of the ink.

More broadly, the work presented herein reinforces the importance of multiphysics considerations in engineering functional devices. Even seemingly minor design choices which a trace width of 1 versus 3 mm, or a sharp corner versus a smooth curve and it can significantly influence crack propagation, percolating network stability, and the onset of electrical failure. This understanding is not only valuable for printed electronics but also applicable to diverse engineering domains where cyclic mechanical loads and complex geometries must be accommodated, such as stretchable interconnects in soft robotics, bio-compatible electronics for healthcare monitoring, and lightweight structures for aerospace applications.

Finally, this book is intended to serve as a useful resource for researchers, practising engineers, and product developers alike. By translating fundamental materials science principles into practical design strategies, this exploration aims to facilitate more informed decisions in the engineering of durable, high-performance electronic devices. The results presented here highlight the value of an integrated, multiscale approach that couples experimental characterisation with rigorous analysis to produce meaningful, real-solutions.

As we look to the future, the ongoing miniaturisation of electronics, combined with society's growing demand for flexible, lightweight, and low-power devices, will require continuous innovation in materials, patterns, and manufacturing techniques. We are optimistic that the insights gained through this book will contribute to this progress and help pave the way for further breakthroughs in printed electronics. It is our sincere hope that this book will encourage new collaborations, foster deeper explorations into the science and engineering of functional inks, and inspire new generations of researchers and practitioners to tackle the exciting challenges that lie ahead.

Durian Tunggal, Melaka, Malaysia — Mohd Azli Salim
Maha Sarakham, Thailand — Chonlatee Photong
Durian Tunggal, Melaka, Malaysia — Norhisham Ismail

Acknowledgements Writing a book is never a solitary endeavour. It is a long and often difficult process that demands time, commitment, and the goodwill of many. The completion of this book would not have been possible without the invaluable support, encouragement, and contributions of numerous individuals and organisations who have guided, inspired, and stood by us throughout this journey.

First and foremost, we would like to extend our sincere gratitude to our academic mentors, colleagues, and friends who have shared their knowledge, expertise, and constructive criticism during the work, writing, and reviewing phases. Their kindness with their time and thoughtful feedback greatly enhanced the depth and severity of the final manuscript. We also appreciate the enlightening conversations and informal debates that challenged our thinking and sharpened the arguments presented in this book.

Special thanks are due to the various research and MTUN institutions and laboratories that facilitated access to key equipment and resources. These practical provisions enabled us to conduct the experimental work that forms the backbone of this book. We also acknowledge the technicians, administrative staff, and research students who offered their invaluable practical help and logistical support at every stage of our research.

On a personal note, we owe a great debt of gratitude to our families, who provided unwavering encouragement and understanding as we embarked on this book. Their patience during long hours spent at the computer, and their kind words during the more challenging phases of this process, sustained us more than they could know. Without their emotional support and sacrifices, this book would never have been completed.

Finally, we wish to acknowledge the readers which are students, academics, researchers, and practitioners for whom this book is intended. It is our sincere hope that this book will provide valuable knowledge, foster further research, and contribute meaningfully to ongoing developments in this field. Knowing that this book might help others explore new ideas and create new solutions is what motivated us throughout this journey.

To everyone who played a role, directly or indirectly, in bringing this book to fruition, thank you. We are humbled by the collaborative spirit that has shaped this book, and we trust that it will inspire continued dialogue and discovery for years to come.

Competing Interests The authors have no competing interests to declare that are relevant to the content of this manuscript.

About This Book

Over the past two decades, printed and flexible electronics have transformed from rigid, fragile devices into stretchable, wearable technologies capable of withstanding complex mechanical stresses. This book presents a comprehensive investigation into the mechanical, electrical, and morphological behaviour of graphene nanoplatelet (GNP) and carbon black (CB) conductive inks under cyclic loading. Through systematic analysis of various trace geometries including straight, curved, square, and zigzag patterns, and it reveals how design choices influence durability and performance. Incorporating microstructural insights using SEM and EDX, the investigation uncovers key structure–property relationships that govern long-term reliability. With its focus on pattern engineering, filler behaviour, and multi-scale analysis, this work offers practical strategies for advancing sustainable, high-performance electronic materials. Essential reading for researchers, engineers, and developers, this book bridges fundamental materials science with real-applications in wearable sensors, soft robotics, and beyond, by providing critical guidance for the future of flexible and printed electronics.

Contents

About the Authors

Mohd Azli Salim, Associate Professor Ir. Ts. Dr., received his Bachelor of Engineering (Honours) in Mechanical Engineering with a specialisation in Structures and Materials in 2006, followed by a Master of Mechanical Engineering in Control Engineering in 2009. He was awarded his Doctor of Philosophy in Mechanical Engineering, majoring in Vibration Analysis, in 2016. He currently holds TRIZ Level 3 Practitioner Certification and is a Certified Instructor for both TRIZ Level 1 and 2, as well as the TRIZ Master Class. In addition, he is a Professional Technologist registered with the Malaysia Board of Technologists in Manufacturing and Industrial Technology. He is also a Chartered Engineer with the Institution of Mechanical Engineers, UK, and is professionally registered with the Engineering Council, United Kingdom. In 2021, he was conferred the title of Professional Engineer in Mechanical Engineering by the Board of Engineers Malaysia and is recognised as an ASEAN Chartered Professional Engineer. Since January 2023, he has served as an Evaluator and Panel Member for the Engineering Accreditation Council (EAC) under the Board of Engineers Malaysia. Currently, he is Associate Professor of Flexible Nanoelectronics at the Fakulti Teknologi dan Kejuruteraan Mekanikal (FTKM), Universiti Teknikal Malaysia Melaka (UTeM). He serves as Head of the Advanced Academia-Industry Collaboration Laboratory (AiCL) and Principal Researcher at the Green and Efficient Energy Technology Research Group (GReeT). His research interests include finite element analysis, structural dynamics, control engineering, vibration and acoustic analysis, condition monitoring, advanced materials, nanotechnology, and nano-composites. He is among the key pioneers of stretchable conductive ink research at UTeM and the founder of electromagnetic energy harvesting (EEH) studies on composite materials. He was awarded "Most Cited Paper" by Praise Worthy Prize Publications indexed in Scopus. He currently supervises 18 Ph.D. (which are 4 of them graduated) candidates and 14 M.Sc. students, all working within the domains of flexible nanoelectronics, vibration, and structural dynamics. To date, he was published more than 13 research books since 2017. He has been appointed External Examiner for Higher Degree Programmes at Universiti Brunei Darussalam for the research title Thermo-reflective Polarising Roof Materials, and he is also an international editor in Thailand indexed by Scopus. Dr. Azli is also currently a Visiting Professor at Institut Teknologi Sepuluh Nopember (ITS), Indonesia, and is a registered mentor for Chartered Engineers in the Southeast Asia and Oceania region through the Institution of Mechanical Engineers. Most recently, he has been appointed as a Subject Matter Expert in Research and Publication by Politeknik Ungku Omar (PUO), Ipoh, Perak, for the term 2024–2026.

Chonlatee Photong, Associate Professor Dr. and had been a development engineer at Sony Device Technology (Thailand) Co., Ltd., Pathum Thani, for 3 years (2002–2004) and a maintenance engineer at Seagate Technology (Thailand), Nakorn Ratchasima, for 2 years (2005–2006). He received his master's degree of Power Electronics and Drives in 2007 and his doctoral degree in Electrical and Electronic Engineering in 2013 from the University of Nottingham, United Kingdom. After his doctoral graduation, he has started working as a lecturer in Electrical and Computer Engineering, Faculty of Engineering, Mahasarakham University, Thailand, and experienced working as an editor-in-chief of *Mahasarakham International Journal of Engineering Technology* and *Engineering Access* since 2014, a co-editor-in-chief of *Community University Engagement Journal* since 2022. He had been an assistant to the President for Public Relation and International Affairs at Mahasarakham University during

2015–2018. He is currently the Deputy Dean for Administration, Graduate School, Mahasarakham University, Thailand. He is currently the head of the Electric Automotive and Energy Technology Research Unit. His research expertise is power electronics and battery technologies for renewable energy and electric motor drives. He has more than 100 published research works, 15 research funds and several awards covering the innovation and research studies on a current source inverter for solar and renewable energy conversion and generation, as well as, motor control, drives and applications. He is a senior member of the IEEE PES and IES-Thailand Section.

Norhisham Ismail, hold a Doctor of Philosophy (Ph.D.) in Mechanical Engineering with a specialization in Nanomaterials from Universiti Teknikal Malaysia Melaka (UTeM), completed in 2023. His doctoral research focused on the formulation and characterization of electrically and mechanically conductive graphene nanoplatelet ink patterns on flexible substrates. This work contributed to improving material performance in flexible electronics and thermal management applications. Previously, his earned a Master's degree in Mechanical Engineering (Automotive) from UTeM and a Bachelor's degree in Engineering Technology (Mechanical—Automotive) from UniKL Malaysian Spanish Institute. His academic background provided me with strong foundations in material science, heat transfer, and mechanical systems, along with practical experience in simulation and design tools such as ANSYS, MATLAB, and CATIA. Currently, he worked as a Research and Development Engineer, where focused on developing and testing innovative nanomaterial formulations to contributing to engineering innovation through material development and technical education.

Chapter 1
Introduction to Nanocomposite Conductive Inks for Flexible Electronics

1.1 Introduction

Nanotechnology holds immense promise in addressing persistent challenges in building materials. Innovations such as nano-engineered concrete and sensor skins for robotic systems exemplify this potential [1, 2]. These materials support the development of structural health monitoring systems, wearable communication devices, and other applications requiring thin, lightweight, conformable, and durable architectures [3].

At the nanoscale, particles exhibit markedly different properties due to their enhanced surface area and surface energy. This size-dependent behaviour has propelled novel directions in electronics research, diverging from the traditional pursuit of faster and smaller planar which is two-dimensional devices. One significant evolution in this field is the emergence of printed electronics, which have found applications across various sectors including healthcare, aerospace, smart packaging, thermos-chromic displays, RFID tags, and batteries [4–6].

According to Huang and Zhu [7], printed electronics are crucial in advancing flexible and, more recently, stretchable electronic systems. Initially conceived as a cost-effective alternative to traditional silicon-based electronics [8], printed electronics have matured into a robust technology, enabling integration into standard manufacturing processes due to the thinness and flexibility of the printed layers. The main goals of printed electronics are twofold, which are to offer economically viable alternatives by substituting conventional electronic components and layers, and to develop multifunctional materials with enhanced capabilities. One strategy to realise these objectives involves using a two-component formulation to control ink viscosity, facilitating self-assembly based on the affinity of the ink's constituents [9].

Stretchable electronics, as noted by Suikkola et al. [10], represent a critical technology for minimising user disruption in wearable electronic applications. A central challenge in this field is engineering conductive and printable inks or electrodes that retain functionality under mechanical deformation [11]. The dynamic nature of

M. A. Salim et al., *Nanotechnology in Conductive Inks*, Engineering Materials,
https://doi.org/10.1007/978-981-95-5708-0_1

stretchable platforms, which are regularly subjected to cyclic motion, necessitates the design of materials capable of pervasive and self-effacing sensing and display functionalities [12]. The pervasive sensing refers to the capability of acquiring real-time data from a wide array of sources, while unobtrusive sensing pertains to non-intrusive monitoring of user behaviour or environmental conditions. Achieving this necessitates materials that can endure compression, torsion, and conformation to complex geometries, which can attribute that are crucial for healthcare, entertainment, and energy-related applications [13].

The mechanical and electrical behaviours of these systems are typically evaluated through the characterisation of conductive inks. Both metallic and non-metallic nanoparticle-based inks are assessed in terms of resistivity and conductivity. The percolation threshold theory is frequently employed to explain the conduction mechanisms. This theory presumes that a critical loading of conductive filler is required to establish conductive pathways throughout the matrix [14]. In response to the limitations of conventional tensile testing for thin films and nanostructures, many works have increasingly adopted depth-sensing nanoindentation techniques. These enable precise measurements of mechanical properties at the micro- and nanoscale using controlled contact geometries [15]. Furthermore, optimising performance under mechanical loading can be explored by examining fatigue behaviour across diverse printed geometries. Many investigations have analysed how patterns such as straight lines, sine waves, semi-circles, serpentines, zigzags, and horseshoes influence stress distribution and strain dissipation [16]. Finite element analysis (FEA), including metrics such as Von Mises stress and principal elastic strain, has been utilised to predict potential failure regions and evaluate stretchability [17].

Electronic printing encompasses several methods including screen printing, inkjet printing, and flexographic printing [18]. Conductive ink serves as the cornerstone of these technologies, offering the ability to print conductive paths onto flexible substrates. Typically, the inks are deposited, followed by a mild thermal process to evaporate solvents and promote particle sintering [19]. These inks can supplant traditional printed wiring in compact, low-cost circuit designs. Stretchable conductive inks must remain functional under deformation, but several practical limitations persist, including durability under harsh environments and compatibility with specific substrates.

Materials utilised in conductive inks range from metal-based (for example silver, copper), to carbon-based [for instance carbon black (CB), graphene, carbon nanotubes (CNTs)], and nanoparticle-based composites [20]. While silver offers excellent conductivity, its cost is a barrier. Cheaper alternatives such as copper, aluminium, and nickel are susceptible to oxidation, forming insulating layers. Carbon-based materials, particularly graphene, offer stability under extreme conditions and possess tuneable electrical characteristics. Organic compounds generally lack stretchability which a major constraint for applications like prosthetics, AI systems, wearable sensors, and soft robotics.

Graphene-based conductive inks, once thermally reduced and printed onto polymer substrates, have demonstrated consistent electrical conductivity across repeated bending cycles. The thinness of such circuits is essential, as the bending strain is inversely related to film thickness. Rigid microelectronic devices mounted on flexible substrates can endure deformation thanks to engineered interconnectors, often using stretchable metals in innovative structural designs. The interplay between film thickness and ductility directly impacts the mechanical resilience of these systems. This work will specifically investigate graphene and CB as primary fillers in conductive inks. These fillers serve both as electrical conductors and structural reinforcements, helping to minimise polymerisation shrinkage. Graphene, in particular, is selected for its exceptional electrical and physical properties which a high surface area (2630 m^2/g), a high Young's modulus (1 TPa), excellent thermal conductivity (5000 W/m K), high optical transparency (97.7%), high electron mobility (2.5×10^5 cm^2/V s), electromagnetic interference (EMI) shielding capability, and overall flexibility and transparency [21].

The fabrication of high-performance conductive inks involves optimising ink formulation, printing processes, substrate interactions, and curing parameters. The relationship between the filler, binder, and hardener is crucial, as this determines ink stability and conductivity. Following the formulation stage, the ink is deposited onto a glass substrate and thermally cured under controlled conditions. A comprehensive characterisation of the printed films by using techniques such as Four-Point probe measurements, nanoindentation, and image-based analysis, can provides critical insights into their electrical and mechanical behaviour. Building upon these characterisation results, further explorations will evaluate different printed trace geometries on stretchable substrates. These patterned films will undergo various testing to assess their electrical performance, mechanical integrity, and morphological features under deformation. The outcomes of this investigation aim to inform the design and optimisation of next-generation stretchable electronic systems suitable for a broad range of real-applications.

1.2 Advancing the Future of Conformable Electronics

Stretchable conductive ink (SCI) represents a transformative advancement in the field of printed electronics. While still a relatively novel concept, its applications are both current and far-reaching, addressing the demand for highly deformable yet electrically stable materials in a wide range of next-generation technologies. Materials capable of maintaining high conductivity under tensile strains exceeding 100% are essential for emerging systems such as flexible displays, neuro-prosthetics, cardiac stimulation implants, soft robotics, and other curvilinear electronic configurations [3, 22].

Stretchable electronics (SE) that can endure bending, twisting, and stretching without compromising device integrity are especially valuable in biomedical domains. These include electroencephalogram (EEG) and electrocardiogram (ECG) electrodes, photodetectors, wearable therapeutic patches, and implantable diagnostic or therapeutic devices. In addition to medical use, SEs have demonstrated potential in highly deformable photovoltaic cells, artificial skins equipped with sensory arrays, and actuators for artificial muscles, components essential for humanoid robotics and responsive interfaces. The field of SE is also closely aligned with the evolution of the Internet of Things (IoT), wearable electronics (WE), and flexible displays (FD). These domains demand electronics that are lightweight, unobtrusive, and mechanically compliant. A SCI technologies enable these requirements through their ability to be seamlessly integrated into wearable systems, smart textiles, and biocompatible patches. A compelling example is the potential application of stretchable electronics for quick monitoring which are flexible, skin-conformable patches worn by expectant mothers could enable non-invasive, real-time monitoring of high-risk pregnancies. Likewise, performance monitoring of elite athletes could be significantly enhanced using Lycra-based garments embedded with stretchable sensors that map physiological data during training or competition.

Conductive inks can be printed onto a broad range of lightweight and flexible substrates, including polyester, polycarbonate, polyurethane, and even paper. The use of such substrates enhances creative freedom in device architecture and simplifies the production process. These flexible, low-cost platforms also encourage sustainability through easier recyclability and material efficiency. Despite their universality, most consumers remain unaware of the advanced technologies that underpin modern printed electronics. For instance, soft sensors integrated into textiles or consumer goods often consist of deformable conductive materials embedded within non-stretchable matrices. With SCI, these sensors can now be made stretchable and adaptable, capable of being laminated directly onto biological or mechanical surfaces.

One promising area of application is in energy scavenging. The SCI-printed flexible thermoelectric sensors could be wrapped around engine components to harvest waste heat. Similarly, stretchable photovoltaic modules could be deployed on curved or moving surfaces, unlocking applications in aerospace and automotive systems where rigidity has traditionally limited energy integration.

The SEs have also been adopted in optoelectronic components such as light-emitting diodes (LEDs) and biologically inspired cameras, as well as in soft, multifunctional skins for robotics. These robotic skins, which mimic the sensory functions of human skin, are capable of pressure, temperature, and deformation sensing, powered by underlying SCI-printed networks. The use of conductive ink has further revolutionised the development of flexible electrodes and sensor pads. Traditional rigid electrodes are being rapidly replaced by flexible, disposable counterparts that are more comfortable, lower in cost, and better suited for integration into everyday garments and devices. For example, Volf et al. [23] demonstrated how conductive inks can serve as transducers by converting mechanical pressure into electrical signals with high sensitivity and repeatability. These transducer capabilities make conductive inks ideal for touch-sensitive surfaces, pressure mapping arrays, and intelligent control interfaces.

Another outstanding use case is in smart packaging. The integration of polymer thick film (PTF) circuitry into product packaging enhances supply chain traceability and product lifecycle management. These PTF-based circuits can include embedded sensors, QR codes, and near-field communication (NFC) features. Stretchable labels, increasingly popular in consumer marketing, are not only aesthetic but also functional, offering interactive features or real-time condition monitoring of the product within. Therefore, SCI technologies represent a convergence of material science, electronics, and mechanical design. They have unlocked new paradigms in medical diagnostics, energy systems, wearable technology, robotics, and consumer goods. Their ability to function reliably under dynamic deformation makes them indispensable in realising the next generation of smart, adaptive systems.

1.3 Formulation and Durability Considerations in Conductive Inks

The development of conductive inks with high electrical conductivity necessitates thoroughly optimisation of both material composition and printing architecture. Despite the technological promise of conductive inks, a significant drawback lies in the lack of standardised formulation procedures. Presently, ink development is predominantly driven by empirical methods which an approach heavily reliant on trial and error. This methodology has led to considerable variability in the electrical and mechanical performance of printed inks, impeding reproducibility and scalability. By overcoming these challenges, systematic enhancement in ink formulation protocols and printing techniques is essential. Factors such as ink rheology, filler dispersion, substrate compatibility, and curing mechanisms must be finely tuned to ensure consistent and reliable properties. Particular attention must also be paid to ink-substrate interfacial behaviour, as adhesion quality significantly impacts performance under strain.

Graphene-based conductive inks represent a novel class of materials that, despite their extraordinary electrical and mechanical properties, are still underexplored in commercial ink formulation. Graphene's high carrier mobility, mechanical flexibility, and chemical stability render it an attractive candidate for next-generation flexible electronics. However, comprehensive explorations that validate its performance, particularly in comparison to more established conductive fillers such as carbon black where it is remained limited. Therefore, it is scientifically applicable to conduct a comparative analysis between graphene and CB-based inks, evaluating parameters such as conductivity, stretchability, fatigue life, and interfacial adhesion.

The foundational objective of conductive ink development is to offer a cost-effective and mechanically robust alternative to traditional printed circuit boards (PCBs). Conventional PCBs are often fabricated using multistep photolithographic processes that are labour-intensive, expensive, and yield boards that are brittle and susceptible to mechanical failure under stress. In contrast, printed electronics, when developed using optimised conductive inks, present a lightweight, flexible, and lower-cost solution that can be rapidly prototyped and integrated into curved or dynamic surfaces. However, validating the electrical and mechanical durability of conductive inks under static and dynamic loading conditions is imperative. This validation should encompass a variety of printed patterns such as straight lines, sine waves, serpentine, and zigzag configurations, and span a range of trace widths and thicknesses. Each pattern introduces distinct stress concentrations and deformation profiles, which affect crack propagation, delamination, and resistivity under load. The mechanical behaviour of these patterns should be thoroughly analysed using methods such as finite element analysis (FEA), strain mapping, and in situ mechanical–electrical testing.

The issue of cyclic loading fatigue is particularly relevant. As SEs are inherently designed for dynamic applications such as wearable systems, soft robotics, and bio-integrated devices, they are frequently subjected to repetitive mechanical strain. Conductive materials must maintain functional integrity under thousands of cycles of stretching, bending, and torsion. Unfortunately, many formulations suffer from micro-crack formation, delamination, or filler disconnection under prolonged cyclic loading, ultimately leading to an increase in resistivity or complete failure.

Moreover, performance degradation can also arise due to plastic deformation and loss of toughness. Prolonged exposure to mechanical fatigue may compromise the filler-binder matrix, leading to mechanical instability and reduced electrical performance. For emerging inks, especially those incorporating high aspect-ratio fillers like graphene or CNTs, understanding how the filler network responds to repeated deformation is essential. The network exhibits one of two responses which are it either undergoes self-healing or reorientation, or it becomes fragmented and suffers irreversible compromise.

By addressing these concerns, a comprehensive mechanical–electrical reliability work is proposed. This should include: (i) static load testing for resistance stability under various strain levels, (ii) cyclic fatigue testing over extended durations, (iii) microscopic imaging [for example Scanning Electron Microscopy (SEM)] to visualise crack evolution and filler dispersion, (iv) comparative assessment with standard materials (for example CB), and (v) characterisation of changes in electrical performance using Four-Point probe methods and impedance spectroscopy. By generating a robust dataset across different ink compositions and printed geometries, researchers and engineers will be better positioned to rationally design conductive inks tailored for specific applications, ensuring both performance and longevity. Ultimately, the commercial and technological viability of printed SEs hinges on the repeatable and predictable performance of the conductive inks that underpin them.

1.4 Rationale and Intended Impact

The primary aim of this book are fundamental, as they serve as a framework for guiding the experimental approach and analytical processes. This book specifically focuses on the development of graphene nanoplatelet (GNP) conductive traces, building upon baseline formulations of conductive ink. The overall plan is to enhance the electrical and mechanical performance of printed conductive patterns using GNP and CB composites.

By achieving this, the following specific targets have been established: (i) to formulate a new composite ink blend incorporating GNP and CB, and to characterise its electrical, mechanical, and morphological properties through standard techniques, (ii) to investigate and understand the performance of various printed pattern geometries and widths composed of GNP and CB-based conductive ink under static loading conditions, (iii) to apply cyclic mechanical testing in order to assess the durability, repeatability, and performance of different patterns and widths of GNP and CB conductive inks under repetitive loading scenarios and (iv) to evaluate the overall effectiveness of different conductive ink configurations by analysing changes in electrical, mechanical, and morphological properties before and after testing.

This book operates within defined experimental parameters and technical constraints. The scope is focused on developing and evaluating GNP and CB-based conductive inks for application in flexible and SE, which are: (i) the initial phase involved the formulation of a conductive ink composite by blending both fillers with a bisphenol-A epoxy resin (DGEBA) and a polyetheramine-based curing agent (Huntsman D230), with careful attention paid to dispersion and ink homogeneity, (ii) a comparative analysis was conducted to evaluate the electrical and mechanical performance of both conductive inks against established baseline formulations. Key metrics included conductivity, flexibility, and structural stability, (iii) the conductive ink patterns for both static and cyclic tests were printed using a stencil with a uniform thickness of 1 mm, ensuring consistency in trace deposition across all samples, (iv) the SEM was used to assess both surface and cross-sectional morphology of the printed ink layers. Then, the Energy Dispersive X-ray (EDX) analysis was employed to confirm the presence and distribution of carbon-based components within the cured matrix, (v) cyclic loading tests were carried out using a custom-designed test rig. While suitable for the intended experiments, this apparatus introduced limitations, such as fixed deformation speeds and constraints on sample dimensions and (vi) functional validation of the conductive ink traces was conducted via in-situ resistance monitoring during both static and cyclic tests. This was complemented by measurements of bulk resistance before and after loading cycles to assess performance degradation.

1.5 Potential Benefits for Sustainable Development Goals

Following the successful completion of this book, a novel formulation of conductive ink based on carbonaceous nanomaterials, specifically GNP and CB was developed, characterised, and evaluated in terms of its electrical, mechanical, and morphological properties. The formulation exhibits considerable potential for real-implementation due to its low resistivity, excellent thermal conductivity, and high tolerance to mechanical stress. These attributes make it a promising alternative to traditional metal-based inks, which are often expensive, rigid, and less sustainable. This innovation addresses the increasing demand for flexible, low-cost, and environmentally sustainable electronics that can be integrated into next-generation technologies. The ability to print circuits directly onto flexible substrates (for textiles and paper) opens up opportunities for smart packaging, wearable medical sensors, and lightweight electronics, especially in resource-constrained environments.

Moreover, the work provides critical insights into how different printed trace geometries and widths influence the functional reliability of conductive inks under mechanical loading. This includes cyclic bending, twisting, and tensile stress conditions often encountered in stretchable and wearable devices. These insights are particularly relevant for the development of human-interactive electronics, implantable biomedical devices, and conformable sensor arrays. The book also demonstrates the feasibility of using carbon-based materials as sustainable alternatives to metal-based fillers, offering cost-effectiveness, material abundance, and recyclability, thus contributing directly to the global effort to create greener and more sustainable technologies. Several original and impactful contributions to scientific and technological knowledge in the field of printable and stretchable electronics and been discovered in this book.

The work introduces a novel dual-method approach which are mechanical mixing and ultrasonication for optimising the dispersion of GNP and CB nanomaterials within an epoxy matrix. This method ensures homogeneity, printability, and improved curing dynamics. It addresses the lack of standardisation in formulation processes by providing a systematic understanding of how material ratios and processing conditions affect ink performance. This innovation promotes resource-efficient manufacturing and contributes to SDG 9: Industry, Innovation and Infrastructure, by supporting the development of sustainable industrial processes and resilient infrastructure through advanced materials engineering. Besides, this work also validates the use of GNP and carbon black as sustainable, high-performance alternatives to silver and copper fillers. These carbon-based materials offer comparable electrical properties while being non-toxic, corrosion-resistant, and abundantly available and making them more suitable for mass production and environmentally conscious design. This directly aligns with SDG 12: Responsible Consumption and Production, as it encourages the use of materials that reduce environmental impact while maintaining high functionality.

Moreover, it provides a deep analysis of how different trace geometries affect signal propagation, impedance matching, heat dissipation, and resistance to mechanical fatigue. These findings serve as practical design guidelines for engineers creating flexible and durable circuit layouts, particularly for wearable and implantable electronics. This supports SDG 3: Good Health and Well-Being, by enabling the development of more reliable biomedical monitoring devices, and SDG 11: Sustainable Cities and Communities, through the integration of flexible electronics into smart infrastructure. The book also explores the long-term stability and degradation mechanisms of conductive inks under repetitive mechanical deformation. By simulating real-loading conditions, it identifies the thresholds beyond which performance degradation occurs, and it establishes parameters for designing robust, fatigue-resistant inks. This work thus lays the foundation for life-cycle engineering of conductive materials, contributing to the goals of SDG 13: Climate Action, through durable, low-energy, and long-life electronics that reduce waste and minimise energy usage over time.

Besides, the outcomes of this book are closely aligned with several United Nations Sustainable Development Goals (SDGs), demonstrating the broader global relevance and impact of the work: (i) SDG 3—Good Health and Well-Being, development of robust, flexible, and biocompatible conductive inks can enhance the performance of wearable biosensors, EEG/ECG monitors, and implantable medical devices, thereby supporting improved health monitoring and diagnostics, (ii) SDG 9—Industry, Innovation and Infrastructure. The work promotes innovation in materials science and printable electronics, helping to build sustainable industrial capabilities and resilient electronic systems, (iii) SDG 11—Sustainable Cities and Communities. Integration of flexible electronics in smart cities for energy harvesting, pollution monitoring, and infrastructure tracking is enabled by sustainable ink formulations, (iv) SDG 12—Responsible Consumption and Production. The replacement of rare, expensive, and non-recyclable metal inks with abundant and environmentally friendly carbon materials supports sustainable production practices, and (v) SDG 13—Climate Action by developing long-lasting, energy-efficient, and recyclable electronic materials, this book supports mitigation strategies that reduce carbon emissions associated with electronic waste and manufacturing.

In summary, this book contributes both theoretically and practically to the evolving field of SEs and printed electronics. It not only addresses critical gaps in material formulation and mechanical reliability but also aligns with the global imperative to create sustainable, inclusive, and innovative technologies. The findings offer significant value to academia, industry, and policymakers alike, and serve as a robust foundation for future advancements in flexible, scalable, and sustainable electronic systems.

References

1. Oke, A., Aigbavboa, C., Semenya, K.: Review of the application of nanotechnology for sustainable construction materials. In: Advances in Engineering Research (AER), Volume 102. Conference: Second International Conference on Mechanics, Materials and Structural Engineering (ICMMSE) (2017)
2. Majidi, C.: Soft-matter engineering for soft robotics. Adv. Mater. Technol. **2018**, 1800477 (2018)
3. Rogers, J.A., Someya, T., Huang, Y.: Materials and mechanics for stretchable electronics. Science **327**, 1603–1607 (2010)
4. Russo, S., Dosio, A., Graversen, R.G., Sillmann, J., Carrao, H., Dunbar, M.B., ... Vogt, J.V. (2014). Magnitude of extreme heat waves in present climate and their projection in a warming world. J. Geophys. Res. Atmospheres **119**(22), 12–500.
5. Cano-Raya, C., Denchev, Z.Z., Cruz, S.F., Viana, J.C.: Chemistry of solid metal-based inks and pastes for printed electronics–A review. Appl. Mater. Today **15**, 416–430 (2019)
6. Mason, J., Carreira, L., Collier, R., Felice, C., Isganitis, L., Lauro, S., Mason, P., Shellhammer, L.: Highly Conductive Custom Inks for Aerospace Applications (2020)
7. Huang, Q., Zhu, Y.: Printing conductive nanomaterials for flexible and stretchable electronics: a review of materials, processes, and applications. Adv. Mater. Technol. **4**(5), 1–41 (2019)

8. Dahiya, S., Katakojwala, R., Ramakrishna, S., Mohan, S.V.: Biobased products and life cycle assessment in the context of circular economy and sustainability. Mater. Circ. Econ. **2**(1), 7 (2020)
9. Matsuhisa, N., Kaltenbrunner, M., Yokota, T., Jinno, H., Kuribara, K., Sekitani, T., Someya, T.: Printable elastic conductors with a high conductivity for electronic textile applications. Nature communications **6**(1), 7461 (2015)
10. Suikkola, J., Björninen, T., Mosallaei, M., Kankkunen, T., Iso-Ketola, P., Ukkonen, L., Vanhala, J., Mäntysalo, M.: Screen-printing fabrication and characterization of stretchable electronics. Sci. Rep. **6**, 25784 (2016)
11. Liang, J., Wang, C., Wang, Y., Xu, Z., Lu, Z., Ma, Y., ... Liu, J.: All-inorganic perovskite solar cells. J. Am. Chem. Soc. **138**(49), 15829–15832 (2016)
12. Sekitani, T., Nakajima, H., Maeda, H., Fukushima, T., Aida, T., Hata, K., Someya, T.: Stretchable active-matrix organic light-emitting diode display using printable elastic conductors. Nat. Mater. **8**, 494–499 (2009)
13. Tang, Y., Zhou, K., Palacios, T.: Mechanical and electrical properties of graphene-based conductive ink for inkjet printing flexible electronics. J. Mater. Sci. **52**(6), 3261–3272 (2017)
14. Kumar, V., Sharma, A., Kaur, P., Sidhu, G.P.S., Bali, A.S., Bhardwaj, R., Thukral, A.K., Cerda, A.: Pollution assessment of heavy metals in soils of India and ecological risk assessment: A state-of-the-art. Chemosphere **216**, 449–462 (2019)
15. Hu, Z.: Characterization of materials, nanomaterials, and thin films by nanoindentation. In: Microscopy Methods in Nanomaterials Characterization, pp. 165–239 (2017)
16. Ab Wahid, A.K., Salim, M.A., Ali, M., Masripan, N.A., Dai, F., Saad, A.M.: Measurement of optimal stretchability graphene conductive ink pattern by numerical analysis. Def. S&T Techn. Bull. **14**(1), 43–54 (2021)
17. Spazzin, A.O., Costa, A.R., Correr, A.B., Consani, R.L.X., Correr-Sobrinho, L., dos Santos, M.B.F.: Effect of bar cross-section geometry on stress distribution in overdenture-retaining system simulating horizontal misfit and bone loss. J. Biomech. **46**(12), 2039–2044 (2013)
18. Khan, K., Khan, H., Lu, Y., Ihsanullah, I., Nawab, J., Khan, S., Shah, N.S., Maryam, A.: Evaluation of toxicological risk of foodstuffs contaminated with heavy metals in Swat, Pakistan. Ecotoxicol. Environ. Safety **108**, 224–232 (2014)
19. Yang, S., Wang, Y., Liu, P., Cheng, Y. B., Zhao, H. J., Yang, H.G.: Functionalization of perovskite thin films with moisture-tolerant molecules. Nat. Ener. **1**(2), 1–7 (2016)
20. Mokhlis, M., Salim, M.A., Masripan, N.A., Saad, A.M., Omar, G.: Electrical performances of graphene materials with different filler loading for future super conductor. Def. S&T Tech. Bull. **12**(2) (2019)
21. Wei, J., Vo, T., Inam, F.: Epoxy/graphene nanocomposites: processing and properties: a review. Mater. Sci. Technol. **31**(9), 1059–1071 (2015)
22. Fan, Z., et al.: Toward the development of printable nanowire electronics and sensors. Adv. Mater. **21**, 3730–3743 (2009)
23. Volf, M., Diviš, J., Havlík, F.: Thermal, moisture and biological behaviour of natural insulating materials. Ener. Proc. **78**, 1599–1604 (2015)

Chapter 2
Materials, Printing Techniques and Testing of Conductive Inks

2.1 Theoretical Basis of Flexible and Stretchable Electronics

Thin films and conductive inks are becoming increasingly prominent in the field of modern electronics, particularly in applications where mechanical flexibility, weight reduction, and manufacturability are critical. Their adoption marks a significant paradigm shift from traditional rigid and bulky electronics toward soft, flexible, and even stretchable systems. This transformation is not merely aesthetic or ergonomic and it reflects a deeper evolution in materials science, device engineering, and user-centric design. According to Aggarwal et al. (2017), Roy Bjorlin, Director of Commercial and Strategic Strategies at Sun Chemical, observed a surging interest in integrating electronics into everyday items. These include wearables, smart packaging, interactive labels, and other Internet of Things (IoT) devices. This growing trend illustrates the convergence of consumer electronics with ubiquitous computing, where sensors and circuits must conform to the physical constraints of the objects they inhabit be it textiles, plastic films, or organic surfaces.

Traditional electronic systems, predominantly silicon-based, are fabricated using subtractive manufacturing techniques such as photolithography and etching. While these methods produce high-performance microelectronics, they are inherently limited by the rigidity of silicon and the fragility of interconnects on brittle substrates like glass or boards. Consequently, these systems are poorly suited for applications requiring bending, stretching, or integration into curved surfaces.

In contrast, thin films defined as materials with thicknesses in the nanometre to micrometre range, enable the deposition of active and passive electronic components on lightweight, flexible substrates. These films can be produced using various low-temperature techniques such as vacuum deposition, sputtering, spin coating, or solution processing, making them compatible with polymeric substrates such as polyethylene terephthalate (PET), polyimide (PI), or thermoplastic polyurethane (TPU). Conductive inks play a central role in enabling printed electronics, where circuitry

M. A. Salim et al., *Nanotechnology in Conductive Inks*, Engineering Materials,
https://doi.org/10.1007/978-981-95-5708-0_2

is formed by printing rather than etching. These inks are typically composed of conductive fillers such as silver nanoparticles, carbon nanotubes (CNTs), graphene, or copper flakes, dispersed within a carrier medium. When printed onto a substrate, they form conductive pathways after curing, allowing for the fabrication of resistors, capacitors, antennas, and interconnects in a cost-effective and scalable manner.

2.2 Historical Development of Thin Film Electronics

The origins of thin-film silicon electronics trace back to the early 1990s, where they were initially implemented on glass substrates as active switching matrices for liquid crystal display (LCD) pixels. These thin-film transistors (TFTs) rely on field-effect transistor (FET) theory, where electrical conduction in a semiconductor channel is modulated by a gate voltage applied through a dielectric. Their evolution into flexible forms involves depositing similar materials, for example amorphous Si, IGZO, onto flexible polymeric substrates instead of glass, utilising techniques such as low-temperature chemical vapour deposition (CVD) or sputtering.

The introduction of printed electronics emerged soon after as a promising low-cost and additive-manufacturing alternative to conventional lithography-based silicon electronics. The foundational principle here is that functional inks, containing active materials such as semiconductors, conductors, or insulators, can be deposited using printing methods which are inkjet, screen, or gravure printing onto flexible substrates, leading to electronics that are lightweight, rollable, and conformable.

As substrates become thinner and more compliant, printed electronics enable seamless integration into existing packaging production lines. Consumer Packaging Groups (CPGs) have begun to adopt these innovations to enhance product interactivity and supply chain intelligence. Smart labels embedded with RFID or NFC circuits, printed temperature sensors, and freshness indicators enable data capture and enhance the consumer experience.

This transformation is underpinned by Moore's Law and More-than-Moore paradigms. While traditional microelectronics follow Moore's Law, scaling down transistors for better performance printed and stretchable electronics (SE) expand functionality by integrating sensors, displays, and communication interfaces into unconventional form factors, not necessarily focusing on higher computation, but on distributed intelligence and adaptability. The potential applications for flexible and SEs span various sectors, including industrial automation, automotive interfaces, biomedical diagnostics, aerospace systems, thermal management, and smart textiles. Each of these sectors presents multi-billion-dollar market opportunities. For instance, the global lighting market was estimated to reach USD 150 billion by 2020 equivalent to the annual revenues of the television market [1]. Saleh et al. [2] projected the global market for flexible electronics to hit USD 87.21 billion by 2024, with the Asia-Pacific region alone expected to grow from USD 5 billion in 2013 to USD 30 billion by 2024. Additionally, IDTechEx forecasted that flexible hybrid electronics

(FHE) which integrate printed components with silicon ICs, would surpass USD 3 billion by 2030.

A significant challenge to widespread adoption is the cost of flexible electronics, especially as they continue to be evaluated against cost performance benchmarks set by traditional silicon manufacturing. Geographical location plays a major role, with Asia becoming a global hub for electronics manufacturing due to economies of scale and infrastructure. Malaysia, in particular, hosts several major electronics companies such as Intel, Infineon, AMD, Seagate, Osram, Motorola, Honeywell, BBraun, and Agilent.

Recognising the potential of nanotechnology in this sector, Malaysia established NanoMalaysia Berhad under the Ministry of Science, Technology and Innovation (MOSTI) in 2011 to spearhead the commercialisation of nanotech innovations, following its classification as a strategic economic pillar during the National Innovation Council meeting in 2009 [3]. Flexible circuits refer to thin, lightweight interconnect systems capable of bending and conforming to irregular surfaces. These are typically fabricated using PI or PET substrates via roll-to-roll or screen printing methods. The theory of mechanical bending suggests that the bending strain experienced by a film of thickness on a substrate of radius. Lower values of thickness and higher radius enable greater flexibility without exceeding the fracture strain of the material.

The SEs are more challenging, as they must accommodate tensile strain while maintaining electrical continuity. This is achieved through either: (i) structural engineering of conventional materials, for example serpentine wiring layouts or kirigami-inspired cuts [4], or (ii) new materials with intrinsic stretchability, such as elastomeric conductors [5]. In this context, stretchable conductive inks (SCI) are crucial. These inks must combine high conductivity with mechanical compliance, allowing expansion and contraction with minimal resistance change. According to Zhang et al. (2018), these inks demonstrate minimal piezo resistive effect by maintaining stable performance even under strain. The conductivity of SCIs is determined by the percolation theory, where a continuous network of conductive fillers must form to allow charge transport. Conductive nanofillers such as carbon nanotubes (CNTs), silver nanowires (AgNWs), and graphene are chosen for their high aspect ratio and surface area, which reduce the percolation threshold. For example, in a polymer matrix, the critical volume fraction for percolation can be reduced by aligning or dispersing high-aspect-ratio fillers effectively.

Mechanical stability is critical in wearable and biomedical applications. Embedding conductive fillers in elastomers such as PDMS or Ecoflex produces composites capable of withstanding large deformations. The Young's modulus and Poisson's ratio of the matrix, as well as the filler–matrix interaction, influence the stretchability and durability of the final ink. For instance, AgNW–PDMS composites leverage both filler conductivity and matrix elasticity for use in soft sensors and implantable devices. Fatigue performance under repeated bending or strain cycles is assessed using the gauge factor (GF). A low GF under repeated strain indicates robust electrical stability, a key parameter for long-term use in flexible displays and physiological monitors.

2.3 Properties and Functional Applications

Conductive ink is an essential material in the advancement of flexible and SE technologies. It is formulated to produce patterned traces that are capable of conducting electricity, thus enabling the development of electronic devices that are both lightweight and mechanically adaptable. These inks are composed of conductive fillers such as metallic nanoparticles, conductive polymers, metal complexes, or carbon-based substances, dispersed within a polymer matrix. Together, these constituents determine the ink's electrical performance, mechanical resilience, thermal stability, and resistance to environmental degradation. Since the early 2000s, significant research efforts have been directed towards enhancing conductive ink formulations to meet demanding application requirements. The key performance parameters include high electrical conductivity, durability under mechanical stress, resistance to oxidation and corrosion, thermal stability, and long-term environmental reliability [6]. The versatility of conductive inks makes them ideal for a wide array of uses, including printed circuit interconnects, battery electrodes, electromagnetic shielding, antenna fabrication, electrostatic discharge mitigation, and advanced smart packaging [7, 8]. Figure 2.1 shows the application and conductive ink market in various fields.

Conductive inks can be broadly classified into three categories based on their primary filler materials which are metal-based, carbon-based, and polymer-based inks. Metal-based inks, particularly those utilising silver, copper, or gold nanoparticles, are highly valued for their exceptional electrical performance. However, their

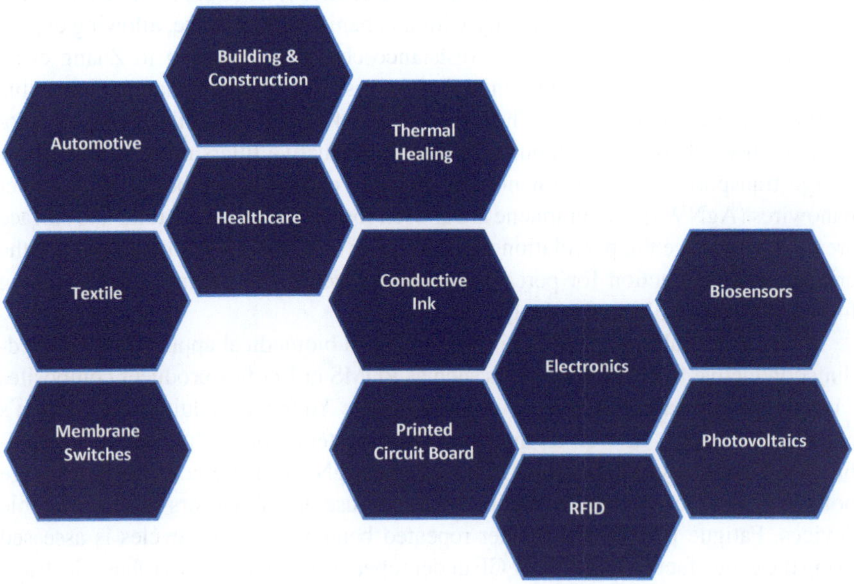

Fig. 2.1 The application and conductive ink market in various fields

widespread adoption is sometimes limited due to high material costs, susceptibility to oxidation, and stability issues. In response, investigators have developed hybrid materials combining different types of fillers, such as silver nanowires with nanoparticles in an elastomer matrix, which show great promise for use in flexible and stretchable electronics [9].

An essential concept in understanding how conductive inks function is the idea of the percolation threshold. This concept describes the critical point at which the concentration of conductive filler particles becomes sufficient to form a continuous conductive path through the otherwise insulating polymer matrix. Below this threshold, the conductive particles remain isolated, and the ink remains non-conductive or only weakly conductive. Once the threshold is surpassed, a network forms that enables the movement of electrons across the material, significantly increasing its conductivity [10]. This transition is a fundamental aspect of the design and optimisation of conductive inks, as it dictates how much filler is required to achieve functional conductivity.

This behaviour is visually represented in diagrams such as Fig. 2.2, which illustrates how filler distribution affects electrical performance. In a poorly distributed system, the lack of connectivity results in weak or negligible conductivity. However, with appropriate dispersion and concentration, the filler forms a cohesive network that allows for efficient electron flow. The careful tuning of filler content not only improves conductivity but also influences printability, viscosity, and adhesion characteristics factors that are essential for practical manufacturing processes.

The role of the polymer binder is equally significant. It provides mechanical integrity, ensures homogeneous dispersion of conductive particles, and contributes to the flexibility and thermal resistance of the ink. Thermoplastic and thermosetting polymers are commonly used, with thermoplastics offering ease of reprocessing and flexibility, while thermosets provide high thermal and chemical resistance. Examples of common binder materials include polyimide, polystyrene, polyurethane, and silicone-based resins [11]. These polymers not only influence the mechanical properties of the final printed trace but also determine how well the ink adheres to different substrates. Figure 2.3 shows the conductive fillers reinforced polymer composites percolation zone.

Composite conductive materials those composed of a polymer matrix reinforced with conductive fillers are particularly valued for their unique combination of properties. These include high tensile strength, stiffness, durability, and resistance to cracking, all while maintaining a low weight. The superior performance of composites arises from effective stress transfer between the matrix and filler materials. This relationship is reinforced by the orientation and interaction of particles at the microscopic scale, enabling the final product to maintain both mechanical and electrical functionality under strain [12–14].

One of the major advantages of using conductive inks in place of traditional bulk metal conductors is the ability to adapt their properties by adjusting the composition. The percentage of filler, type of polymer binder, and even processing conditions can be customised to meet the specific requirements of an application [15]. For example, a formulation intended for wearable electronics may prioritise flexibility

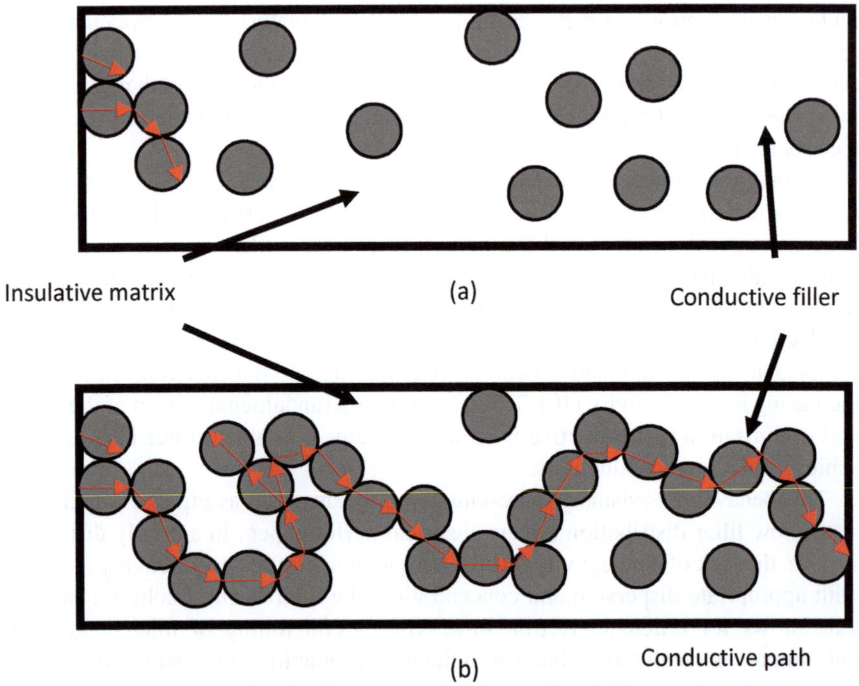

Fig. 2.2 Formation of conductive fillers in the formulation, with **a** depicting a poor distribution of filler and **b** depicting an acceptable distribution

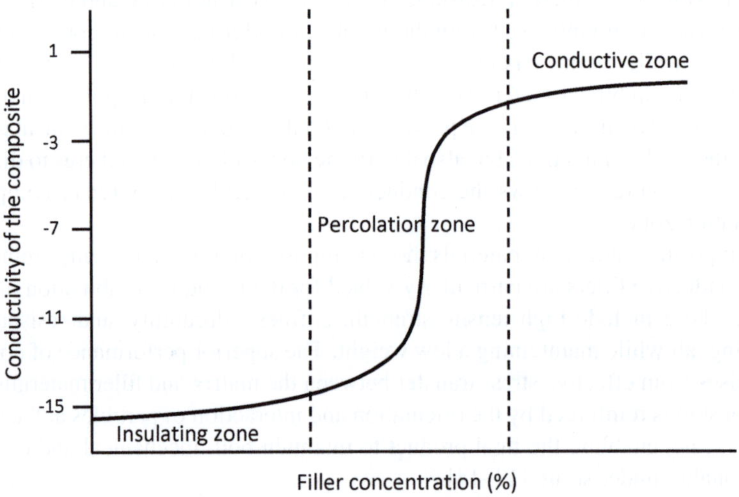

Fig. 2.3 The conductive fillers reinforced polymer composites percolation zone

and biocompatibility, whereas an ink for printed antennas may focus on maximising electrical conductivity and thermal dissipation.

The selection of conductive fillers also plays a crucial role in determining ink performance. Four primary material groups are typically employed: carbon-based, ceramic, metal-based, and hybrid materials [16]. Among these, carbon-based fillers such as carbon black (CB), CNTs, graphite, and especially graphene are particularly attractive. Graphene, in particular, exhibits a remarkable combination of properties including excellent electrical conductivity, chemical resistance, high mechanical strength, and a large surface area. Its two-dimensional structure allows it to form effective percolative networks at relatively low filler loadings, which is beneficial for maintaining flexibility and reducing production costs [17].

Another important consideration in ink formulation is the interplay between viscosity and surface tension. These rheological properties determine whether an ink is suitable for specific printing techniques such as inkjet, screen, gravure, flexographic, or direct-writing methods [18]. An ink that is too viscous may block nozzles or result in uneven deposition, whereas one with insufficient viscosity may spread uncontrollably, leading to poor resolution and performance. Therefore, the development of a successful conductive ink requires a delicate balance between electrical, mechanical, and fluidic properties.

From an application standpoint, the range of conductivity required varies significantly depending on the intended use. For instance, materials used for electrostatic discharge protection typically require low to moderate conductivity, while electromagnetic interference shielding and antenna applications demand much higher conductivity values. These variations highlight the need for a versatile range of ink formulations to address the broad spectrum of functional requirements in modern electronics [19]. Figure 2.4 shows the range of conductivity of conducting polymers and conductive polymer composites.

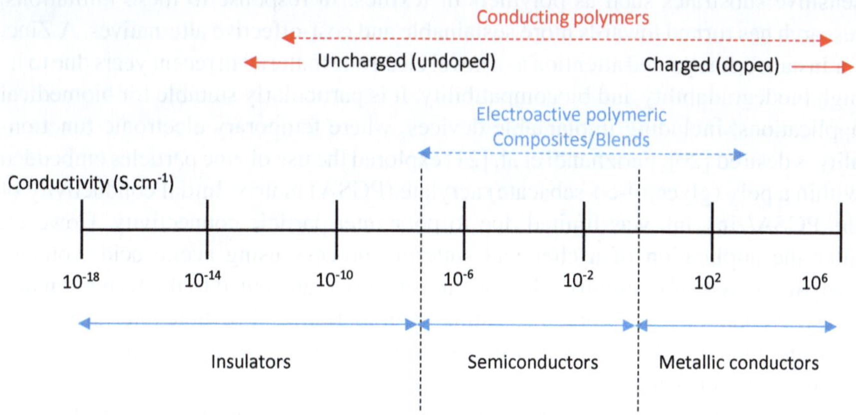

Fig. 2.4 The range of conductivity

Therefore, the development and application of conductive inks represent a multidisciplinary challenge that spans materials science, polymer chemistry, electrical engineering, and manufacturing technology. By understanding and controlling the fundamental factors that govern conductivity, mechanical performance, and printability, researchers and engineers are better positioned to design next-generation electronic components that are not only functional but also scalable, cost-effective, and environmentally sustainable.

2.4 Materials of Conductive Inks

Conductive inks are typically categorised into three main groups based on their composition: metal-based, conductive polymer-based, and carbon-based inks. Each category is distinguished by the nature of its conductive filler, with distinct advantages and limitations depending on the intended application and processing conditions. Metal-based conductive inks are widely explored and commercially utilised due to their superior electrical conductivity. Within this category, there are two principal forms which are metal nanoparticle dispersions and metal compound precursor solutions. Among the various conductive nanomaterials, metallic systems remain the preferred choice owing to their inherent high conductivity. Gold and silver nanoparticles have historically been used in printed and flexible electronics, valued for their chemical stability and excellent electrical performance. These metals have found particular utility in advanced applications such as bioelectronics, wearable sensors, and flexible circuitry.

However, the use of likeness metals poses significant cost and processing challenges. High material expense and the requirement for high-temperature sintering to form conductive pathways limit their scalability, particularly for temperature-sensitive substrates such as polymers or textiles. In response to these limitations, research has turned towards more sustainable and cost-effective alternatives. A Zinc, for instance, has gained attention as a novel electrode material in recent years due to its high biodegradability and biocompatibility. It is particularly suitable for biomedical applications, including implantable devices, where temporary electronic functionality is desired [20]. Farizhandi et al. [21] explored the use of zinc particles embedded within a poly (glycerol-co-sebacate) acrylate (PGSA) matrix. Initial conductivity of the PGSA/zink ink was limited due to poor inter-particle connectivity. However, after the application of a chemical sintering process using acetic acid, conductivity improved substantially. This enhancement was attributed to the formation of a physical conductive network, suggesting the importance of particle interface chemistry and matrix–filler interaction in achieving low percolation thresholds and high electrical performance.

Conductive polymers represent another significant category of conductive ink materials. These polymers inherently possess the ability to transport charge carriers through their conjugated π-electron systems. The examples include poly (3,4-ethylenedioxythiophene)-poly (styrenesulfonate) (PEDOT: PSS), polythiophene, polyaniline (PAN), and polystyrene sulfonate. These materials are particularly attractive for applications requiring flexibility, transparency, and solution-processability.

The PEDOT: PSS, for instance, is widely recognised for its excellent film-forming ability, environmental stability, and compatibility with printing technologies. It has found widespread use in organic photovoltaics, OLEDs, and touch screens. The charge transport in such polymers is facilitated by delocalised electrons that can hop or tunnel across chains, and their performance can be tuned through doping, morphology control, and composite formulation. Although their conductivity is generally lower than that of metal-based inks, conductive polymers provide a compelling balance between processability and functionality.

Carbon-based inks incorporate a range of allotropes and derivatives of carbon, such as graphene, CNTs, CB, fullerenes, and graphite. These materials offer a combination of excellent electrical, thermal, and mechanical properties while being more cost-effective than metals. Carbon-based fillers are often employed in electromagnetic interference (EMI) shielding, electrostatic discharge (ESD) protection, and sensor applications.

The versatility of carbon arises from its ability to adopt various hybridised states (sp, sp^2, and sp^3), resulting in a diverse set of allotropes with unique structural and electrical properties. For example, diamond has a tetrahedral sp^3 structure, making it an electrical insulator. In contrast, graphite consists of layers of sp^2-bonded carbon atoms, where each carbon is bonded to three others in a hexagonal arrangement. The fourth valence electron is delocalised, enabling the material to conduct electricity. The layers in graphite are held together by weak van der Waals forces, which allow them to slide over one another, imparting lubricating properties. When these layers are reduced to a single atomic plane, the material is termed graphene. Graphene exhibits extraordinary properties, including high intrinsic electron mobility, thermal conductivity, mechanical strength, and optical transparency. Its high aspect ratio and two-dimensional morphology facilitate efficient percolation at lower filler loadings compared to spherical or irregular particles.

Fullerenes (0D) and CNTs (1D) further expand the structural landscape of carbon. Fullerenes consist of closed-cage carbon molecules, while CNTs are essentially rolled-up graphene sheets, exhibiting either metallic or semiconducting behaviour depending on their chirality and diameter. These nanomaterials can form continuous conductive networks within a polymer matrix, enhancing conductivity without compromising flexibility. The CB, composed of over 97% amorphous carbon, is another common filler in conductive inks. It is characterised by its particulate nature, typically forming aggregates and agglomerates. Its relatively low cost and ease of dispersion make it suitable for high-volume applications where moderate conductivity is acceptable. However, its conductivity is often lower than that of metallic or graphene-based fillers.

Graphene, owing to its exceptional properties, holds substantial promise for the next generation of electronic materials. Its high surface area, mechanical robustness, and thermal stability make it a superior candidate for use in sensors, composite materials, and energy storage systems [22]. In addition, its sensitivity to deformation makes it ideal for use in strain sensors that can detect stretching, bending, and torsional motion. This capability is particularly relevant in the field of wearable electronics and human–machine interfaces.

Due to its planar structure and delocalised π-electron system, graphene facilitates efficient electron transport even at low loadings. It also enhances the mechanical integrity of the host polymer, improving resistance to cracking and delamination under mechanical stress. Furthermore, its compatibility with various functionalisation techniques allows for modifying surface chemistry to improve dispersion and matrix adhesion. The selection of conductive fillers whether metal-based, polymeric, or carbonaceous must be aligned with the specific functional and processing requirements of the intended application. Metal-based inks offer unparalleled conductivity but face challenges in cost and thermal processing. Conductive polymers provide design flexibility and environmental compatibility. Carbon-based materials, particularly graphene, offer a promising middle ground with their unique balance of performance, cost, and mechanical properties. Understanding the structure property relationships of these materials, as well as their interactions within the ink matrix, is essential for advancing the development of high-performance conductive inks for next-generation electronics.

2.5 Graphene-Based Conductive Inks in Advanced Printed Electronics

Graphene, first isolated by Geim and Novoselov [23], has revolutionised the scientific landscape of nanoscience and condensed matter physics, owing to its remarkable electrical, physical, and chemical properties. The discovery sparked a global research surge, with significant interest from academic and industrial communities leading to exponential growth in publications and applications of graphene. This two-dimensional (2D) material has since emerged as a ground-breaking candidate in fields ranging from nanoelectronics to biomedical devices. In printed electronics, graphene-based conductive inks have gained traction as an applicable alternative to traditional metallic inks, conductive polymers, and carbon-based inks. Their adoption addresses several limitations associated with conventional materials, including high production costs, limited flexibility, and environmental or biological hazards. The unique structural and functional properties of graphene, including high carrier mobility, superior thermal conductivity, and exceptional mechanical strength, render it highly suitable for flexible, lightweight, and durable electronics. Figure 2.5 illustrates the graphene methodologies development.

Graphene production methods have evolved considerably since its discovery. Techniques such as chemical vapour deposition (CVD), arc discharge, epitaxial growth, and the unzipping of CNTs have been widely explored. Each method offers distinct advantages and limitations, particularly concerning material purity, defect levels, and scalability. A commonly used technique involves the oxidation of graphite

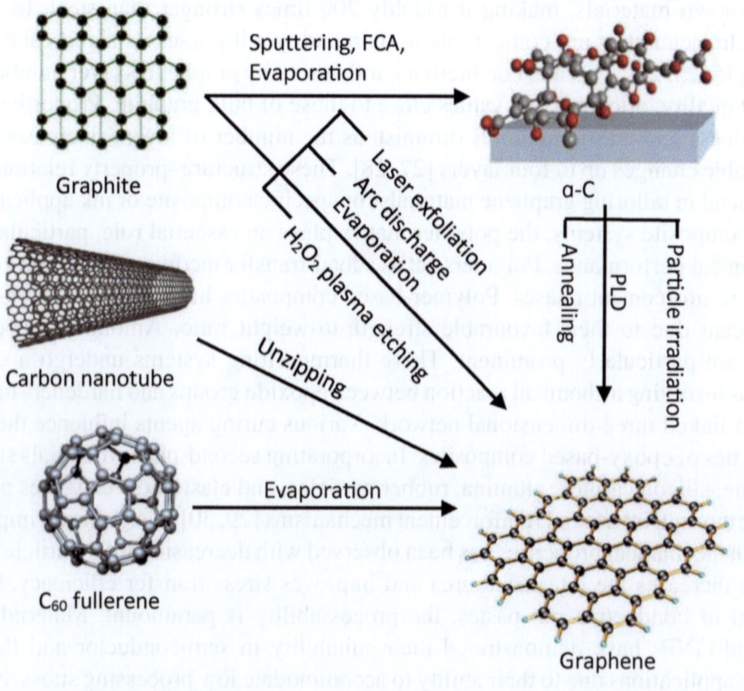

Fig. 2.5 The development of graphene materials

to produce graphite oxide (GO), followed by its reduction and mechanical exfoliation to obtain reduced graphene oxide (rGO) [24]. The GO, which is electrically insulating and hydrophilic, is easily dispersed in aqueous media, making it suitable for ink formulation. However, the presence of oxygen-containing functional groups requires reduction steps to restore conductivity.

Further advances in graphene chemistry have enabled the functionalisation of graphene sheets, enhancing their dispersion in organic solvents and polymer matrices. Functionalised graphene sheets (FGS) offer improved compatibility with various resins and substrates, which is critical for the fabrication of homogenous composites. Graphene materials are available in various formats, such as single-layer sheets deposited on substrates or as graphene nanoplatelets (GNPs) in powder, paste, or liquid dispersions. Structurally, graphene comprises a single layer of sp2-hybridised carbon atoms arranged in a hexagonal lattice, conferring outstanding strength and electrical properties. Interestingly, Allen et al. [15] demonstrated that monolayer graphene exhibits intrinsic ripples in its structure, deviating from the ideal flat 2D form.

Graphene remains the first true 2D atomic crystal known for its ultra-high thermal and electrical conductivities. Its mechanical robustness is highlighted by a Young's modulus of approximately 1 TPa and an intrinsic tensile strength that far surpasses most known materials, making it roughly 200 times stronger than steel. Its resistance characteristics are comparable to diamond, yet it possesses significantly less mass; [15, 25, 26]. Thermal conductivity, influenced by graphene's layer number and crystal quality, often reaches values close to those of bulk graphite. Properties such as hardness and elastic modulus diminish as the number of layers' increases, with noticeable changes up to four layers [27, 28]. These structure–property relationships are crucial in tailoring graphene materials for specific composite or ink applications.

In composite systems, the polymer matrix plays an essential role, particularly in mechanical performance. Polymers act as a load-transfer medium, distributing stress across reinforcement phases. Polymer-based composites have become increasingly significant due to their favourable strength-to-weight ratio. Among them, epoxy resins are particularly prominent. These thermosetting systems undergo a curing process involving a chemical reaction between epoxide groups and hardeners to form a cross-linked three-dimensional network. Various curing agents influence the final properties of epoxy-based composites. Incorporating second-phase materials such as graphite, silicon carbide, alumina, rubber particles, and elastomers enhances performance through interfacial reinforcement mechanisms [29, 30]. A significant improvement in mechanical properties has been observed with decreasing filler particle sizes, as this increases the interfacial area and improves stress transfer efficiency. In the context of conductive ink pastes, the processability is paramount. Materials like CB and GNPs have demonstrated their suitability in semiconductor and flexible device applications due to their ability to accommodate low processing stress, ensure substrate compatibility, and enable high-speed curing. These properties are vital for achieving high throughput and reliable printing in industrial settings.

Base resins used in conductive ink pastes, especially epoxy systems, can be classified by solvent presence. For instance, Su [31] explored the formulation of silver-based inks using both epoxy and polyamide resins. Solvent-based systems can facilitate curing but may introduce processing limitations such as short tack-free times and void formation. Conversely, non-solvent systems offer better stability but may lack rapid cure kinetics. Thermoplastic additives in binders can also serve as toughening agents, improving fracture resistance and reducing brittleness.

In recent developments, binders are being optimised to assist in automatic performance layouts. These include pressure- and heat-sensitive adhesives that reduce manual intervention during assembly, improving dimensional control and consistency [32]. Essential binder properties include sufficient cohesion to maintain filler distribution and structural integrity during processing and end-use conditions. Epoxy systems typically involve two components which are the resin and the hardener. The resin exhibits higher viscosity and provides the bulk of mechanical properties, while the hardener governs the crosslinking density and glass transition temperature. Post-curing, the composite undergoes a transition to a rigid and thermally stable structure. When nanoparticles are introduced into such matrices, several nanoscale interactions occur. These interactions yield improved thermal, optical, and mechanical performance, depending on particle type, distribution, and interfacial adhesion [33].

Applications include semiconductor packaging, optical encapsulants, and advanced structural composites. Nanoparticles such as silica, alumina, and carbon derivatives enhance thermal conductivity, dielectric behaviour, and mechanical reinforcement. These improvements are evidenced by increases in the glass transition temperature and impact strength, making them ideal for electronic, automotive, and aerospace sectors. The integration of graphene and other nanofillers into epoxy-based conductive inks and composites represents a major advancement in printed electronics. Their superior multifunctional performance, coupled with emerging fabrication and formulation technologies, is transforming the landscape of flexible, wearable, and high-performance electronic devices. Understanding the underlying material science and engineering principles remains essential for optimising these systems for future industrial applications.

2.6 Influence of Trace Ink Patterns on Circuit Performance

The influence of trace ink shape or pattern on the performance of electronic circuits is a critical area of investigation in the field of printed and SEs. The configuration of conductive traces, those narrow paths of ink that transmit electrical signals can significantly affect the overall functionality and efficiency of a circuit. These patterns govern not only electrical characteristics such as resistance, capacitance, and impedance but also influence thermal dissipation, mechanical resilience, and spatial optimisation. As such, their design must be approached with a strong understanding of both theoretical principles and practical considerations.

At the core of this discussion lies the relationship between geometry and electrical properties. Conductive traces serve as the essential channels for signal transmission and power delivery within a circuit. Their shape, width, and thickness directly influence the resistance and impedance along the path. According to electromagnetic theory, resistance increases with longer paths and narrower widths, while impedance particularly in high-frequency applications is heavily affected by both the shape of the trace and its relation to surrounding dielectric materials. Therefore, the design of a trace must take into account not only the direct current (DC) characteristics but also the alternating current (AC) behaviour, especially in radio frequency (RF) and high-speed digital circuits.

In high-frequency circuits, signal integrity becomes a dominant concern. Unintended alterations in impedance due to abrupt trace width changes, sharp angles, or inconsistent routing can cause signal reflections, phase distortion, and electromagnetic interference (EMI). These issues lead to crosstalk and degradation of the signal, which can compromise data integrity. Investigations such as those conducted by Liu et al. [34] and Alresheedi et al. [35] have shown that smoother curves and controlled impedance paths significantly enhance signal fidelity. The use of serpentine, spiral, or meandering traces helps in managing electromagnetic effects and allows for better integration of inductive or capacitive functions directly into the trace layout.

Thermal management is another crucial aspect influenced by trace shape. As electronic circuits operate, especially under high current loads, they generate heat. The ability of a trace to dissipate this heat is dependent on its cross-sectional area and its interaction with the substrate. Wider and thicker traces provide a greater surface area for heat dissipation and reduce the likelihood of thermal hotspots, thus improving the reliability and longevity of the device [36]. The importance of effective thermal dissipation is especially evident in power electronics, where failure to manage heat can result in significant performance degradation or even device failure.

Beyond electrical and thermal considerations, mechanical integrity is a prominent concern in the context of flexible and stretchable electronics. When circuits are subjected to bending, twisting, or stretching, common in wearable technologies, the trace shape becomes a determinant of mechanical durability. Conventional straight-line traces may fracture or delaminate under mechanical strain, whereas curved or serpentine traces distribute stress more evenly and allow for greater elongation without compromising electrical connectivity. Larson et al. [37] and Kumar et al. (2020) have both demonstrated the enhanced mechanical endurance of optimised trace shapes, noting that geometric flexibility directly contributes to long-term device stability under dynamic conditions.

The choice of trace patterns also influences manufacturability and design efficiency. While intricate patterns may offer superior electrical and mechanical properties, they can introduce challenges in the fabrication process. Complex geometries may lead to inconsistencies during printing, particularly in screen printing or inkjet deposition, where resolution limitations and ink spreading behaviours must be managed. Kim et al. [38] highlighted the potential for defects when overcomplicating trace paths, reinforcing the need for a balance between performance and manufacturability.

Trace design also plays an important role in space utilisation. Modern electronics demand ever-smaller form factors, with increased component density. Effective use of space requires precise routing strategies. Customised trace geometries such as meanders or zigzag configurations can optimise the layout, reduce the footprint, and facilitate dense integration without sacrificing performance. Work by Zhang et al. (2017) using optimisation algorithms for zigzag traces illustrates how algorithmic design approaches can be employed to iteratively refine trace parameters for minimal loss and maximal efficiency.

Moreover, trace thickness is a determining factor not only for electrical performance but also for structural and manufacturing robustness. Thicker traces generally exhibit lower resistance, enabling better current handling and reduced power loss. They also enhance the mechanical strength of the circuit, providing robustness against handling and flexure. However, they require more ink material and may reduce the achievable resolution of printed features. This introduces trade-offs that must be carefully balanced during the design process, particularly when working with substrates of varying flexibility or thermal expansion characteristics.

In practical applications, the synergy between trace design and overall circuit architecture is paramount. For instance, Wang et al. [39] explored how meandering traces can improve the flexibility of printed electronics while maintaining signal integrity. Their findings underscore the complex interplay between mechanical and electrical performance metrics and the need for a holistic design strategy. The shape, pattern, and thickness of trace inks are not merely aesthetic or incidental considerations and they are fundamental design parameters that impact every critical performance metric of a printed electronic circuit. From electrical and thermal characteristics to mechanical resilience and manufacturability, the trace geometry plays a decisive role in defining the capability and reliability of the final device. An investigation continues to advance, supported by both experimental and computational tools, engineers and designers are increasingly equipped to exploit these design freedoms to develop more efficient, durable, and miniaturised electronic systems. Understanding and optimising trace ink design is thus a cornerstone in the development of next-generation flexible and stretchable electronic technologies.

2.7 Printing Techniques and Substrate Considerations

Printed electronics is an interdisciplinary field that merges traditional printing techniques with the fabrication of electronic components, enabling the creation of innovative devices or enhancing existing systems. The process involves the controlled deposition of functional materials such as conductive, semi conductive, or dielectric inks onto various substrates, including paper, plastic, textiles, and even flexible films. This method has been employed to produce an array of devices, ranging from simple circuit interconnects to complex sensors, flexible displays, radio-frequency identification (RFID) tags, photovoltaic cells, and energy storage components [40].

A fundamental distinction in printed electronics lies in the classification of printing methods into two major categories which are conventional (contact-based) and non-impact (digital) printing techniques. Figure 2.6 provides a comprehensive comparison of these methods, illustrating their operational principles and typical specifications.

Conventional printing techniques rely on the use of a physical printing plate or stencil to transfer ink onto the substrate. These methods include lithography, flexography, gravure, and screen printing. In these processes, the plate or screen carries the image to be printed, and the ink is applied to selected regions, producing a patterned deposition on the substrate surface. The principle underlying these techniques is based on selective adhesion or transfer, where ink is deposited only on the areas intended for printing, thus generating contrast between conductive and non-conductive regions. Among these, screen printing has emerged as one of the most widely adopted methods in printed electronics due to its compatibility with high-viscosity inks, scalability, and ability to deposit thick ink layers. It is particularly suitable for fabricating large-area and low-resolution structures such as electrodes and interconnects.

Non-impact printing technologies, on the other hand, do not require a physical image carrier. These digital printing methods such as inkjet printing and electrophotography rely on computer-controlled nozzles or electrostatic mechanisms to deposit functional materials directly onto the substrate. This allows for greater design flexibility, high-resolution patterning, and minimal material waste. Inkjet printing, in

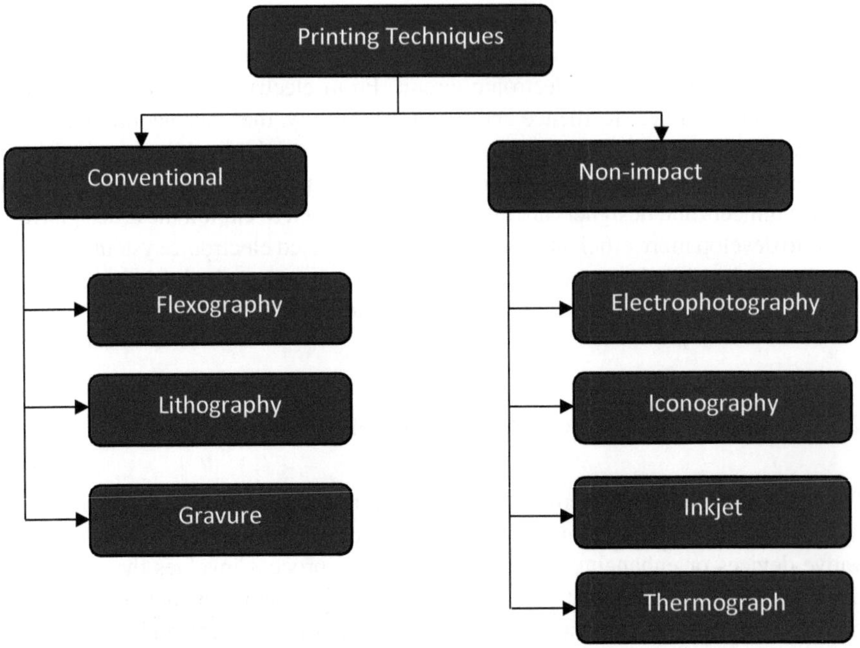

Fig. 2.6 Overview of printing technology

particular, has seen rapid advancements over the past decade, largely due to its non-contact nature, which is ideal for sensitive substrates and rapid prototyping.

As described by Huang et al. [41], the development of non-impact digital printing technologies has been instrumental in enabling flexible and SEs. However, these technologies also present unique challenges. Achieving consistent and uniform ink deposition is essential to ensuring reliable electrical performance. The interaction between the ink and the substrate surface including wettability, spreading behaviour, and curing characteristics and plays a decisive role in defining the quality of the printed features.

Furthermore, as noted by Mendez et al. (2019) and Wahid et al. [42], ink formulations for conductive printing must meet stringent rheological and surface energy requirements to form defect-free patterns on uneven or dynamically deforming substrates. Successful deposition requires a careful balance of ink viscosity, surface tension, and solvent evaporation rate to prevent issues such as coffee-ring effects, satellite droplets, or incomplete coverage.

The substrate is a critical component in printed electronics, serving not only as the physical foundation of the device but also influencing the functional behaviour of the printed layers. For flexible electronics, substrates must conform to stringent mechanical, electrical, and thermal criteria. Flexibility is one of the most fundamental requirements, enabling the device to bend, twist, or stretch without compromising functionality. This is typically achieved by employing thin, laminar substrates composed of organic or inorganic materials integrated with elastomeric backings. The substrate must also be non-conductive to avoid interfering with the printed conductive pathways. It should possess suitable surface energy to ensure proper ink adhesion and spreading, as well as exhibit thermal stability to withstand processing temperatures. Moreover, its surface morphology including roughness and porosity, must be tightly controlled to facilitate consistent conductivity, even at minimal ink film thicknesses.

Traditional rigid substrates such as glass have favourable barrier properties and dimensional stability, making them suitable for applications requiring high precision. However, they are brittle and susceptible to micro-cracking under mechanical stress. Although ultra-thin glass can be manufactured to mimic the flexibility of plastic films, its mechanical fragility remains a limitation for applications requiring repeated deformation. Plastic and paper substrates, by contrast, offer lightweight, flexible, and low-cost alternatives. Paper, in particular, presents advantages such as foldability, biodegradability, and excellent optical properties, including high reflectivity and contrast ratio. These attributes make paper a promising candidate for disposable or transient electronic devices. However, paper's fibrous structure and inherent porosity require surface treatments or coatings to render it suitable for high-resolution printing.

Plastic substrates are widely used in flexible electronics due to their mechanical robustness, processability, and barrier tunability. Polymers such as PI, PET, polyethylene naphthalate (PEN), and polycarbonate (PC) are common choices. The thermal and mechanical performance of these substrates is largely determined by their molecular structure, including whether they are amorphous or semi-crystalline.

A key material property for these polymers is the glass transition temperature, defined as the temperature at which the polymer transitions from a rigid, glassy state to a soft, rubbery state. This property is particularly significant in printed electronics, where substrates must endure both the thermal conditions of ink curing and the mechanical deformations encountered during operation. Below the transition temperature, the polymer behaves rigidly, whereas above it, the molecular chains gain mobility, contributing to flexibility but potentially leading to dimensional instability.

For instance, PET, with a transition temperature of approximately 78°C, offers an optimal balance between thermal stability and mechanical flexibility, making it a popular choice for applications requiring low-cost, transparent, and durable substrates. It retains its mechanical integrity and optical clarity across a broad temperature range, which is crucial for applications such as flexible displays and wearable sensors. However, achieving long-term reliability also necessitates consideration of barrier properties. Plastics are inherently permeable to oxygen and moisture, which can degrade the performance of printed conductive materials. By mitigating this, barrier layers or coatings often composed of inorganic oxides or multilayer polymer stacks—are applied to enhance environmental stability.

The integration of appropriate printing technologies with well-engineered substrates is foundational to the advancement of printed and flexible electronics. A deep understanding of material interactions, process limitations, and functional requirements is essential for engineers, academics, and researchers to develop robust, scalable, and application-specific solutions. As the field evolves, the synergy between materials science and printing technology will continue to play an important role in shaping next-generation electronic systems.

2.8 Load Conditions in Stretchable and Wearable Electronics

In the context of mechanical and structural engineering, static load refers to a force or combination of forces that are applied gradually to an object or structure and remain constant over time. These forces are introduced slowly and maintained without fluctuation, allowing a system to stabilise under load and enabling a more straightforward analysis of stress and deformation. Static loading is necessary in determining the strength, stiffness, and failure thresholds of materials and assemblies under predictable conditions. For example, the structural load exerted by stationary vehicles on a car park constitutes a static load, as the weight does not shift dynamically once the vehicles are at rest [43].

In contrast, a dynamic load often termed a live load is characterised by its temporal variability. These loads may fluctuate in magnitude, direction, or point of application, introducing a more complex interplay of inertial and damping forces within the structure. Dynamic loading scenarios include walking or running on a wearable device, wind pressure on a building, or vibrations induced by machinery. Dynamic loads can be further categorised into periodic and non-periodic types. Periodic loading repeats at regular time intervals, such as the cyclic movement of a piston or motor. Non-periodic loading, however, may arise from irregular or stochastic sources, such as earthquakes, impacts, or human motion [43–45].

Understanding the mechanical loading conditions whether static or dynamic is especially important in the development and optimisation of flexible and SE systems. Such systems often integrate conventional electrical properties with soft substrates and materials that are mechanically compliant. These advanced devices must sustain their functionality under various mechanical deformations, including bending, twisting, and stretching. Emerging materials and device architectures have made it possible to fabricate electronic circuits that can conform to arbitrary geometries while maintaining reliable performance [42].

The theory of strain–stress response in flexible electronics helps to explain the impact of mechanical loading on electronic performance. Strain develops when mechanical loads deform a structure, while stress is the internal reaction of the material to that deformation. When a device is subjected to bending, the outer surface undergoes tensile strain, while the inner surface is compressed. The degree of bending strain is inversely proportional to the thickness of the flexible material. This foundational concept explains why ultra-thin electronics especially those constructed with nanomaterials like graphene, demonstrate superior flexibility and resilience under repeated loading cycles.

Indeed, Huang et al. [46] demonstrated that thermally reduced graphene-based conductive inks printed on polymer substrates can maintain high electrical conductivity even after enduring numerous bending cycles. This durability is attributed to the two-dimensional structure of graphene, which provides exceptional flexibility and strength at the atomic level. However, the integration of SEs into practical applications particularly in wearables demands more than simple bending tolerance. These systems must survive repeated, complex, and multi-directional deformations caused by body motion without degradation in electrical performance. For instance, stretchable interconnects and electrodes embedded in clothing or skin patches experience frequent and localised changes in strain, which can result in delamination, cracking, or mechanical fatigue if not properly engineered.

According to Park et al. [9], one of the primary failure mechanisms in wearable electronics under dynamic loading is the differential strain between the substrate and the conductive film. As the body moves, the soft elastomeric substrate stretches and contracts, but the thin conductive layer often composed of metallic or carbon-based materials may not accommodate the same level of deformation, leading to eventual cracking or failure. This mismatch highlights the importance of understanding interfacial adhesion, material compatibility, and structural design when fabricating stretchable conductors.

Therefore, mechanical deformability testing is essential to evaluate the real-reliability of SEs. Standardised mechanical tests such as cyclic bending, uniaxial stretching, and torsion are often adapted to simulate operational conditions. These tests not only determine a device's fatigue life but also monitor changes in electrical resistance, capacitance, and overall functionality during deformation. As part of best practice, test setups should be integrated with resistance measurement instruments to enable real-time monitoring of electrical properties during mechanical strain cycles.

Importantly, mechanical characterisation must go hand-in-hand with electrical evaluation. For instance, the coupling between mechanical strain and electrical resistance in stretchable conductors can be described through gauge factor analysis and crack propagation theory. These approaches allow for the prediction of electrical degradation under various deformation modes, thereby guiding the design of robust and reliable stretchable systems. The distinction between static and dynamic loading is fundamental to the engineering of next-generation wearable and stretchable electronics. The ability of these devices to endure diverse mechanical stresses while maintaining electrical performance is a product of thoughtful material selection, structural design, and advanced testing protocols. As this field evolves, a deeper understanding of load-induced failure modes and material responses will be crucial to realising high-performance, mechanically adaptive electronic systems for biomedical, consumer, and industrial applications.

2.9 Characterisation of Mechanical Deformation

The first essential step in realising stretchability in electronic devices is achieving sufficient bendability. Bending serves as a preliminary evaluation of a material's mechanical compliance and is often a gateway property for more complex deformations such as stretching and twisting. Fundamentally, bending strain in a layered structure is inversely proportional to its thickness. Therefore, any sufficiently thin material can achieve a high degree of bendability without structural failure [47]. This principle is rooted in classical plate and shell theory, where the strain distribution across the thickness of a bent structure varies linearly, with the neutral axis experiencing no strain. As a result, reducing the overall thickness allows materials with relatively low strain tolerance to bend safely without exceeding their fracture limits.

Several standard methods are used to characterise bending performance in Ses. According to Mohammed [47], one of the simplest experimental techniques is the fixed-radius bend test, where a sample is curved to a specific radius, and its electrical resistance is measured at different levels of bending strain. In the variable-radius bend test, the sample is positioned between two parallel plates, and the curvature is adjusted by modifying the distance between the plates. A variable-angle bend test involves holding the sample between a fixed and a co-planar plate, then varying the angle to induce controlled bending. The sliding plate test bends the sample into a U-shape and applies cyclic movement via a reciprocating clamp to evaluate flexibility. Meanwhile, the multi-mode bend test applies a load to the centre of a supported sample, inducing bending until a predefined angle or failure point is reached. In the free arc test, grips hold the edges of the sample while repeated linear motions generate bending.

Empirical investigations have shown that electronic films, even those built from traditionally rigid materials like silicon, can tolerate significant deformation when engineered in ultra-thin geometries. Lu and Wang [48] demonstrated that a 100 nm-thick single-crystal silicon film on a 20 mm-thick polymer substrate can undergo bending to a radius of 1 cm while experiencing only 0.1% strain far below the 1% fracture threshold of silicon. This performance is made possible by the significant strain gradient across the substrate and conductive film, with tensile stresses developing on the convex surface and compressive stresses on the concave side.

Mohammed [47] further observed that decreasing the bending radius increases the local stress and strain in the conductor-substrate system, particularly affecting materials with low fracture toughness. Up to now, conductive films formulated from silver pastes demonstrated remarkable stability, retaining their electrical properties even under tight curvatures down to a 1 mm bending radius. Such resilience under deformation can be attributed to the percolation pathways formed by metallic particles, which remain intact due to effective filler dispersion and strong interfacial adhesion with the polymer matrix. Typical failure modes observed during bending tests include metal layer rupture, film cracking, delamination from the substrate, and loss of adhesion or cohesion within the film itself [49]; [9]. These phenomena are governed by the mechanical mismatch between layers, the modulus contrast, and the interface energy. Hence, bending tests provide essential insights into mechanical integrity, fatigue behaviour, and long-term performance under cyclic loads.

Mechanical performance in stretchable conductors is further influenced by multiple factors, including the modulus of the substrate, the volume fraction and morphology of the conductive filler, and the particle-particle interactions within the ink matrix. A soft elastic substrate acts as a mechanical buffer, distributing the applied stress and preventing catastrophic failure. Even when micro-cracks form in the ink film, the substrate may preserve the macroscopic integrity of the device. Over time, however, accumulated damage leads to a gradual reduction in stiffness and conductive performance. Kraft et al. [50] calculated the lifetime of conductive films by tracking the degradation in stiffness across strain cycles, linking this to fatigue-induced damage propagation in the composite structure.

In addition to bending, uniaxial tensile testing is crucial for assessing the stretchability of printed conductors. This test simulates stretching by applying tension to a sample either until failure or across multiple cycles to evaluate durability. A single-cycle tensile test increases strain continuously until electrical failure, defined either by a complete loss of conductivity or by resistance exceeding a critical threshold (for example 30× the baseline). In cyclic stretch tests, smaller strains, typically 10–20%, are applied repeatedly to simulate long-term use. During these tests, resistance is recorded before, during, and after stretching to assess both elastic recovery and permanent deformation.

Stretchable substrates typically viscoelastic polymers exhibit time-dependent behaviour such as stress relaxation and hysteresis. By capturing accurate data, it is recommended that hold times be incorporated between cycles, allowing the polymer matrix to recover. This viscoelastic recovery is governed by the molecular chain mobility of the polymer and is strongly influenced by temperature, strain rate, and polymer architecture.

Designing conductive traces on stretchable substrates requires careful geometrical planning. For most conductive materials are inherently brittle or rigid, conductive lines must be designed to accommodate deformation. One effective approach is the use of meander-shaped interconnects, which distribute strain across the length of the trace and allow elongation through unfolding rather than material stretching. Mohammed [47] observed that interconnects with embedded LEDs retained full functionality even at 100% strain, and performance returned to baseline upon recovery. However, a sharp rise in resistance at 130% strain resulted in the failure of LED illumination, indicating a threshold beyond which the conductive pathways degrade irreversibly.

Another critical to date underexplored area in stretchable electronics is torsion, or twisting. Flexible and wearable electronics are often subjected to torsional forces during use such as when wrapped around limbs or embedded in garments. Despite its relevance, torsional deformation remains less thoroughly characterised than bending or tensile loading. Torsion introduces shear stress along the axis of rotation and can lead to complex stress distributions, particularly in layered or anisotropic structures. Horgan and Chan [51] examined the torsional behaviour of inhomogeneous materials and demonstrated that material tailoring could shift the location of peak shear stress within a structure, allowing engineers to avoid premature failure at boundaries. Similarly, Lee and Liu [52] measured the shear modulus of fibre-reinforced composites using plate-twist and torsion tube methods, providing valuable data on anisotropic responses. Soldatos [53] used variational mechanics to explore large-deflection torsion of orthotropic plates, further expanding our understanding of complex laminate responses under twist.

Torsion tests typically involve anchoring one end of a sample and rotating the other, thereby applying torque about the sample's axis. Variations of the test include applying opposing torques at both ends or measuring the shear modulus, torsional yield strength, fatigue life, and ductility. These parameters differ from those captured in tensile testing but are equally critical for assessing service conditions in flexible electronics [54]. Torsion tests are usually classified into failure testing, which continues until the sample fractures such as proof testing, which subjects the sample to a sustained torque for a defined period, and operational testing, which simulates actual service conditions.

By understanding the response of SE components to bending, stretching, and torsion is fundamental to their integration into reliable, long-lasting systems. These tests reveal critical information about material fatigue, structural design limitations, and interfacial bonding. As flexible electronics continue to evolve, rigorous mechanical testing rooted in sound physical theory, will be central to bridging the gap between laboratory innovation and real-application.

2.10 Towards Integrated Mechanical, Electrical and Functionality

Previous work on conductive ink systems has largely concentrated on optimising the performance of either individual materials or hybrid composites, often limited to ink formulation and basic conductivity characterisation. However, such explorations frequently neglect the broader implications on mechanical robustness, electrical reliability, and advanced functional attributes such as self-healing capabilities. Consequently, the present work aims to address this critical gap by holistically investigating the electrical, mechanical, and self-healing performance of conductive inks under diverse deformation scenarios, with specific attention given to trace design and pattern morphology.

Several influential works have laid the groundwork for the development of stretchable and flexible conductive materials. For example, Park et al. [9] fabricated a highly stretchable silver conductor comprising silver flakes dispersed within a polyester matrix. Utilising infrared (IR) curing techniques, they printed the conductive ink onto a polyurethane elastomer substrate. This configuration demonstrated exceptional performance, which a low sheet resistance of 46 milliohms per square, a resistivity of 3.68×10^{-7} ohm-metre, and superior stretchability reaching up to 150%. Furthermore, the conductors sustained 10,000 cycles of cyclic stretching at 40% strain and endured bending radii down to 1 mm without mechanical or electrical failure. These metrics highlight the critical balance between electrical conductivity and mechanical deformability, both of which are influenced by interfacial adhesion, filler distribution, and substrate compatibility.

In the domain of fatigue performance, Chen et al. [55] explored nanocrystalline gold and gold-titanium films deposited on flexible polyimide substrates. Their findings revealed that even minor deviations in the resistance-strain relationship by approximately 5% from theoretical predictions, served as early indicators of microcrack formation. The cyclic mechanical testing demonstrated that electrical degradation occurred due to localised plasticity and crack initiation within the thin film, especially under high cycle fatigue conditions ($N = 10^5$ cycles). Interestingly, the failure strain of the gold/titanium composite was significantly higher than that of pure gold films, indicating the reinforcing role of titanium in enhancing strain tolerance and suppressing crack propagation. This highlights the critical importance of material synergy in hybrid multilayer systems to delay fatigue-induced failure.

In a separate investigation, Mohammed [47] investigated the role of trace geometry specifically straight-line and sine-wave patterns on the stretchability of silver-based conductive inks. Samples subjected to cyclic stretching at strains up to 50% exhibited a pronounced increase in resistance over 100 cycles, from an initial 2.8–141 Ω. This change signifies microstructural rearrangements such as filler reorientation, inter-particle separation, and possible crack development. The elastic modulus of the initial ink was 40 MPa, a value that positions it within the soft polymeric range, suitable for stretchable applications, albeit with limited structural resilience over time. Besides, Chow et al. [56] further extended these findings by subjecting printed silver ink traces of varying widths to cyclic torsion. The results showed that monotonic twisting up to 135° caused negligible changes in resistance, below 0.5%, while 5000 cycles of torsion led to a modest increase of 1.5% in resistance. These findings suggest that torsional deformation, within certain limits, does not critically impair the percolative pathways of conductive fillers in well-anchored films, particularly if the ink adheres uniformly to a flexible in torsion-resistant substrate.

Additionally, Cho et al. [57] developed gravure-printed PEDOT: PSS electrodes exhibiting a sheet resistance of 359 Ω/sq and remarkable mechanical integrity across multiple bending modes. The electrodes demonstrated minimal resistance change (0.08) even when subjected to bending radii below 10 mm. Furthermore, under a 2000-cycle bending fatigue test at a 10 mm radius, the devices exhibited no delamination or visible micro-cracks. During uniaxial stretching up to 4% strain, its value remained as low as 0.004, affirming the potential of PEDOT: PSS as a highly compliant, for stable, conductive polymer for flexible systems. For twisting, the resistance change remained low until angles exceeded 40°, beyond which interfacial delamination between the film and PET substrate was observed, indicating the limits of adhesive and cohesive integrity.

These work, while significant in their contributions, collectively reveal a prevailing a gap in fully understanding how trace pattern geometries and microstructural architecture influence multifunctional performance metrics, especially in contexts where mechanical, electrical, and self-healing characteristics must co-exist and function synergistically. Most investigations have emphasised planar performance metrics which are resistivity, elongation at break, and modulus without adequately considering how trace curvature, loop structures, or serpentine pathways mitigate or exacerbate failure under compound mechanical loads such as simultaneous bending, twisting, and stretching.

From a physics perspective, the performance of these materials is fundamentally governed by the percolation theory, which describes how the connectivity of conductive filler particles determines the macroscopic electrical conductivity. As mechanical deformation induces local strain gradients and filler realignment, the percolative network undergoes dynamic evolution often resulting in increased tunnelling distances, contact resistance, and eventual circuit disruption. In addition, fracture mechanics and continuum damage theory provide insights into how micro-cracks nucleate and grow in polymer matrices under repetitive loading, often initiating at defect sites or filler agglomerates due to stress concentration.

Moreover, the interaction between the substrate's viscoelastic properties and the conductive film plays a critical role in strain dissipation and energy absorption. Viscoelastic relaxation mechanisms allow time-dependent strain recovery but also induce creep and hysteresis effects, influencing cyclic performance outcomes. Hence, accurate modelling and testing of these systems must account for such time-dependent behaviour and interfacial dynamics. In light of the above, the current work seeks to bridge these gaps by systematically analysing how trace design including geometry, dimensionality, and spatial orientation affects not only electrical and mechanical performance but also the initiation and recovery of self-healing responses under dynamic conditions. This integrated approach will better inform the design of next-generation conductive ink systems for emerging applications in flexible, wearable, and stretchable electronics.

References

1. Yim, C., Sandwell, A., Park, S.S.: Hybrid copper-silver conductive tracks for enhanced oxidation resistance under flash light sintering. ACS Appl. Mater. Interf. **8**(34), 22369–22373 (2016)
2. Saleh, R., Barth, M., Eberhardt, W., Zimmermann, A.: Bending setups for reliability investigation of flexible electronics. Micromachines **12**, 78 (2021)
3. Karim, M.E., Munir, A.B.: Nanotechnology: sketching the next big thing in Malaysia context. Nanotechnol. Law Bus. **12**(2), 161–172 (2015)
4. Kim, D.-H., Xiao, J., Song, J., Huang, Y., Rogers, J.A.: Stretchable, curvilinear electronics based on inorganic materials. Adv. Mater. **22**(19), 2108–2124 (2010)
5. Sekitani, T., Someya, T.: Stretchable, large-area organic electronics. Adv. Mater. **22**(20), 2228–2246 (2010)
6. Saidina, D.S., Eawwiboonthanakit, N., Mariatti, M., Fontana, S., Hérold, C.: Recent development of graphene-based ink and other conductive material-based inks for flexible electronics. J. Electr. Mater. **48**(1), 3428–3450 (2019)
7. Chee, W.K., Lim, H.N., Zainal, Z., Huang, N.M., Harrison, I., Andou, Y.: Flexible graphene-based supercapacitors: a review. J. Phys. Chem. C **120**(8), 4153–4172 (2016)
8. Toda, K., Furue, R., Hayami, S.: Recent progress in applications of graphene oxide for gas sensing: a review. Anal. Chim. Acta **878**, 43–53 (2015)
9. Park, J.Y., Lee, W.J., Kwon, B.-S., Nam, S.-Y., Choa, S.-H.: Highly stretchable and conductive conductors based on Ag flakes and polyester composites. Microelectron. Eng. **199**, 16–23 (2018)
10. Potts, J.R., Dreyer, D.R., Bielawski, C.W., Ruoff, R.S.: Graphene-based polymer nanocomposites. Polymer **52**(1), 5–25 (2011)
11. Mohan, V.B., Lau, K.T., Hui, D., Bhattacharyya, D.: Graphene-based materials and their composites: A review on production, applications and product limitations. Compos. Part B Eng. **142**, 200–220 (2018)
12. Hg, P.K., Xavior, M.A.: Graphene reinforced metal matrix composite (GRMMC): a review. Proced. Eng. **97**, 1033–1040 (2014)

13. Stankovich, S., Dikin, D.A., Dommett, G.H., Kohlhaas, K.M., Zimney, E.J., Stach, E.A., Piner, R.D., Nguyen, S.T., Ruoff, R.S.: Graphene-based composite materials. Nature **442**(7100), 282 (2006)
14. Ji, X., Xu, Y., Zhang, W., Cui, L., Liu, J.: Review of functionalization, structure, and properties of graphene/polymer composite fibers. Compos. A Appl. Sci. Manuf. **87**, 29–45 (2016)
15. Allen, M.J., Tung, V.C., Kaner, R.B.: Honeycomb carbon: a review of graphene. Chem. Rev. **110**(1), 132–145 (2009)
16. Mittal, G., Dhand, V., Rhee, K.Y., Park, S.-J., Lee, W.R.: A review on carbon nanotubes and graphene as fillers in reinforced polymer nanocomposites. J. Ind. Eng. Chem. **21**, 11–25 (2015)
17. Shah, R., Kausar, A., Muhammad, B., Shah, S.: Progression from graphene and graphene oxide to high performance polymer-based nanocomposite: a review. Polym. Plast. Technol. Eng. **54**(2), 173–183 (2015)
18. Huang, Q., Zhu, Y.: Printing conductive nanomaterials for flexible and stretchable electronics: a review of materials, processes, and applications. Adv. Mater. Technol. **4**(5), 1–41 (2019)
19. Alemour, B., Badran, O., Hassan, M.R.: A review of using conductive composite materials in solving lightening strike and ice accumulation problems in aviation. J. Aerospace Technol. Manag. **11**(1), e1919 (2019)
20. Kim, J., Harris, N., Tian, J.: The Roles of Carbon Black in Wire and Cable Conductor, Insulation and Jacketing Applications. https://www.birlacarbon.com/the-roles-of-carbon-black-in-wire-cable-conductor-insulation-and-jacketing-applications/ (2021)
21. Farizhandi, A.A., Khalajabadi, S.Z., Krishnadoss, V., Gohari Derakhshandeh, P., Gohari Derakhshandeh, N.: Synthesized biocompatible and conductive ink for 3D printing of flexible electronics. J. Mech. Behav. Biomed. Mater. **110**, 103960 (2020)
22. Olabi, A.G., Abdelkareem, M.A., Wilberforce, T., Sayed, E.T.: Application of graphene in energy storage device: a review. Renew. Sustain. Energy Rev. **135**, 110026 (2021)
23. Novoselov, K.S., Geim, A.K., Morozov, S.V., Jiang, D.E., Zhang, Y., Dubonos, S.V., Grigorieva, I.V., Firsov, A.A.: Electric field effect in atomically thin carbon films. Science **306**(5696), 666–669 (2004)
24. Nogueira de Melo, C.C., Beatrice, C.A.G., Pessan, L.A., Dantas de Oliveira, A., Machado, F.M.: Analysis of nonisothermal crystallization kinetics of graphene oxide-reinforced polyamide 6 nanocomposites. Thermochim. Acta **667**, 111–121 (2018)
25. Novoselov, K.S., Fal′ ko, V.I., Colombo, L., Gellert, P.R., Schwab, M.G., Kim, K.: A roadmap for graphene. Nature 490(7419), 192–200 (2012)
26. Brown, C.: Graphene: the material that is changing the world. In: Everything You Need to Know About Graphene and Carbon Nanotubes [Blog post]. Medium. https://charlotte-innovate.medium.com/graphene-the-material-that-is-changing-the-world-fd32575bccd5 (2021)
27. Jana, N.R., Ray, S.C.: Graphene-based carbon nanoparticles for bioimaging applications. In: Ray, S. (ed.) Applications of Graphene and Graphene-Oxide Based Nanomaterials, pp. 57–84. Elsevier, Amsterdam (2015)
28. Zhang, Y., Pan, C.: Measurements of mechanical properties and number of layers of graphene from nano-indentation. Diam. Relat. Mater. **24**, 1–5 (2012)
29. Bagheri, R., Williams, M.A., Pearson, R.A.: Use of surface modified recycled rubber particles for toughening of epoxy polymers. Polym. Eng. Sci. **37**(2), 245–251 (1997)
30. Kinloch, A.J., Taylor, A.C.: Mechanical and fracture properties of epoxy/inorganic micro- and nanocomposites. J. Mater. Sci. Lett. **22**(20), 1439–1441 (2003)
31. Su, Y.-K.: Nitride-based LEDs and superluminescent LEDs. Compr. Semicond. Sci. Technol. **6**, 28–100 (2011)
32. Bantcev, D., Ganyushkin, D., Terekhov, A., Ekaykin, A., Tokarev, I., Chistyakov, K.: Isotopic composition of glacier ice and meltwater in the arid parts of the Altai Mountains (Central Asia). Water **14**(2), 252 (2022)
33. Holister, P., Weener, J.W., Roman, C., Harper, T.: Nanoparticles. Technology white papers, **3**, 1–11 (2003)
34. Liu, H., Wu, J., Xu, H., Zou, X.: Research on conductive trace ink design for screen printing. IEEE Access **9**, 58890–58899 (2021)

35. Alresheedi, M.T., Li, H., Zhu, Y., Alrashdi, A., Albalawi, A., Xu, L.: Simulation of inkjet printed spiral inductor for wearable electronics. AIP Adv. **10**(2), 025201 (2020)
36. Araki, T.L., Jackson, T.N.: Conductive ink patterns in PCBs for power electronics applications: a review. IEEE Trans. Industr. Electron. **65**(2), 1402–1410 (2018)
37. Larson, C., Peele, B., Li, S., Robinson, S., Totaro, M., Beccai, L., Wood, R.: Highly stretchable electroluminescent skin for optical signaling and tactile sensing. Science **351**(6277), 1071–1074 (2016)
38. Kim, B.K., Kim, C.J., Cho, D.I.: Tailoring conductive-adhesive interconnects for stretchable electronics. IEEE Trans. Comp. Pack. Manufact. Technol. **7**(4), 625–632 (2017)
39. Wang, Q., Wang, Z., Awasthi, M. K., Jiang, Y., Li, R., Ren, X., Zhang, Z. (2016). Evaluation of medical stone amendment for the reduction of nitrogen loss and bioavailability of heavy metals during pig manure composting. Bioresource Technol. **220**, 297–304
40. Kattumenu, R.C.: Flexography printing of silver based conductive inks on packaging substrates. Western Michigan University (2008)
41. Huang, J., Yang, X., Her, S.-C., Liang, Y.-M.: Carbon nanotube/graphene nanoplatelet hybrid film as a flexible multifunctional sensor. Sensors **19**(2), 317 (2019)
42. Wahid, A.K.A., Lim, Y.T., Cheong, J.Y., Lim, S.K.: Driving monitoring system application with stretchable conductive inks: a review. Int. J. Nanoelectr. Mater. **13**, 327–346 (2020)
43. Rajasekaran, S.: Introduction to dynamics. Struct. Dyn. Earthq. Eng. **21**, 1–6 (2009)
44. Maxwell: Introduction to Cyclic (Dynamic) Triaxial Testing. VJ Tech Limited. https://www.vjtech.co.uk/blog/introduction-to-cyclic-dynamic-triaxial-testing (2019)
45. TWI: What is Dynamic Loading? (A Definitive Guide). https://www.twi-global.com/technical-knowledge/faqs/what-is-dynamic-loading (2022)
46. Huang, L., Huang, Y., Liang, J., Wan, X., Chen, Y.: Graphene-based conducting inks for direct inkjet printing of flexible conductive patterns and their applications in electric circuits and chemical sensors. Nano Res. **4**(7), 675–684 (2011)
47. Mohammed, A.A.: Development of a New Stretchable and Screen Printable Conductive Ink (Doctoral dissertation, University of Maryland, College Park) (2017)
48. Lu, N., Wang, P.: Bio-integrated electronics. In: Proceedings of the 2014 IEEE International Conference on IC Design and Technology, pp. 1–5. IEEE (2014)
49. Lu, L., Han, X., Li, J., Hua, J., Ouyang, M.: A review on the key issues for lithium-ion battery management in electric vehicles. J. Power Sourc. **226**, 272–288 (2013)
50. Kraft, O., Schwaiger, R., Wellner, P.: Fatigue in thin films: lifetime and damage formation. Mater. Sci. Eng. A **319–321**, 919–923 (2001)
51. Horgan, C.O., Chan, A.M.: Torsion of functionally graded isotropic linearly elastic bars. J. Elast. **52**(2), 181–199 (1998)
52. Lee, Y.-C., Liu, T.-S.: Torsional analysis and measurement for flexible displays. J. Solid Mech. Mater. Eng. **5**(9), 476–483 (2011)
53. Soldatos, K.P.: A nonlinear transverse shear deformable plate theory allowing a multiple choice of trial displacement and stress distributions. Compos. Eng. **3**(9), 885–897 (1993)
54. Finn, M., Martens, C.J., Zaretski, A.V., Roth, B., Søndergaard, R.R., Krebs, F.C., Lipomi, D.J.: Mechanical stability of roll-to-roll printed solar cells under cyclic bending and torsion. Sol. Energy Mater. Sol. Cells **174**, 7–15 (2018)
55. Chen, H.-L., Luo, X.-M., Wang, D., Schaaf, P., Zhang, G.-P.: Achieving very high cycle fatigue performance of Au thin films for flexible electronic applications. J. Mater. Sci. Technol. **89**, 107–113 (2021)
56. Chow, J.H., Meth, J., Sitaraman, S.K.: Twist testing for flexible electronics. In: Proceedings of the 2019 IEEE 69th Electronic Components and Technology Conference (ECTC), pp. 245–251 (2019)
57. Cho, C.-K., Hwang, W.-J., Eun, K., Choa, S.-H., Na, S.-I., Kim, H.-K.: Mechanical flexibility of transparent PEDOT: PSS electrodes prepared by gravure printing for flexible organic solar cells. Sol. Energy Mater. Sol. Cells **95**, 3269–3275 (2011)

Chapter 3
Fabrication of Nanocomposite Conductive Inks

3.1 Introduction to Experimental Work

Flexible and wearable electronics have emerged as an essential area in contemporary technological advancement, driven by the demand for lightweight, low-cost, and space-efficient alternatives to traditional rigid systems. Among the most significant advantages offered by these emerging systems is their mechanical flexibility, an attribute that sets them apart from conventional electronics fabricated on printed circuit boards (PCBs) or thin films deposited on rigid glass substrates. This flexibility enables these systems to conform to complex surfaces, endure dynamic deformation, and function reliably under mechanical stresses such as bending, folding, stretching, and rolling.

The electrical and mechanical behaviours of flexible electronics have been the subject of extensive work, particularly with respect to the durability and stability of conductive materials under repeated mechanical loading. In recent years, conductive polymer composites and nanomaterial-based inks have formed the basis of many explorations aimed at enabling reliable electrical performance under dynamic conditions. Materials such as graphene nanoplatelets (GNPs), carbon nanotubes (CNTs), and metallic nanoparticles such as silver and copper, have shown promising potential due to their high electrical conductivity and mechanical robustness when embedded in flexible matrices. These systems are now regarded as foundational to the development of next-generation electronics for applications such as biomedical sensors, flexible displays, soft robotics, and energy harvesting devices.

This chapter outlines the experimental methodology employed in the present work, detailing the procedural flow and underlying principles. The work primarily investigates conductive ink previously developed in earlier work by Maizura et al. (2018), building upon this foundation to explore the effects of cyclic mechanical deformation on its electrical performance. The primary objective of this chapter is to examine how different trace geometries and widths influence the electrical and mechanical integrity of conductive inks when printed on flexible and stretchable

M. A. Salim et al., *Nanotechnology in Conductive Inks*, Engineering Materials,
https://doi.org/10.1007/978-981-95-5708-0_3

substrates. Specifically, the conductive ink comprising GNPs as the conductive filler and a polymeric binder system was formulated and printed into four type of pattern shapes with three different line widths. The choice of GNPs is justified by their exceptional electrical conductivity, high aspect ratio, and inherent flexibility, making them ideal candidates for forming percolative conductive networks in a polymeric matrix.

The ink formulation process commenced with the accurate weighing of the filler and binder components, adhering to predefined concentration ratios to ensure consistency and repeatability. The mixture was homogenised using a centrifugal mixing machine, a technique known for its ability to eliminate air bubbles and ensure uniform dispersion of nanomaterials. Following the mixing process, the microstructural distribution of the GNPs within the ink was assessed using image analysis tools. This step is critical for verifying the homogeneity of the conductive network, which directly affects the ink's electrical conductivity and mechanical reliability. Subsequently, the prepared conductive ink paste was deposited onto the flexible substrates using the stencil printing technique. This method, widely used in printed electronics, allows for precise control over the trace geometry and thickness, which are crucial parameters influencing the ink's performance under mechanical stress. Once printed, the samples were cured under controlled thermal conditions to facilitate solvent evaporation and binder cross-linking, thereby solidifying the printed structure.

By simulating the mechanical stresses encountered in real-applications, the printed samples were subjected to three types of cyclic loading which are bending, tensile, and torsional deformation. These mechanical tests were selected to represent a range of common dynamic actions that flexible electronics may experience during operation in various environments, ranging from wearable sensors exposed to repetitive body movements, to flexible circuitry embedded in soft robotics.

Cyclic bending involves repeated flexing of the sample along a neutral axis, imposing alternating tensile and compressive forces across the thickness of the trace. This test is essential for evaluating the ink's resistance to delamination and crack formation. Cyclic tensile loading subjects the sample to repeated elongation and contraction, testing the ink's capacity to accommodate strain without experiencing fracture or conductivity loss. Besides, cyclic torsion introduces shear stresses through twisting motions, thereby assessing the ink's shear modulus and interfacial adhesion to the substrate. Each loading type introduces different modes of mechanical stress, and the work seeks to determine how these stresses influence both the microstructural integrity of the conductive network and the electrical performance of the cured ink. Prior to loading, all samples were characterised to establish baseline measurements of resistance and mechanical strength. These properties were then re-evaluated after exposure to a predetermined number of loading cycles.

The experimental setup enabled rigorous and reproducible characterisation of the printed samples before and after cyclic testing. Electrical measurements focused on detecting changes in resistance, which may indicate micro-crack formation, loss of percolation pathways, or degradation of the filler–matrix interface. Mechanical observations included visual inspection under optical and electron microscopes to identify structural damage, as well as possible delamination or substrate failure.

Conducting the experiments under controlled laboratory conditions ensured environmental stability such as temperature and humidity, which is an important for minimising external variables that could influence the results. This systematic approach enhances the validity of the findings and supports their application to broader engineering challenges.

Understanding the behaviour of conductive inks under cyclic mechanical stress has direct implications for the reliability and longevity of flexible electronic systems in real-applications. Materials and devices used in aerospace, automotive, renewable energy, and healthcare sectors must endure prolonged exposure to mechanical fatigue without compromising performance. The findings from this work contribute to the fundamental understanding of how mechanical loading affects nanocomposite-based conductive inks, thereby informing better design strategies for flexible and stretchable electronics.

Future investigations may expand upon this work by exploring alternative nanomaterial fillers, varying polymer matrices, or introducing multi-layer structures. Additionally, advanced simulation techniques and in-situ monitoring methods could further elucidate failure mechanisms at the micro- and nanoscale. Identifying such mechanisms is essential for advancing predictive models and enhancing material formulations.

3.2 Materials Selection and Description

This section discusses the key materials utilised in this investigation, with each material selected based on its intrinsic properties and its functional role in the development of flexible and stretchable conductive ink systems. The primary materials include GNPs, carbon black (CB), a polymer binder (epoxy resin), a curing agent (polyetheramine), and a selection of flexible substrates. A comprehensive understanding of each material's characteristics is essential, as these properties directly influence the electrical performance, mechanical reliability, and processability of the resulting composite system.

The GNPs in flake powder form, were used as one of the principal conductive fillers in this work. These nanoplatelets possessed an average lateral dimension of approximately 5 µm. The GNPs are widely recognised for their unique two-dimensional structure, which consists of several layers of graphene stacked into thin, platelet-like shapes. This morphology offers a high aspect ratio, large surface area, and exceptional electron mobility, resulting in enhanced electrical conductivity when incorporated into polymer matrices. Beyond electrical performance, GNPs also impart improved thermal conductivity, gas barrier properties, and mechanical reinforcement to the host material. Compared to other carbon-based nanostructures such as carbon nanofibres (CNFs) and CNTs, GNPs offer advantages in terms of safer handling, lower cytotoxicity, and reduced chemical reactivity, particularly when compared to graphene oxide. These characteristics make GNPs a promising choice for use in polymeric

Table 3.1 Details description of GNPs

Description	
Particle size	5 μm
Surface area	500 m^2/g
Density	0.03–0.1 g/cm^3
Appearance	Powder
Color	Black

systems intended for flexible and wearable electronics, where stability and processability are critical. The use of GNPs in this investigation serves to create a percolated conductive network within the ink, enabling charge transport even under mechanical deformation. Table 3.1 shows the details description of GNPs used in this work.

The CB was employed as a secondary conductive filler. It is a widely used industrial carbon material produced via thermal-oxidative decomposition of hydrocarbon feedstock's, typically through the furnace black process. This method yields fine, spherical carbon particles with a high surface area and structure, essential characteristics for forming conductive pathways in composite systems. The selection of CB was based on several compelling factors. First, it is a cost-effective alternative to nanocarbon materials while still offering respectable conductivity. Second, it exhibits excellent dispersibility in aqueous and organic media, which facilitates homogeneous ink formulation. Third, it is inherently non-toxic, which is advantageous for applications requiring biocompatibility or human interface.

Due to its high specific surface area and aggregate structure, this material enables the formation of an effective conductive network at relatively low filler loadings, thereby preserving the mechanical flexibility of the host polymer. Importantly, it has a relatively low impact on the mechanical integrity of the matrix, allowing the composite to retain its elasticity and conformability, traits essential for stretchable electronic systems. Table 3.2 outlines the key physicochemical properties of the materials as provided in the manufacturer's datasheet.

The polymeric binder used in this study is Araldite® EP 300 A. In conductive ink formulations, the binder plays a crucial role in maintaining cohesion among filler particles, ensuring adhesion to the substrate, and preventing defects such as cracking or delamination during or after processing. Epoxy-based thermosetting resins are particularly favoured in printed electronics due to their excellent thermal stability,

Table 3.2 Details description of carbon black

Description	
Particle size	30–90 nm
Surface area	500 m^2/g
Density	1.16 g/cm^3
Appearance	Powder
Color	Black

Table 3.3 Details description of epoxy

Description	
Maximum temperature	204 °C
Density	1.3 g/cm^3
Appearance	Liquid
Color	Off-white

strong adhesion, and ability to form durable cross-linked networks upon curing. In contrast to thermoplastic binders, which soften and deform upon heating, thermoset epoxies provide permanent structural integrity, making them ideal for applications requiring long-term mechanical stability. During the curing phase, reactive epoxide groups in the resin react with amine hardeners to form a highly cross-linked, three-dimensional matrix.

Previous works such as Wang et al. [1] have demonstrated the adaptability of epoxy systems in flexible electronics, particularly through modifications that enhance flexibility and enable low-temperature or room-temperature curing. These developments are instrumental in preventing thermal degradation of underlying flexible substrates during processing. Table 3.3 summarises the essential characteristics of the epoxy resin based on the product datasheet.

The Jeffamine® D-230, a polyetheramine curing agent was selected for its favourable reactivity, low viscosity, and its ability to produce mechanically resilient cured networks. This amine-based compound acts as a hardener by reacting with the epoxide functional groups of the epoxy binder, initiating the cross-linking reaction that results in a thermoset structure.

Polyetheramines such as Jeffamine D-230 are preferred in flexible electronic systems because they yield coatings and adhesives with excellent toughness, impact resistance, and adhesion. Their low vapour pressure and colourless nature make them compatible with aesthetic and environmentally sensitive applications. Furthermore, the flexible ether linkages within the polyetheramine backbone contribute to the elasticity and flexibility of the final composite. Table 3.4 presents the physical and chemical properties of Jeffamine D-230, which influence not only the curing kinetics but also the mechanical performance and environmental resistance of the ink system.

The choice of substrate plays a critical role in determining the mechanical compatibility, adhesion behaviour, and electrical testing accuracy of conductive inks. In this

Table 3.4 Details description of curing agent

Description	
Flash point	121 °C
Density	0.948–0.999 g/cm^3
Appearance	Liquid
Color	Colorless to pale yellow liquid with slight haze

chapter, three types of substrates were utilised which are glass, polyethylene terephthalate (PET), and thermoplastic polyurethane (TPU), where each offering distinct benefits.

Microscope glass slides with dimensions of 2.54 × 7.62 cm and a thickness ranging from 0.10 to 0.12 cm were employed during the early stages of the investigation. Besides, glass substrates provide a stable and inert base that resists deformation under thermal or mechanical stress. Their excellent insulating properties and dimensional stability allow for accurate baseline electrical characterisation. However, due to their rigidity, glass substrates serve primarily as a control platform rather than as a candidate for stretchable applications.

The PET was selected for its widespread use in flexible electronic applications, owing to its favourable balance of mechanical strength, thermal resistance, and optical transparency. Chemically, it is a polyester derived from the polymerisation of ethylene terephthalate units, yielding a structure that resists mechanical deformation while maintaining flexibility. Its surface morphology and compatibility with various coating techniques make it suitable for large-area, roll-to-roll printed electronics. The crystallinity of this material, particularly when aligned, enhances tensile properties while maintaining light weight and processability.

A TPU represents a high-performance substrate material for stretchable electronics. Comprising soft and hard segment domains, derived from polyols and isocyanates, a TPU combines flexibility, toughness, and excellent abrasion resistance. Its elasticity remains stable across a broad range of hardness grades, making it especially well-suited to applications where frequent mechanical deformation occurs. Furthermore, its high surface energy promotes strong adhesion with printed inks, enhancing reliability under cyclic strain. Table 3.6 provides the material specifications of TPU used in this work.

3.3 Formulation and Preparation of Conductive Ink

This section outlines the procedural workflow undertaken from the initial formulation of the conductive ink to the completion of mechanical and electrical testing. The successful development of high-performance conductive inks, particularly those incorporating graphene GNPs, relies on careful attention to material formulation, dispersion, patterning, and curing. Each step in the process plays an important role in determining the final properties of the printed conductive traces, including their electrical conductivity, mechanical durability, and surface integrity. By producing high-quality samples for investigating the behaviour and potential of graphene-based inks in flexible electronics, the experimental methodology followed three primary stages which are formulation and mixing, printing, and curing. These stages are essential to ensure a homogeneous distribution of conductive fillers, the formation of robust conductive networks, and good interfacial bonding with the substrate.

The formulation phase serves as the foundation for developing a conductive ink capable of supporting reliable electrical performance under mechanical deformation.

The aim is to produce a composite ink where the conductive filler concentration meets or exceeds the percolation threshold where it is an essential requirement for achieving continuous pathways for electron transport. In this chapter, two types of conductive fillers were employed which are GNPs and CB. These were individually mixed with a polymeric matrix consisting of Araldite® epoxy resin and Jeffamine® D-230 polyetheramine hardener. The combination of filler and polymer binder must be optimised not only to attain the desired electrical conductivity but also to preserve the mechanical flexibility and processability of the ink.

A well-balanced formulation ensures that the filler particles are sufficiently interconnected to allow electron movement while being adequately encapsulated by the binder for structural integrity. The binder not only serves to hold the fillers in place but also enhances adhesion to the substrate and mitigates cracking during and after the curing stage. An improperly formulated ink, whether due to insufficient dispersion, excessive filler loading, or improper ratios between components, can result in defects such as lumpiness, agglomeration, or phase separation. Such issues compromise both printability and final performance. For example, poor mixing can result in agglomerated clusters of filler material, as shown in Fig. 3.1, which inhibits uniform conductivity and leads to mechanical failure under stress.

The hardener was loaded at 30 wt.% relative to the binder (epoxy resin), a ratio found to be effective in achieving full curing and optimal mechanical behaviour. The epoxy-hardener system must be precisely balanced to facilitate cross-linking reactions that form a robust, thermoset matrix upon curing. Any deviation from this ratio can hinder the curing process, resulting in weak interfacial adhesion or brittle traces.

Table 3.5 outlines the composition and proportions of each component in the baseline formulations for both GNPs and CB inks. These formulations were derived based on prior optimisation explorations to ensure electrical percolation and mechanical compatibility with flexible substrates. The formulation calculations accounted for the total weight of each component, including the filler, epoxy binder, and hardener,

Fig. 3.1 Condition of ink when mixture when loading concentration

Table 3.5 Details list of the main components used

Materials	Descriptions
GNP powder	As conductive fillers, they are added to polymer resin to make it more conductive, allowing electricity and electrons to move more easily through it
CB powder	As conductive fillers, they are added to polymer resin to make it more conductive, allowing electricity and electrons to move more easily through it
Araldite epoxy resin	As a binder. Binder elements are used to hold the conductive filler particles together and act as an influencing agent in the conductive ink synthesis. So, the formation of a polymer film on the surface strengthens the bonding element
Huntsman Jeffamine D230 polytheramine	As a hardener to complete the curing process through cross-linking, and as additives to adjust the processability or functional properties. These provide improved adhesion and drying ease for all substrates

Table 3.6 Formulation composition for GNPs and carbon black conductive ink

No.	Filler		Binder		Hardener, 10% (g)	Total (g)
	(%)	(g)	(%)	(g)		
1	10	0.4	90	3.6	0.36	4.36
2	15	0.6	85	3.4	0.34	4.34
3	20	0.8	80	3.2	0.32	4.32
4	25	1.0	75	3.0	0.30	4.30
5	30	1.2	70	2.8	0.28	4.28
6	35	1.4	65	2.6	0.26	4.26
7	40	1.6	60	2.4	0.24	4.24

to ensure repeatability and consistency in sample preparation. The theoretical underpinning of this process lies in composite material science. When conductive particles are dispersed in an insulating matrix, they begin to form a percolating network once the filler content surpasses a critical volume fraction. This network enables charge carriers to move through the composite, converting it from an insulator to a conductor. However, excessive filler loading can increase viscosity and reduce printability, while insufficient loading leads to poor conductivity. Therefore, precise control of formulation parameters is fundamental to optimising ink performance for printed electronics. Table 3.6 shows the formulation composition of the formulation.

3.4 Sample Preparation and Mixing Procedure

One of the main functions of conductive ink in flexible and printed electronics is to establish effective electrical interconnections between electronic components. Therefore, validating the electrical performance of the prepared conductive ink is a critical step in this work. This validation requires the ink to be formulated, deposited, and cured on suitable substrates before undergoing electrical and mechanical characterisation. Proper sample preparation ensures that the conductive pathways reflect the material's intrinsic performance, free from external artefacts or substrate-related interferences.

By eliminating undesirable effects on electrical readings, all conductive ink samples were prepared on substrates with excellent insulating properties. This ensured that the electrical performance measured was attributable solely to the conductive ink and not influenced by underlying substrate conductivity. Two different sample preparation routes were followed in this investigation which are one involving the deposition of conductive ink on a rigid, inert glass substrate as a control reference, and another involving patterned traces on flexible and stretchable substrates, specifically PET and TPU. These two scenarios allowed for both baseline characterisation and application-relevant testing under flexible conditions.

For the basic or control formulation, a microscope glass slide served as the substrate. The glass slides used in this work had dimensions of 25.4 by 76.2 mm, with a thickness ranging from 1.0 to 1.2 mm. Glass is a commonly used substrate in electronics application due to its excellent thermal stability, chemical inertness, and insulating properties, all of which are crucial for ensuring accurate and repeatable electrical measurements. Furthermore, glass provides a smooth, uniform surface ideal for film deposition and thickness control.

By defining the printing area and ensure consistency in sample dimensions, high-temperature-resistant tape, such as Scotch tape was affixed along the edges of the glass slide. This not only acted as a mask during printing but also allowed control over the ink thickness, providing clear boundaries for conductive trace deposition. As illustrated in Fig. 3.2, the resulting printed samples were 25.4 mm in length, 3 mm in width, and 0.1 mm in thickness. By ensuring the reproducibility of results, electrical readings were taken at three different points on each sample.

The deposition was carried out using the direct screen-printing technique combined with a doctor-blade method. This method is widely adopted in printed electronics due to its simplicity, controllability, and cost-effectiveness. The doctor blade is used to uniformly spread the conductive ink paste across the masked region of the glass slide, ensuring consistent layer thickness. This printing method enables

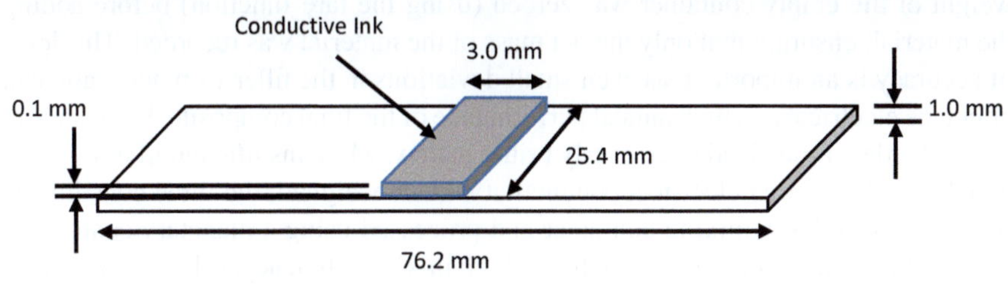

Fig. 3.2 Diagram of conductive ink on glass slide

fine control of line width and thickness, essential for evaluating ink rheology, print fidelity, and film uniformity.

In addition to the baseline glass samples, it was also necessary to fabricate patterned conductive traces on flexible and stretchable polymer substrates, which more closely resemble real-applications of printed electronics. Once the formulation was optimised for stretchability and flexibility, it was used to create patterned traces on PET and TPU substrates. These materials were selected due to their favourable balance of mechanical flexibility, surface processability, and affordability.

Basically, PET, a widely used thermoplastic polymer, is known for its good dimensional stability, high tensile strength, and transparency, making it an excellent choice for flexible electronics. The TPU, on the other hand, offers superior elasticity, abrasion resistance, and toughness, making it ideal for applications requiring mechanical resilience and stretchability. For this purpose, PET and TPU sheets with dimensions of 120.0×40.0 mm and thicknesses of 0.1 and 0.05 mm respectively were prepared. A patterned stencil was placed on each substrate to guide the shape and size of the conductive ink trace, as shown in Fig. 3.3. The stencil provided precise control over the trace geometry during deposition. The final printed traces had lengths of 90.0 mm, widths varying between 1.0 and 3.0 mm, and thicknesses of approximately 1.0 mm. Similar to the glass samples, electrical measurements were taken at three points along each trace to ensure measurement consistency and statistical reliability.

Again, the direct screen-printing method was used to apply the ink through the stencil openings, ensuring accurate and repeatable patterns. The printing process on polymer substrates had to account for surface energy and adhesion properties, which affect ink wetting and spreading. Therefore, substrate surface cleanliness and ink viscosity were carefully controlled to avoid defects such as smudging or incomplete prints.

The first step in preparing conductive ink samples involved precise weighing of all component materials. The main constituents included the conductive filler (either GNPS or CB), Araldite® EP 300 A epoxy resin as the polymer binder, and Jeffamine D-230 polyetheramine as the curing agent. For accurate and consistent formulation, a digital analytical balance was used. This instrument provided high-precision weight readings, essential for formulating small-scale batches with strict control over component ratios.

Each material, whether in powder or liquid form, was weighed individually. The weight of the empty container was zeroed (using the tare function) before adding the material, ensuring that only the net mass of the material was recorded. This level of accuracy is an important, as even small deviations in the filler-to-binder ratio can affect the electrical and mechanical performance of the final composite. For instance, excess hardener may lead to an overly brittle matrix, while insufficient filler may fail to achieve the desired electrical conductivity. Once weighed, the three constituents were transferred to a mixing container and processed using a planetary centrifugal mixer. This type of mixer is known for its high mixing efficiency, achieved by simultaneously rotating and revolving the container. Such motion eliminates air bubbles, enhances filler dispersion, and ensures homogeneous mixing without introducing shear-induced degradation that might occur in conventional mixers.

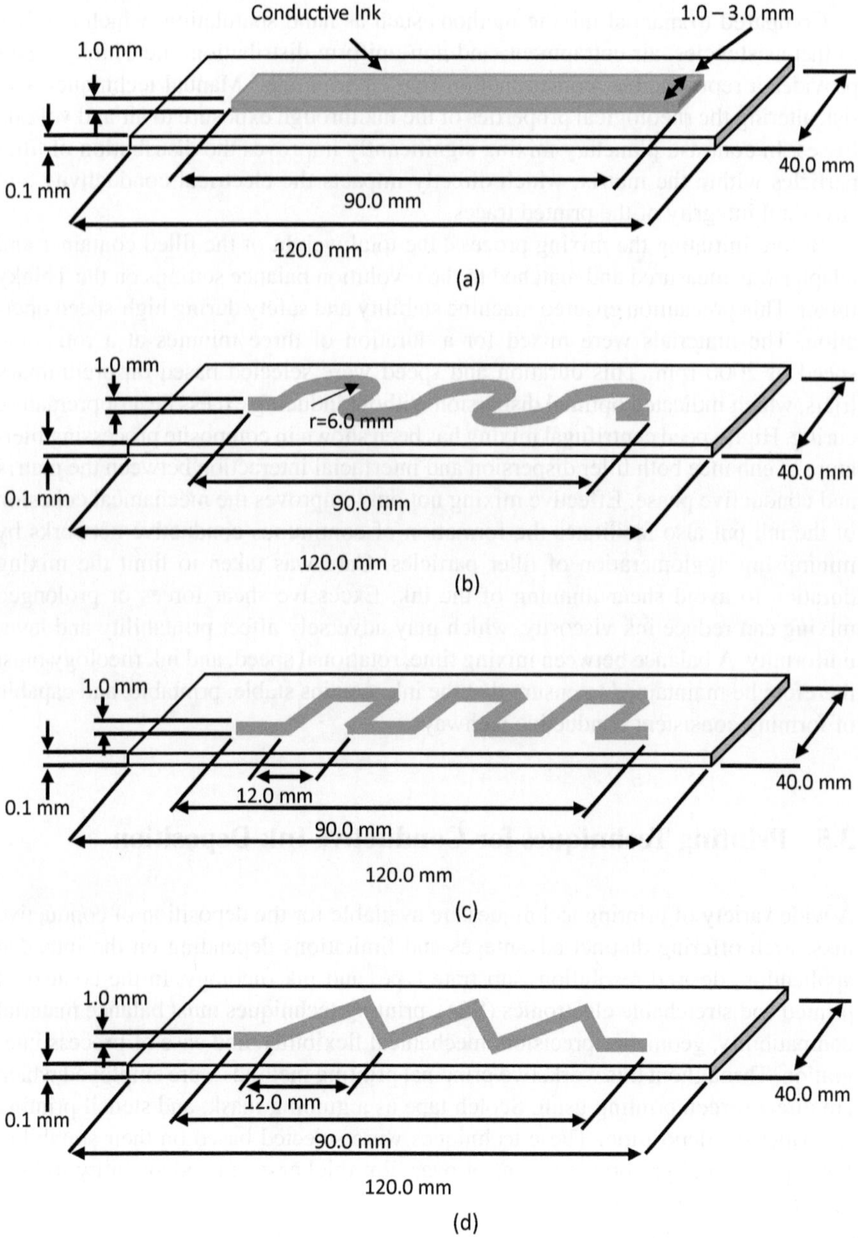

Fig. 3.3 Diagram of patterned conductive ink on flexible substrate; **a** straight-line pattern, **b** curve pattern, **c** square pattern, and **d** zigzag pattern

Compared to manual mixing methods such as hand spatulation, which are flats to inconsistencies, air entrapment, and non-uniform distribution, the Thinky mixer provides a reproducible, contamination-free environment. Manual techniques also risk altering the rheological properties of the ink through exposure to air and volatile losses. In contrast, planetary mixing significantly improves the distribution of filler particles within the matrix, which directly impacts the electrical conductivity and structural integrity of the printed traces.

Before initiating the mixing process, the total weight of the filled container and adapter was measured and matched to the revolution balance settings on the Thinky mixer. This precaution ensured machine stability and safety during high-speed operation. The materials were mixed for a duration of three minutes at a rotational speed of 2000 rpm. This duration and speed were selected based on preliminary trials, which indicated optimal dispersion without inducing excess heat or premature curing. High-speed centrifugal mixing has been shown in composite processing literature to enhance both filler dispersion and interfacial interaction between the matrix and conductive phase. Effective mixing not only improves the mechanical cohesion of the ink but also facilitates the formation of continuous conductive networks by minimising agglomeration of filler particles. Care was taken to limit the mixing duration to avoid shear thinning of the ink. Excessive shear forces or prolonged mixing can reduce ink viscosity, which may adversely affect printability and layer uniformity. A balance between mixing time, rotational speed, and ink rheology must therefore be maintained to ensure that the ink remains stable, printable, and capable of forming consistent conductive pathways.

3.5 Printing Techniques for Conductive Ink Deposition

A wide variety of printing techniques are available for the deposition of conductive inks, each offering distinct advantages and limitations depending on the intended application, desired resolution, substrate type, and ink rheology. In the context of printed and stretchable electronics (SEs), printing techniques must balance material compatibility, geometric precision, mechanical flexibility, and ease of process integration. Throughout this work, two principal printing methods were employed which are direct screen printing using Scotch tape as a guiding mask, and stencil printing for patterned deposition. These techniques were selected based on their suitability for experimental prototyping, control over film thickness, and adaptability to both rigid and flexible substrates.

Direct screen printing using Scotch tape is a traditional yet effective method for depositing conductive films, particularly in laboratory-scale studies. In this technique, Scotch tape is applied to the surface of a rigid substrate, such as a glass slide to delineate the boundaries of the printed area. The tape acts as a physical mask, defining the length and width of the conductive ink trace and providing a simple method for film thickness control. A doctor blade is employed to spread the conductive ink across the masked area. This method involves dragging a blade, typically made of stainless steel or similar rigid material across the surface at a fixed height above the substrate. The ink is deposited in a controlled layer, constrained in height by the tape and in spread by surface tension and viscosity. The doctor blade technique is widely used in both investigation and industrial settings for the preparation of thin films with precise and reproducible thickness.

One of the theoretical benefits of this method is its capability to control the wet film thickness, which is essential in achieving uniform conductivity across the printed area. While solution losses are generally estimated to be around 5% under ideal conditions, practical outcomes vary based on ink rheology, substrate flatness, and blade speed. The simplicity of the method is offset by the requirement for iterative optimisation, especially when balancing film uniformity with minimal material waste. Despite its relative simplicity, this method is particularly useful when working with rigid substrates where precision is more easily maintained. However, for applications requiring greater mechanical compliance or patterned features, alternative printing strategies such as stencil printing are more appropriate.

For applications demanding precise patterning, thickness control, and mechanical flexibility, such as wearable electronics or stretchable sensors, the stencil printing offers distinct advantages. In this method, a stencil with pre-defined apertures is placed directly in contact with the substrate surface. Conductive ink is then deposited through the stencil openings to form the desired pattern. Stencil printing is well-suited for printing on polymer substrates such as PET and TPU, both of which were used in this work. These materials are commonly chosen for their favourable mechanical properties, where PET for its dimensional stability and low cost, and TPU for its elasticity and abrasion resistance. Prior to printing, all substrates underwent surface treatment with acetone, a solvent known for its efficacy in removing organic contaminants, dust, and residual oils. Surface preparation is a critical step in the printing process. The presence of any surface contaminants can negatively affect ink adhesion, spread, and uniformity. Acetone, as a polar aprotic solvent, effectively cleans both polymeric and glass surfaces, promoting better ink-substrate interactions and improved film stability.

As illustrated in Fig. 3.4a, the printing process begins by applying a measured quantity of the conductive ink paste onto the edge of the stencil. A blade or squeegee is then used to sweep the ink across the stencil, as shown in Fig. 3.4b, forcing the ink into the stencil apertures through mechanical shear. The action of the squeegee both fills the stencil gaps and levels the ink surface. After the entire patterned area is filled, the stencil is carefully lifted from the substrate, leaving behind the printed pattern, as depicted in Fig. 3.4c.

This method enables the fabrication of thicker films compared to doctor blade printing, which can enhance durability and electrical robustness in flexible applications. Moreover, stencil printing offers greater repeatability for complex geometries and is more amenable to multilayer deposition strategies, often required in advanced flexible circuits. During the printing process, several mechanical factors influence the quality of ink deposition and pattern fidelity. As the blade or squeegee moves across the stencil, the rolling resistance of the conductive ink paste produces an upward force against the tool. This lift force is governed by multiple variables, including the

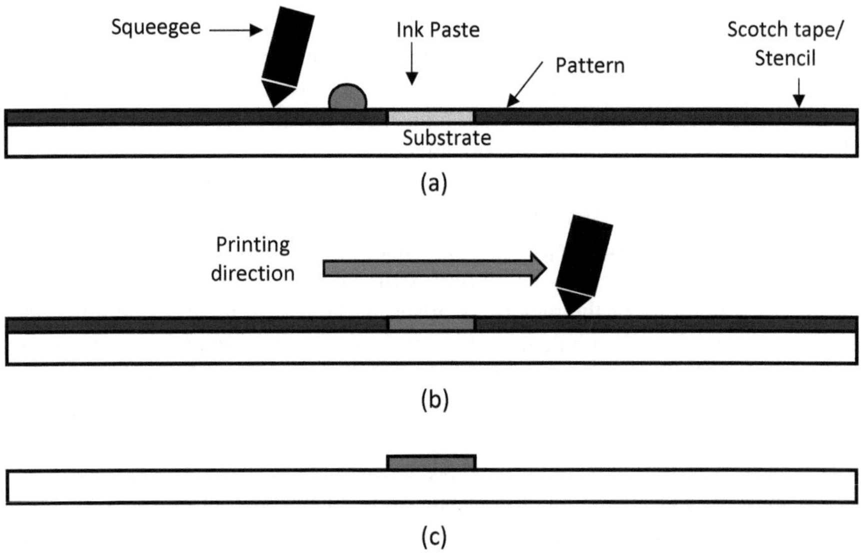

Fig. 3.4 Stencil screen-printing for conductive ink

viscosity of the ink, the speed of the blade movement, and the application angle of the blade or squeegee.

If the downward pressure applied by the blade is insufficient, the ink may not adequately fill the stencil apertures, resulting in incomplete traces or voids. Conversely, excessive pressure or slow movement can force the blade into the stencil gaps, scooping out ink from the filled regions. This can lead to defects such as bleeding, where the ink spreads beyond the intended pattern, or bridging, where neighbouring traces unintentionally connect due to uncontrolled ink spread, as demonstrated in Fig. 3.5.

Therefore, achieving an optimal balance between blade pressure, angle, and speed is essential for high-resolution printing. Empirical findings and prior literature suggest that printing along the length of the stencil aperture, rather than across its width, and can reduce the risk of scooping, and improve paste release behaviour. This is particularly critical for narrow or high-aspect-ratio features, where edge definition plays a crucial role in determining circuit performance. Furthermore, the rheological behaviour of the conductive ink plays a foundational role in printing performance. The ink must exhibit thixotropic properties, allowing it to flow under shear during printing and quickly regain viscosity to maintain shape post-deposition. This shear-thinning behaviour ensures proper aperture filling without slumping or spreading once the shear force is removed.

The selection and implementation of a suitable printing method depends on a complex interplay of substrate characteristics, ink rheology, geometric requirements, and mechanical parameters. While doctor blade printing offers simplicity

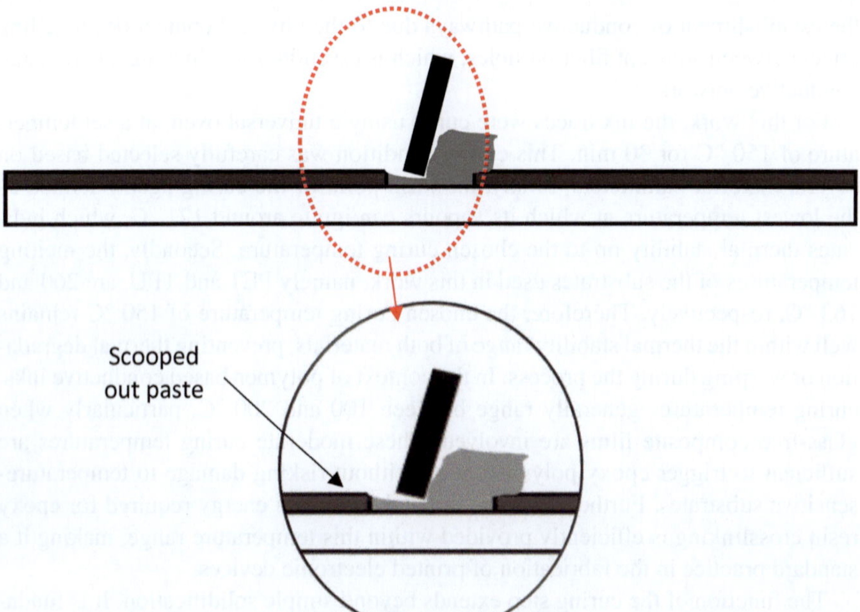

Fig. 3.5 Schematic view of print scavenging or scooped out conductive ink paste

and rapid prototyping advantages, stencil printing provides superior pattern resolution and is more compatible with the mechanical demands of stretchable and wearable electronics. In both cases, careful process optimisation is required to ensure reproducible, high-quality conductive traces suitable for subsequent testing and application deployment.

3.6 Curing Process

The curing process plays an important role in determining the final electrical, mechanical, and structural properties of conductive ink films. It involves two principal mechanisms which are the evaporation of volatile components, such as the curing agent or solvent, and the polymerisation of the resin matrix. These reactions together facilitate the solidification and structural integrity of the ink layer on the substrate. During curing, the evaporation of low molecular weight species reduces the overall volume of the ink film. Concurrently, the thermally activated polymerisation of the epoxy resin occurs, which leads to crosslinking and solidification of the polymer matrix. As the binder component transitions from a liquid or semi-liquid state to a hardened polymer network, the film thickness decreases. This shrinkage causes the conductive filler particles, such as GNPs or CB, to come into closer proximity. The result is the establishment of conductive pathways due to the physical contact or tunnelling effect between adjacent filler particles, which is essential for achieving a percolated conductive network.

For this work, the ink traces were cured using a universal oven, at a set temperature of 150 °C for 30 min. This curing condition was carefully selected based on several material parameters. Firstly, the flash point of the curing agent, defined as the lowest temperature at which its vapours can ignite around 121 °C, which indicates thermal stability up to the chosen curing temperature. Secondly, the melting temperatures of the substrates used in this work, namely PET and TPU, are 260 and 163 °C, respectively. Therefore, the chosen curing temperature of 150 °C remains well within the thermal stability range of both materials, preventing thermal degradation or warping during the process. In the context of polymer-based conductive inks, curing temperatures generally range between 100 and 200 °C, particularly when glass-free composite films are involved. These moderate curing temperatures are sufficient to trigger epoxy polymerisation without risking damage to temperature-sensitive substrates. Furthermore, the thermal activation energy required for epoxy resin crosslinking is efficiently provided within this temperature range, making it a standard practice in the fabrication of printed electronic devices.

The function of the curing step extends beyond simple solidification. It is fundamentally associated with microstructural evolution, which in turn influences both mechanical resilience and electrical conductivity. For example, incomplete curing may leave unreacted monomers or insufficient crosslink density, leading to poor mechanical adhesion and lower conductivity due to irregular filler dispersion. Conversely, excessive curing, particularly at elevated durations give a result in embrittlement of the filler network and micro-crack formation within the polymer matrix, which degrades performance.

By determining the optimal curing period, the investigators systematically investigated the influence of curing time on the electrical resistance of the conductive ink films. The curing durations examined ranged from 10 to 50 min. It was observed that a curing period of 30 min consistently yielded the lowest electrical resistance and the smallest standard deviation, indicating enhanced electrical stability and reproducibility. Extended curing durations beyond this threshold resulted in signs of thermal degradation, particularly in samples with high graphene filler content, where the ink matrix became increasingly brittle. This degradation adversely affected the mechanical stability of the conductive paths and led to inconsistent conductivity measurements.

Following the curing cycle, the conductive ink samples were not removed immediately from the oven. Instead, they were allowed to cool gradually to room temperature within the controlled environment of the oven. This deliberate and gradual cooling process, known as annealing heat treatment, is critical for relieving residual thermal stresses accumulated during rapid heating and crosslinking. Annealing enables molecular rearrangement within the polymer chains, promoting a more uniform and ductile microstructure. This results in improved mechanical flexibility and reduces the likelihood of micro-crack formation or delamination, which are common failure mechanisms in printed electronics.

In polymer composites, annealing is also known to stabilise the conductive network by allowing filler particles to settle into energetically favourable configurations, thereby enhancing long-term conductivity and material robustness. It provides a crucial post-curing step, particularly in flexible and stretchable electronics, where repeated mechanical deformation may expose any microstructural defects induced during rapid cooling or excessive crosslinking. The curing process is an indispensable stage in the fabrication of conductive ink films. The selected temperature of 150 °C and curing duration of 30 min represent a well-balanced compromise between complete polymerisation, optimal filler connectivity, substrate compatibility, and long-term performance. Incorporating an annealing step further ensures that the cured films attain the mechanical resilience and electrical reliability required for advanced flexible electronic applications.

3.7 Design and Preparation of Stencil Patterns

As illustrated in Fig. 3.6, four distinct stencil patterns were developed for this work to explore the influence of printed geometry on the electrical and mechanical performance of conductive ink films. The patterns consisted of one straight-line configuration and three meandering geometries which are curved, square-wave, and zigzag. These configurations were selected to simulate realistic circuit routing patterns commonly found in flexible and stretchable electronics. Meandering structures are frequently employed to accommodate mechanical deformation, enabling stretchability and mitigating strain concentration under bending, folding, or stretching conditions.

Each pattern was fabricated with three different line widths such as 1, 2, and 3 mm. This range was deliberately chosen to investigate the effect of printed trace width and hence the volume of deposited ink on the electrical resistivity of the resulting conductive paths. In particular, variations in line will influence the cross-sectional area of the conductive film, which directly affects current-carrying capacity and resistance. Wider traces generally accommodate higher volumes of conductive ink, promoting more uniform filler dispersion and potentially leading to lower resistivity due to increased conductive pathways. Interestingly, it was observed that printed traces with widths exceeding 2 mm tend to reduce the impact of the stencil thickness on the final height of the printed film. This phenomenon can be attributed to the more favourable paste release and levelling behaviour that occurs when the aperture width allows for more even shear distribution during squeegeeing. Wider also mitigate edge effects and promote consistent ink flow, resulting in more homogenous film thickness and improved pattern fidelity.

Furthermore, all patterns were designed with a target thickness of 1 mm, aligning with the characteristics of thick-film printing techniques such as screen printing, stencil printing, and, to a lesser extent, inkjet printing. While most inkjet printing applications produce thinner films, typically in the micron range, stencil and screen printing are widely used in producing films with thicknesses ranging from several

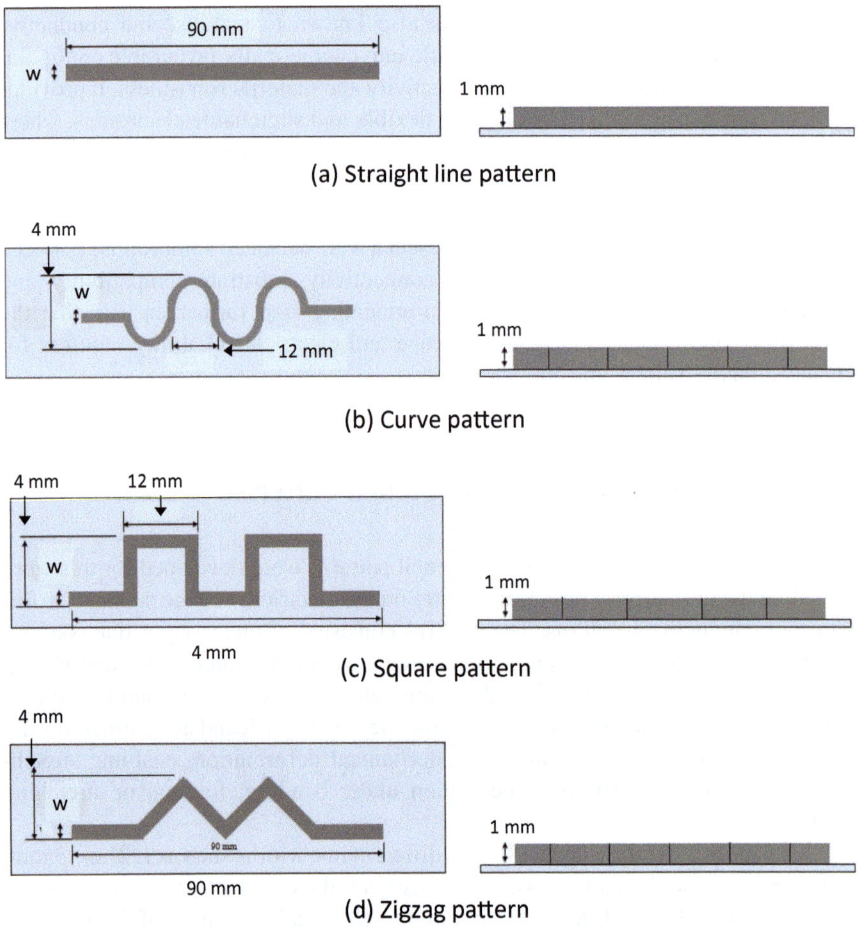

Fig. 3.6 Design of stencil with four different patterns, **a** straight line, **b** curve, **c** square, and **d** zig–zag patterns

tens of microns up to 1 mm. These thick-film approaches are critical in high-power or structural electronics, where increased mechanical robustness, thermal dissipation, and enhanced conductivity are required. The design process was executed using computer-aided design (CAD) software, and the completed pattern layouts were saved in the. igs (Initial Graphics Exchange Specification) format. This standardised file format allows for compatibility with a wide range of precision manufacturing tools, including laser cutters. The CAD files were then directly transferred to the laser cutting machine, facilitating the fabrication of stencil sheets with high dimensional accuracy and repeatability.

Laser cutting presents several distinct advantages in the fabrication of stencils for printed electronics. Unlike traditional stencil manufacturing methods that may involve multiple steps or masks, laser cutting offers a direct-to-tooling approach.

The cutting process is governed by the aperture data, which is digitally adjusted to account for the graze width, the width of material removed by the laser beam. This ensures that the final aperture dimensions are precisely compensated to reflect the original design specifications. As a result, the stencils produced exhibit minimal deviation from the intended pattern, which is essential for maintaining geometric fidelity during the ink deposition process.

An important consideration in laser cutting is the cutting direction. In this work, the laser beam was applied from the underside of the stencil material. This approach is strategic to ensures that any tapering introduced during the ablation process, a common side effect of laser cutting, occurs outward, away from the stencil–substrate interface. This tapering facilitates a cleaner separation of the composite paste from the stencil upon removal, enhancing paste release properties and minimising the risk of edge defects or smudging during the printing process. This detail, while seemingly minor, is a critical parameter in ensuring the accuracy and reproducibility of printed traces, especially when working with viscous composite inks or finely patterned geometries.

Therefore, the design and fabrication of these stencil patterns serve not only as a platform for studying electrical resistivity across different geometries and thicknesses but also demonstrate the importance of precision manufacturing in ensuring high-quality printed electronics. By carefully controlling pattern dimensions, stencil material properties, and manufacturing processes such as laser cutting, researchers and engineers can optimise both the performance and reliability of conductive ink-based devices.

Reference

1. Wang, X., Ma, B., Chen, S., Wei, K., Kang, X.: Properties of epoxy-resin binders and feasibility of their application in pavement mixtures. Constr. Build. Mater. **295**, 123531 (2021)

Chapter 4
Performance Evaluation of Printed Conductive Films

4.1 Introduction

As illustrated in Fig. 4.1, the completed samples of the patterned conductive ink traces were subjected to both static and cyclic mechanical and electrical characterisation to evaluate their fundamental properties and long-term performance. The static test serves as the initial baseline characterisation, performed prior to any mechanical loading. This test provides critical insight into the intrinsic properties of the printed conductive films in their undeformed state, offering a reference point against which subsequent changes, particularly those induced by cyclic stress, can be measured and analysed.

The static characterisation process encompasses a multifaceted evaluation, including measurements of electrical resistance, bulk resistivity, mechanical hardness, elastic modulus, and cross-sectional morphology. These parameters collectively offer a comprehensive understanding of the ink trace's electrical conductivity, structural integrity, and material homogeneity. The measurement of electrical resistance is fundamental to assessing the current-carrying capability of the printed trace. High-performance conductive inks must maintain low resistivity to ensure efficient signal transmission, particularly in high-density or high-frequency electronic applications. Complementing this, bulk resistivity provides a material-specific value that reflects the intrinsic conductive efficiency of the ink composite, accounting for filler dispersion, inter-particle contact, and matrix continuity.

In parallel, mechanical characterisation, including hardness and elastic modulus is essential for understanding how the composite material will perform under physical stress. Hardness testing evaluates the material's resistance to localised deformation or indentation, which is critical for wear resistance in practical applications such as wearable electronics or printed sensors. The elastic modulus, or stiffness, reflects the trace's ability to elastically deform under applied force, directly impacting its mechanical durability and compliance with flexible substrates during bending, stretching, or torsion. Furthermore, the evaluation of the cross-sectional

M. A. Salim et al., *Nanotechnology in Conductive Inks*, Engineering Materials,
https://doi.org/10.1007/978-981-95-5708-0_4

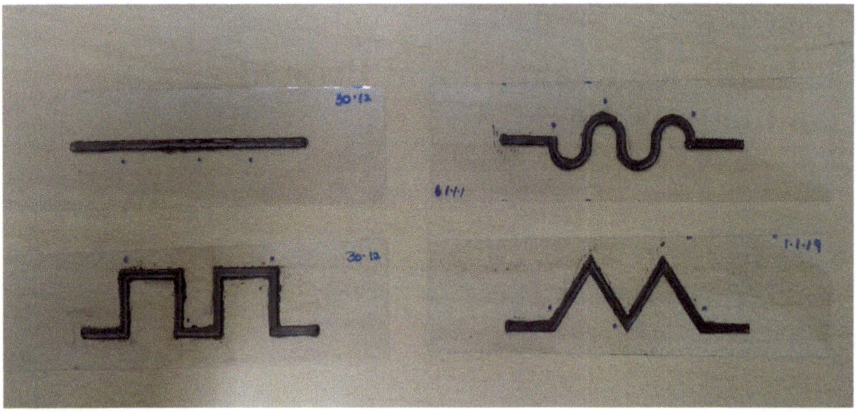

Fig. 4.1 End product of the conductive inks sample

morphology of the printed trace, typically conducted using the Scanning Electron Microscopy (SEM) or optical profilometry, provides visual and quantitative insight into the internal microstructure of the film. Morphological analysis reveals key information such as the distribution and agglomeration of conductive fillers, interfacial adhesion between the ink and substrate, porosity, and film uniformity for all of which significantly influence the overall performance and reliability of the printed circuit.

By performing these static tests at zero loading cycles, it is established a crucial baseline dataset that enables the quantitative assessment of degradation or property changes resulting from mechanical cycling. This initial data is imperative for benchmarking purposes and supports the investigation into the structural stability and functional robustness of conductive ink formulations under practical usage conditions. Therefore, the static characterisation phase provides essential foundational data on both the electrical and mechanical performance of the conductive ink traces in their as-printed condition. This serves as a cornerstone for evaluating how these properties evolve under dynamic cyclic loading, which simulates real-operational stresses encountered in flexible and wearable electronics.

4.2 Characterisation Methods and Procedures

This section describes the characterisation methods and experimental procedures adopted in this book to assess the physical and electrical performance of the conductive ink samples. Two primary categories of characterisation were conducted which are electrical characterisation, and mechanical characterisation.

The electrical conductivity of a circuit material is a fundamental property that directly governs its ability to conduct current. In composite conductive ink systems, this property is determined by the internal distribution, connectivity, and volume fraction of conductive fillers embedded within the insulating polymer matrix. Variations in the weight percentage of these conductive fillers such as graphene nanoplatelets (GNPs) or silver particles, potentially lead to variations in the sheet resistance and overall electrical performance of the printed films. Simultaneously, mechanical properties are equally critical in determining the long-term structural integrity, functional stability, and application-specific reliability of the ink traces. For instance, in flexible or wearable electronics, materials are regularly subjected to deformation, bending, and stretching. In such cases, electrical stability alone is insufficient, where the robust mechanical behaviour becomes paramount. Thus, mechanical characterisation is necessary to complement the electrical data, providing a comprehensive overview of the multifunctional performance of the materials.

4.3 Electrical Characterisation

By evaluating the electrical resistivity of the printed conductive ink films, sheet resistance measurements were carried out using a Four-Point probe system. This characterisation was performed at room temperature with a fixed probe spacing of 1 mm. The objective of this method is to assess the inherent electrical resistance of the printed conductive paths and compare their performance across different formulations and process conditions.

In the measurement setup, the sample is carefully placed on the base plate of the machine. The system is configured to operate in auto-range mode, which automatically adjusts the current and voltage levels applied to the sample, typically settling at around 10 nA current and 100 mV voltage for standard measurements. This automated control ensures that the measurement remains within an optimal range, especially for samples with a wide variation in resistance. However, during the testing process, fluctuations in resistance values are commonly observed, particularly in high-resistance samples. These instabilities are largely attributed to the resistance–capacitance (RC) time constant delay, which arises due to the time required for the current to reach a steady-state level within the sample. This delay can affect the reliability of the readings, especially if measurements are taken prematurely or inconsistently. Consequently, it is crucial to allow the readings to stabilise before recording the final values. Only data points obtained after stabilisation are considered valid and averaged to determine the representative sheet resistance.

In order to account for deviations introduced by the geometry of the sample, especially when the sample dimensions are close to the probe spacing or edge boundaries, a correction factor (G) must be applied. While the base equation used to calculate sheet resistance assumes an infinite sheet or a sample substantially larger than the probe spacing, practical samples often do not meet this criterion. When the sample is either too narrow or too short, the electric field lines are distorted by the edges, leading to an overestimation of the resistance. The correction factor adjusts the result to more accurately reflect the true resistance value, compensating for edge effects and geometric constraints.

The G factor can be obtained from standard reference tables based on the sample's width, length, probe spacing, and thickness. In this work, the sample parameters yielded width/probe spacing = 1 and length/width = 4, resulting in a correction factor $G = 0.9994$, as highlighted in Table 4.1. This correction ensures that the measured resistance reflects the true surface properties of the material rather than being skewed by the limitations of the testing geometry.

By ensuring consistency and accuracy, the sheet resistance measurements were applied across three experimental scenarios. For the initial plan, which involved baseline comparison, three data points were taken per sample, with three repeated readings at each point (see Fig. 4.2 for sample marking). In the second and third scenarios, it is focused on static and cyclic mechanical loading, the same procedure was followed (refer to Fig. 4.3 for markings), allowing for robust statistical comparison across different test conditions.

From a methodological perspective, Four-Point probe sensing is the preferred technique for evaluating the electrical resistivity of conductive inks, particularly when compared to the traditional Two-Point terminal method. In the Two-Point method, both the current and voltage are applied through the same pair of contacts, resulting in measurement interference from contact resistance (RC) and spreading resistance (RSP). These additional resistive components can significantly distort the actual resistance values, especially for thin films or materials with relatively high intrinsic resistance. In contrast, the Four-Point probe method overcomes these limitations by separating the current supply and voltage measurement pathways. Specifically, the

Table 4.1 The correction factor (G) table

Width/probe spacing	Circle	Square	Rectangular length/width = 2	Rectangular length/width = 3	Rectangular length/width = 4
1	N/A	N/A	N/A	0.9988	0.9994
1.25	N/A	N/A	N/A	1.2467	1.2248
1.5	N/A	N/A	1.4788	1.4893	1.4893
1.75	N/A	N/A	1.7196	1.7238	1.7238
2	N/A	N/A	1.9475	1.9475	1.9475
2.5	N/A	N/A	2.3532	2.3541	2.3541
3	2.2662	2.4575	2.7000	2.7005	2.7005
4	2.9289	3.1127	3.2246	3.2248	3.2248
5	3.3625	3.5098	3.5749	3.5750	3.5750
7.5	3.9273	4.0095	4.0361	4.0362	4.0362
10	4.1716	4.2209	4.2357	4.2357	4.2357
15	4.3646	4.3882	4.3947	4.3947	4.3947
20	4.4364	4.4516	4.4553	4.4553	4.4553
32	4.4791	4.4878	4.4899	4.4899	4.4899
40	4.5076	4.5120	4.5129	4.5129	4.5129
Infinity	4.5324	4.5324	4.5325	4.5325	4.5324

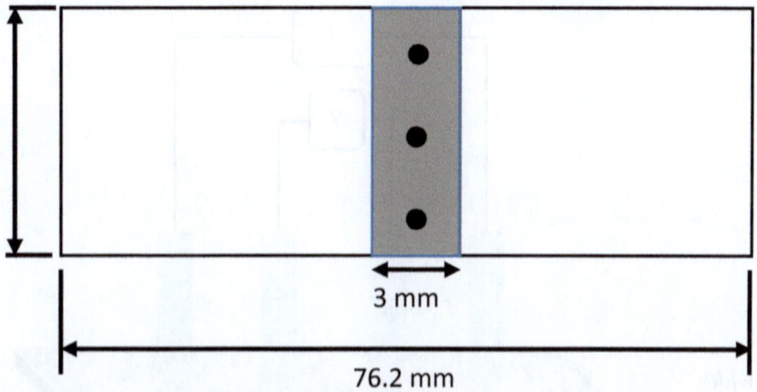

Fig. 4.2 Schematic of printed conductive ink on a glass slide with 3 marking points

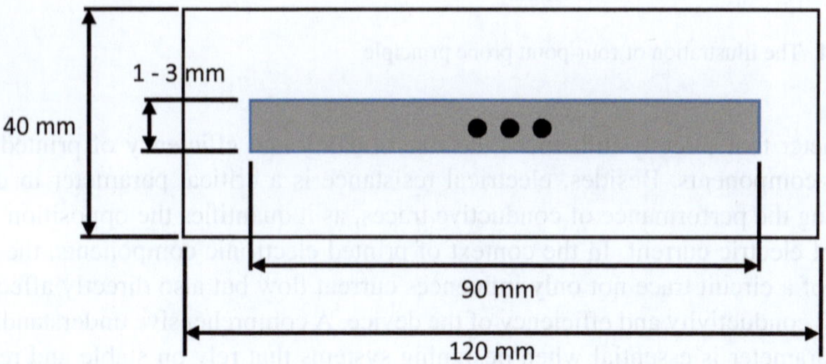

Fig. 4.3 Schematic of printed conductive ink on a flexible substrate with 3 marking points

outer two probes deliver current to the sample, while the inner two probes measure the voltage drop across a precisely defined region. The voltage-sensing probes do not carry current, the measured voltage is not affected by the contact or spreading resistances, thereby enabling highly accurate and repeatable measurements.

The operating principle of this method is further illustrated in Fig. 4.4, where the direction of current flow is shown to be perpendicular to the surface of the conductive film. For thin films, it is generally assumed that electrical current flows predominantly across the surface. However, in reality, some degree of current penetration into the bulk material does occur. Despite this, the surface conduction model remains a useful approximation, especially when comparing relative performance across different thin-film systems. The thin surface layer is typically considered negligible in thickness, allowing for simplification in theoretical analysis and experimental interpretation.

Overall, the sheet resistance characterisation using the Four-Point probe method provides a reliable and non-destructive means of evaluating the electrical performance of printed conductive films. By accounting for geometric effects and ensuring stable measurements, this method enables precise determination of a key electrical

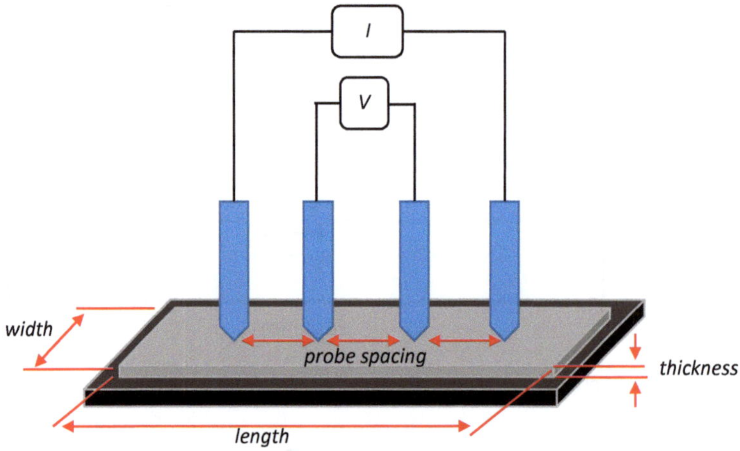

Fig. 4.4 The illustration of four-point probe principle

parameter that directly influences the functionality and efficiency of printed electronic components. Besides, electrical resistance is a critical parameter in understanding the performance of conductive traces, as it quantifies the opposition to the flow of electric current. In the context of printed electronic components, the resistance of a circuit trace not only influences current flow but also directly affects the overall conductivity and efficiency of the device. A comprehensive understanding of this parameter is essential when designing systems that rely on stable and reliable electrical performance.

The bulk resistance of the conductive ink traces was determined by measuring the electrical resistance from one end of the printed pattern to the other. Bulk resistance refers specifically to the resistance offered by the entire volume or bulk of the material, independent of any contributions from surface leakage or superficial effects. This is distinct from surface resistance, which pertains primarily to the resistive behaviour on the surface of thin films or coatings. The bulk resistance of the conductive ink trace was measured using a digital multimeter capable of measuring resistances up to 300 MΩ. The measurements were performed initially under static conditions to establish a benchmark reference for each sample. This baseline data would later serve as a comparison for subsequent measurements taken during cyclic mechanical loading, enabling an evaluation of the material's electrical degradation over time. During each measurement, the positive and negative probes of the multimeter were positioned at the respective ends of the trace, as illustrated in Fig. 4.5, and three repeated readings were taken at each point to ensure accuracy and reproducibility. All resistance measurements were conducted in accordance with the IEEE Standard 118-1978, which outlines best practices for precision resistance testing.

The geometry of the trace namely its length, width, and cross-sectional area has a fundamental influence on the resistance value. A longer trace typically exhibits higher resistance due to the increased path length for electron flow, whereas a wider

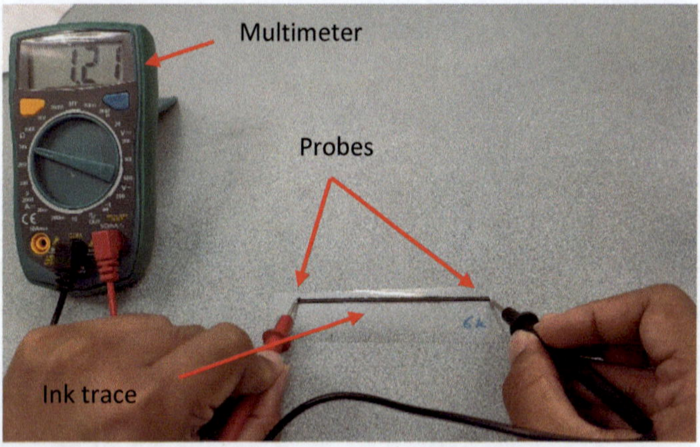

Fig. 4.5 Resistance measurement of conductive ink trace at static loading using digital multimeter

or thicker trace provides a larger cross-sectional area, reducing the overall resistance. Beyond geometry, the inherent properties of the material, specifically its bulk resistivity, are critical. Bulk resistivity is a material-specific constant that indicates how strongly the material resists current flow internally, independent of its physical dimensions.

Understanding the relationship between resistance and resistivity is essential in the context of printable conductors. Although resistivity is a fundamental material property, resistance is influenced by both the material and the geometry of the printed structure. Hence, by analysing and comparing the bulk resistivity across different ink formulations, pattern designs, and trace dimensions, one can determine the most effective configuration for achieving desired resistance levels. This is particularly important in applications where performance, power efficiency, or sensitivity is closely tied to electrical conductivity, such as in sensors, antennas, and interconnects for flexible and stretchable electronics.

Following the static test phase, the samples were subjected to cyclic mechanical loading, simulating repeated operational stress. During this stage, the electrical resistance was monitored in-situ, that is, measured continuously or at fixed intervals throughout the cyclic process. This approach allows for real-time assessment of electrical degradation and damage accumulation as the trace is subjected to repeated deformation. As depicted in Fig. 4.6, the measurement probes were clipped to both ends of the printed pattern and remained in place for the duration of the cyclic test. Resistance readings were recorded at intervals of every 1000 loading cycles, beginning at 1000 and continuing in increments up to 10,000 cycles.

The benefit of in-situ measurement lies in its ability to detect progressive changes in resistance that signal fatigue, microstructural changes, or the onset of failure within the conductive network. As loading cycles increase, micro-cracks or debonding between conductive fillers and the matrix may occur, leading to reduced connectivity and higher resistance. In many cases, a sharp increase in resistance is a precursor

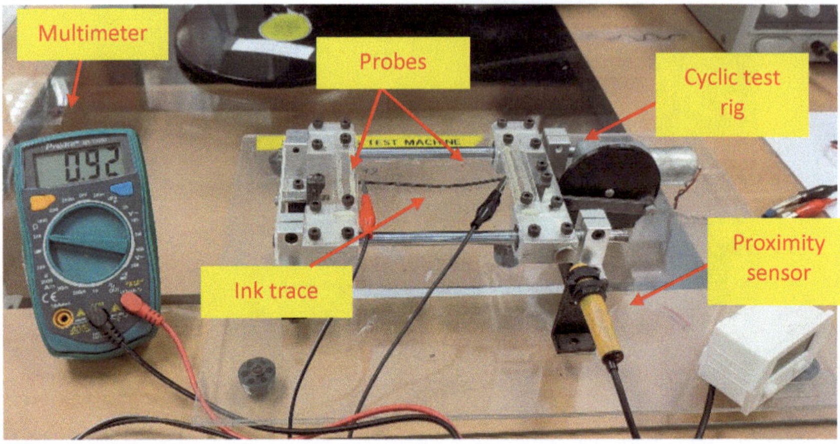

Fig. 4.6 In-situ process of bulk resistance measurement during cyclic loading test

to catastrophic failure. This is confirmed when the multimeter is no longer able to detect a measurable resistance, indicating open circuit failure of the trace.

Bulk resistance is traditionally measured using the Two-Point probe method, particularly when characterising conductive ink traces over extended lengths. In this method, a voltage is applied across the two ends of a material specimen, and the resulting current flow is measured. Resistance is then determined based on the voltage-to-current ratio, following Ohm's law. A conceptual illustration of this process is shown in Fig. 4.7, where a resistor-like device is used to simulate the trace under test.

The simplicity and accessibility of the Two-Point method make it widely used, especially for quick diagnostics and in-process monitoring. However, it is important to note that the method has inherent limitations, particularly when measuring low-resistance samples. One of the primary sources of measurement error is residual

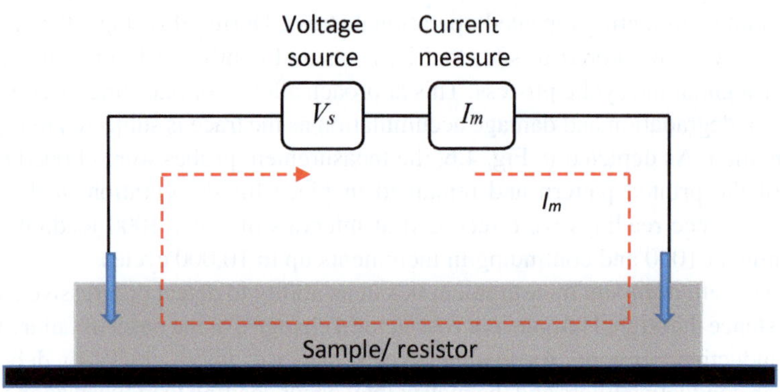

Fig. 4.7 The illustration of two-point probe principle

resistance, which includes contributions from the contact resistance between probes and sample, the intrinsic resistance of the measurement cables, and the interface impedance. For low-resistance materials, these external contributions may be comparable to or even exceed the actual resistance of the sample, leading to significant measurement inaccuracies.

Other error sources include offset voltages, thermoelectric voltages (thermo-EMFs) generated by temperature differences between materials, and instrument drift, all of which may further degrade the precision of the measurement. Such effects are particularly pronounced when working with thin-film materials, where the intrinsic resistance of the sample may be extremely small. In these cases, the parasitic resistances of the probe system can disproportionately influence the result, leading to overestimation of resistance.

For these limitations, the Two-Point method was employed in this work as a practical approach for monitoring resistance during cyclic testing, where its primary function was to track changes rather than determine absolute values. By continuously recording the bulk resistance throughout the cyclic loading phase, it was possible to observe the evolution of the electrical response and identify the point at which performance degradation became significant. While bulk resistance measurements under static conditions provide a snapshot of the initial electrical performance of the printed trace, in-situ monitoring during cyclic loading offers critical insights into the durability and long-term functionality of the conductive ink. The combined use of static and dynamic resistance testing ensures a comprehensive understanding of how printed materials behave under operational stresses, which is essential for designing reliable and resilient electronic systems.

4.4 Mechanical Characterisation

The nanoindentation test serves as a powerful and precise technique for the mechanical characterisation of thin films and micro-scale materials. In this work, nanoindentation was utilised to investigate the mechanical behaviour of conductive ink, specifically in terms of hardness and elastic modulus, which are crucial parameters in evaluating the durability and mechanical resilience of printed conductive traces under both static and dynamic conditions. Throughout the testing process, the position of the indenter tip relative to the sample surface was monitored with high precision using an optical, non-contact depth sensor. This allowed for real-time tracking of penetration depth during the loading and unloading phases of indentation. By plotting the applied load against the corresponding displacement of the indenter, a load–displacement curve was generated. These curves serve as a direct representation of the mechanical response of the material and form the foundation for extracting intrinsic properties such as elastic modulus, stiffness, and hardness.

In accordance with ASTM E2546-15, nanoindentation was performed on all sample types using a Dynamic Ultra Micro Hardness Tester equipped with a three-sided pyramidal Berkovich diamond indenter. This specific indenter geometry is widely accepted in nanoindentation testing due to its ability to produce consistent and interpretable impressions, even at shallow penetration depths. The Berkovich tip also facilitates the distinction between elastic and plastic deformation during testing, making it ideal for characterising thin-film systems.

Prior to indentation, each sample was positioned on the stage and aligned under the optical microscope to ensure accurate spacing between the sample surface and the indenter tip. The test parameters were configured using the software, which allowed for controlled loading and displacement monitoring. The present depth load–unload mode was selected, which is specifically designed for testing thin films. In this mode, a target test force was set, and the system adjusted to maintain the desired loading conditions within a predefined maximum force of 19.6133 mN. This value was carefully selected to remain well within the micro-load testing range, as defined by ISO 14577, which places an upper limit of 2 N for micro-indentation testing. Importantly, previous work has shown that forces below 150 mN are generally sufficient to characterise critical mechanical properties of polymer-based and nanomaterial-enhanced films, such as those used in conductive inks. Lower force levels minimise substrate effects, reduce the risk of over-penetration, and are particularly suited for assessing soft or viscoelastic films, ensuring that the indentation remains within the intended depth and mechanical resolution.

The nanoindentation equipment consists of two functional modules which are the optical side, used for precise alignment and visual inspection, and the indenter side, which houses the actuation and sensing systems required to perform the indentation. Once the setup was complete and the sample accurately positioned, the indentation procedure was initiated. During the loading phase, the indenter penetrated the sample, inducing a combination of elastic and plastic deformation. The extent of penetration, together with the known geometry of the Berkovich tip, enabled an indirect estimation of the contact area at full load, which is essential for computing the material's hardness.

The load was applied gradually to the sample surface until a maximum depth was achieved. A hold time of 5 s at peak load was incorporated into the test protocol. This holding period is critical in mitigating time-dependent deformation phenomena such as creep, especially relevant when testing viscoelastic materials. In the absence of a hold segment, the resulting unloading curve may display an artefact known as the nose effect, which can skew results by artificially reducing the perceived stiffness and introducing errors in the calculation of the elastic modulus. For this reason, a hold time in the range of 5–30 s is often recommended for soft or time-sensitive materials, as it allows for a more stable transition into the unloading phase.

Upon unloading, only the elastic portion of the deformation was recovered, while the plastic deformation remained as a permanent indentation, as illustrated in Fig. 4.8. This behaviour enables the clear separation of elastic and plastic contributions, which is essential for accurate modelling of the mechanical response. The unloading curve itself is typically used to extract the elastic modulus, while the permanent depth and calculated contact area are used to determine hardness. Together, these two parameters provide a comprehensive view of the ink's mechanical strength and its ability to withstand external stress without undergoing irreversible deformation.

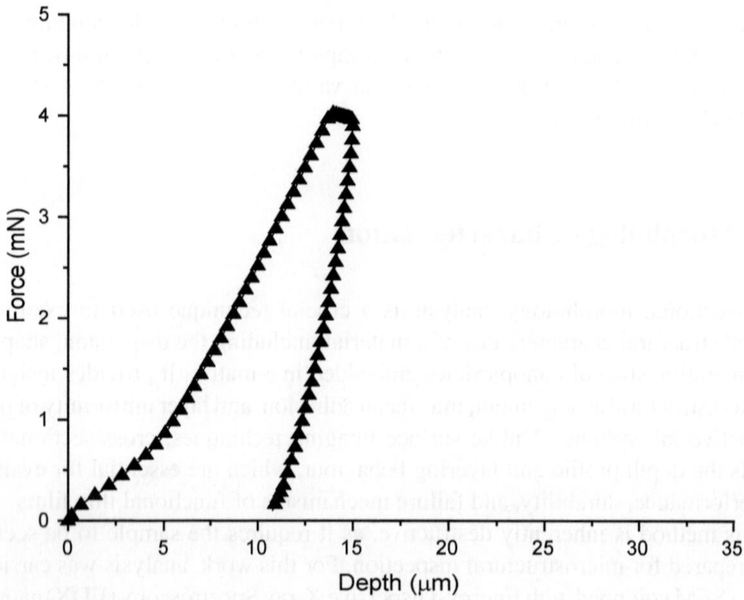

Fig. 4.8 Loading and unloading curve of nanoindentation graph

The samples evaluated in this work were classified into three categories: (i) ink samples deposited on glass slide substrates, (ii) patterned samples on flexible substrates under static conditions, and (iii) patterned samples on flexible substrates subjected to cyclic loading. For the first two categories, indentation was performed once on each specimen to capture the baseline mechanical properties. However, for the third category, which involved cyclic mechanical testing, a more detailed evaluation protocol was followed.

Specifically, nanoindentation was carried out on the cyclic samples at intervals of every 1000 cycles, beginning from 1000 cycles and continuing up to 10,000 cycles. This periodic assessment allowed for the monitoring of mechanical degradation over time as the samples experienced repeated loading. By comparing the mechanical characteristics at each interval, the impact of mechanical fatigue on hardness, modulus, and overall integrity of the ink layer could be systematically assessed.

By ensuring the consistency and maintain the reliability of the experimental process, the electrical resistance of each sample was re-evaluated following every indentation cycle during the cyclic loading tests. This double-checking step was crucial to confirm that the indentation did not introduce unintended damage or alter the sample's conductive behaviour prior to the continuation of further mechanical cycling. In summary, nanoindentation provides a quantitative and reproducible means of assessing the local mechanical performance of conductive ink films. The integration of in-situ optical monitoring, precise load control, and standardised test conditions ensures the generation of meaningful data. When performed alongside electrical characterisation, this method supports a holistic understanding of how printed conductive materials respond to complex operational conditions, making it an invaluable tool in the development and validation of stretchable, flexible, and durable electronic systems.

4.5 Morphology Characterisation

Cross-sectional morphology analysis is a crucial technique used to examine the internal structural characteristics of a material, including the dispersion, shape, and agglomeration state of nanoparticles embedded in a matrix. It provides insight into the microstructural arrangement, interfacial adhesion, and layer uniformity of printed conductive ink systems. Unlike surface imaging techniques, cross-sectional SEM reveals the depth profile and layering behaviour, which are essential for evaluating the performance, durability, and failure mechanisms of functional thin films.

This method is inherently destructive, as it requires the sample to be sectioned and prepared for microstructural inspection. For this work, analysis was carried out using a SEM equipped with Energy-Dispersive X-ray Spectroscopy (EDX) capability for elemental analysis. One of the primary limitations of SEM analysis lies in the requirement for operation under a vacuum environment. This constraint restricts the analysis of materials that are volatile, outgas under vacuum, or are otherwise sensitive to low-pressure conditions. Nonetheless, for conductive or metal-containing films such as those loaded with silver nanoparticles or GNPs, SEM remains a powerful and widely used characterisation tool due to its ability to generate high-magnification, high-contrast images of surface and subsurface features.

By conducting the SEM analysis, a sandwiched specimen containing the conductive ink layer was first sectioned and then mounted on a sample holder. The sample was introduced into the SEM's vacuum chamber, where electron beam imaging was performed. Imaging was conducted using a Secondary Electron (SEN) detector at an accelerating voltage of 15 kV, a commonly employed setting that strikes a balance between image clarity and sample safety. Accelerating voltages significantly higher than this threshold can lead to surface charging, degradation, or even structural alteration of delicate samples prior to successful imaging.

The principle of SEM imaging is grounded in the interaction between a focused primary electron beam and the atoms within the sample. In the SEN mode, inelastic scattering events lead to the emission of low-energy secondary electrons from the top few nanometres of the surface. These emitted electrons are then collected by a detector to form a detailed image of the sample's topography. The intensity of the signal at each point is synchronised with the corresponding pixel in the image display, building a spatial distribution map of electron interactions, hence revealing structural features with nanometre-scale resolution.

Given the destructive nature of cross-sectional SEM, a new set of specimens was prepared for this specific analysis. The sample must be physically cut open to expose the cross-section, and this step must be performed with precision to preserve the integrity of the structure. By enhancing the quality of imaging and to avoid charging artefacts, a conductive coating was applied to the sample surface. In this work, a thin platinum layer, approximately 10 nm thick, was sputter-coated onto the sample using an auto fine coater. Platinum is commonly used due to its fine grain size and excellent conductivity, which helps to stabilise the image and prevent electron build-up on insulating regions. Nevertheless, it is important to note that this coating process can contaminate the sample, rendering it unsuitable for further electrical or chemical measurements.

The preparation of specimens for cross-sectional SEM imaging follows a series of steps derived from metallographic procedures, widely adopted in materials science and engineering. The first step involves the identification of the region of interest (ROI), followed by careful sectioning of the sample to a manageable size for mounting and processing. Sectioning is typically achieved using a low-speed diamond saw to avoid introducing excessive heat or mechanical stress that could alter the sample microstructure. Once sectioned, the sample was cold-mounted to facilitate safe handling and accurate surface preparation. Cold mounting is particularly useful for delicate specimens or composite materials, as it avoids the thermal and mechanical stresses associated with hot mounting. In this process, a two-part epoxy system comprising an epoxy resin and a hardener was used. The components were mixed in a precise ratio of 3 parts resin to 1-part hardener. The materials used for cold mounting in this work differ from those used in the GNPs baseline conductive ink formulation. This distinction is crucial, as mounting materials must provide mechanical support without interfering with the material under study.

The mounting process began by selecting a clean, dry mounting cup, into which the specimen was placed. The epoxy components were then accurately measured and thoroughly mixed, ensuring homogeneity and avoiding the inclusion of air bubbles that can compromise the mechanical stability of the mount. After pouring the mixture over the sample, it was allowed to cure at room temperature, during which time the mixture hardened to encapsulate the specimen securely. Proper mixing and curing times are essential to produce a defect-free mount with consistent hardness and adhesion. An example of the mounted sample is shown in Fig. 4.9, depicting the specimen both before and after the mounting process.

Once the mount was cured, the specimen underwent mechanical grinding and polishing to expose a flat and well-defined cross-section suitable for SEM observation. The grinding process began with coarse grinding using silicon carbide (SiC) abrasive papers of decreasing grit size (180, 320, and 600), with water as a coolant. Water serves to dissipate heat, reduce thermal expansion effects, and remove grinding debris, thereby preventing deformation of the specimen and preserving structural features. Subsequent fine grinding was conducted using finer SiC abrasives up to 1200 grit, providing a smooth, scratch-free surface ideal for high-resolution SEM imaging. This step is critical in eliminating mechanical artefacts from previous grinding stages, thus ensuring the microstructural features observed are representative of the true material behaviour.

Finally, the specimens were polished using a polishing cloth and DIAMAT polycrystalline diamond suspension, applied in decreasing particle sizes of 6, 3, and 1

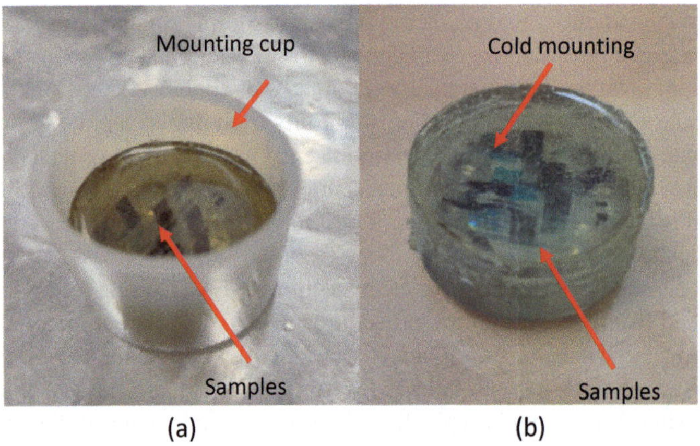

Fig. 4.9 Conductive ink sample during the mounting procedure. **a** Sample in mounting cup, **b** final product of cold mounting sample

μm. Polishing is essential for removing any remaining surface scratches and for achieving a mirror-like finish that enhances contrast and clarity during SEM examination. In conclusion, cross-sectional SEM analysis is an important component in the characterisation of nanocomposite conductive inks. It allows the visualisation of nanoparticle distribution, interfacial bonding, and layer thickness, offering insights that are not possible through surface imaging alone. Though destructive, this technique when properly prepared, and provides high-resolution structural data that are essential for correlating microstructural features with mechanical, electrical, and functional performance of printed electronic materials.

4.6 Mechanical Property Evaluation

The mechanical reliability of conductive ink is a critical performance parameter in the development of flexible and stretchable electronic (SE) devices. Applications such as wearable sensors, foldable displays, electronic skins, and biomedical patches require conductive inks that can endure continuous mechanical deformation without significant degradation in electrical or structural integrity. As such, mechanical testing specifically cyclic bending, tensile, and torsion tests forms an essential part of the characterisation process for materials intended for use in flexible electronics.

By supporting these evaluations, a custom-built, multipurpose mechanical test rig was developed in this work. The apparatus was designed in accordance with international testing standards, particularly those outlined by the Association Connecting Electronics Industries (ACEI) in IPC 9204 [1]: "Guidelines on Flexibility and Stretchability Testing for Printed Electronics." These standards were strictly adhered to during the design, fabrication, and validation of the mechanical test systems. The key advantage of the in-house test rigs is their simplicity, portability, and adaptability, making them suitable for both laboratory and field-based testing environments.

The cyclic bending test evaluates the material's response to repeated flexural deformation. According to the IPC 9204 [1]: "Bending Test Principles for Printed Electronics", this test simulates practical usage conditions where materials experience folding, rolling, or flexing over extended periods. A printed conductive trace or ink, when subjected to cyclic bending, undergoes alternating compressive and tensile stresses along the outer and inner surfaces of the bend, respectively. The repeated bending can induce fatigue damage, manifesting as micro-cracking, delamination, or interfacial failure within the printed layers. Over time, such defects accumulate, leading to a gradual increase in electrical resistance or even complete loss of conductivity. Therefore, evaluating the number of cycles a specimen can endure before failure provides essential insight into the fatigue resistance, mechanical durability, and long-term performance stability of the printed electronics.

The conceptual design of the bending test rig involves dual-end clamping, where the specimen is fixed at both ends and subjected to controlled deformation by a mechanical actuator. The movement is facilitated by a wheel-and-connector system, illustrated in Figs. 4.10 and 4.11, which compresses the sample in a cyclical manner. This setup allows for precise control of bending radius, speed, and repetition, enabling quantitative assessment of mechanical endurance under bending loads.

The cyclic tensile test, as defined in IPC 9204 [1] guidelines, is used to evaluate the behaviour of stretchable printed electronics under repetitive elongation. This test is particularly relevant for applications involving body movement, fabric integration, or elastic substrates, where tensile deformation is a dominant mechanical stress.

During cyclic tensile loading, the conductive ink is stretched longitudinally, placing the material under uniaxial tension. Each elongation cycle introduces strain within the ink matrix and at the ink-substrate interface, potentially leading to the formation of micro-voids, cracks, or debonding. The mechanical and electrical integrity of the ink must be preserved during these cycles to ensure functionality in stretchable devices.

The tensile test rig in this work shares a similar mechanical arrangement with the bending rig, utilising a wheel and connector mechanism (Figs. 4.12 and 4.13).

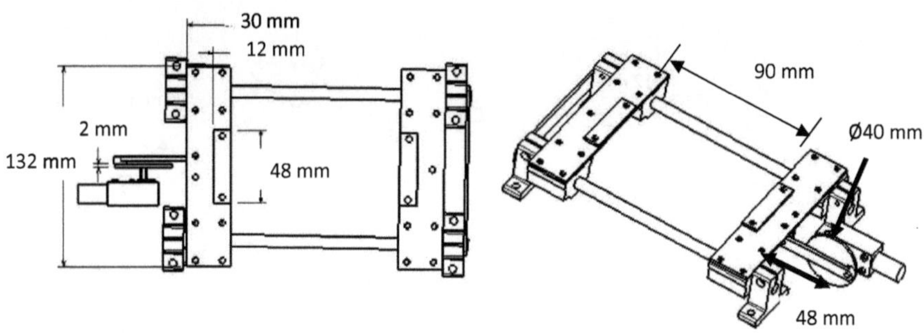

Fig. 4.10 Wheel and connector system in the bending cyclic test rig

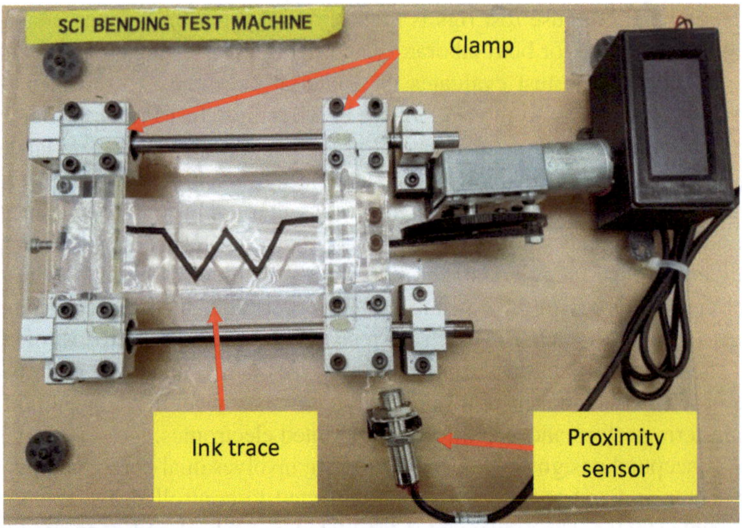

Fig. 4.11 Test-rig model for cyclic bending

However, unlike the compressive action in bending, the tensile rig separates the clamps to apply axial tension to the specimen. The system is calibrated to apply elongation up to 5 mm, a value selected to avoid reaching the necking point, where plastic deformation would lead to irreversible failure after a single cycle. Maintaining the load within the elastic limit ensures that the sample can repeatedly return to its original dimensions without structural damage.

Torsional loading introduces a rotational shear stress along the length of a specimen, replicating scenarios where the device may experience twisting such as during installation around curved surfaces or repetitive twisting motions in wearable applications. The cyclic torsion test, governed by Sect. 4.8 of IPC 9204 [1], determines a material's capacity to endure such cyclic rotational deformation. The test rig for torsion employs a motor generator system coupled with a rack and pinion gear assembly, as illustrated in Figs. 4.14 and 4.15. This configuration enables controlled

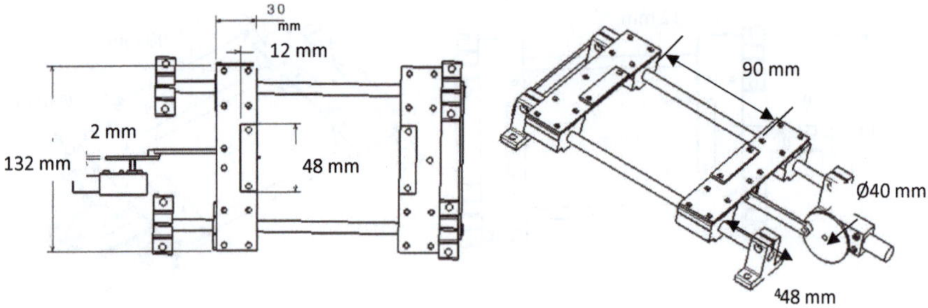

Fig. 4.12 Wheel and connector system in the tensile cyclic test rig

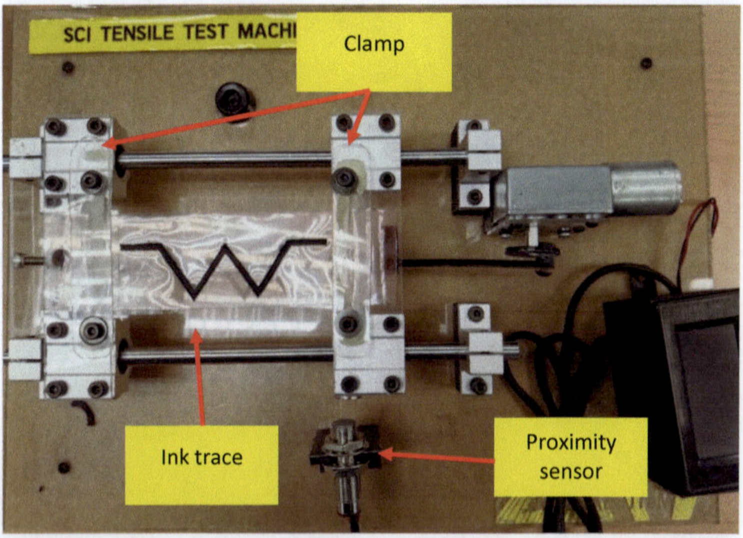

Fig. 4.13 Test-rig model for cyclic tensile

application of cyclic rotational force, typically up to 360° per cycle, followed by a return to the initial position to complete the full cycle. The rack and pinion system translates rotational motion into linear displacement and vice versa, allowing precise angular control during testing.

Under torsional stress, the material experiences shear strain distributed radially from the centre of the cross-section. Over successive cycles, this strain can initiate shear banding, material fatigue, or even catastrophic torsional fracture. Monitoring

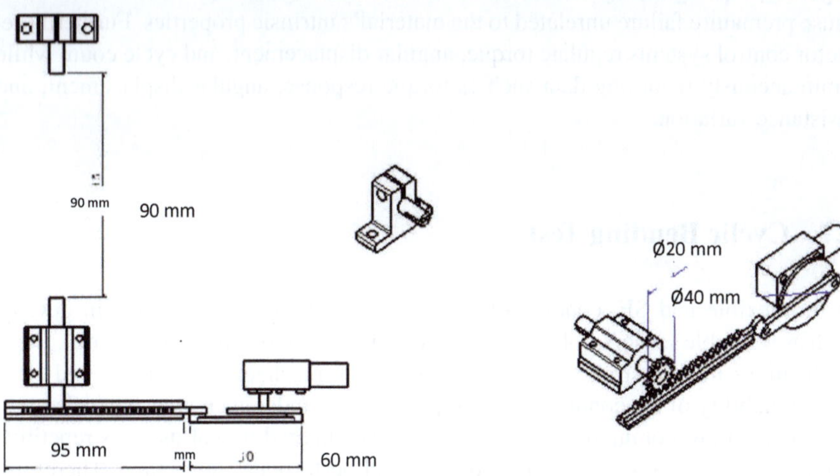

Fig. 4.14 Rack and pinion system in the torsion cyclic test rig

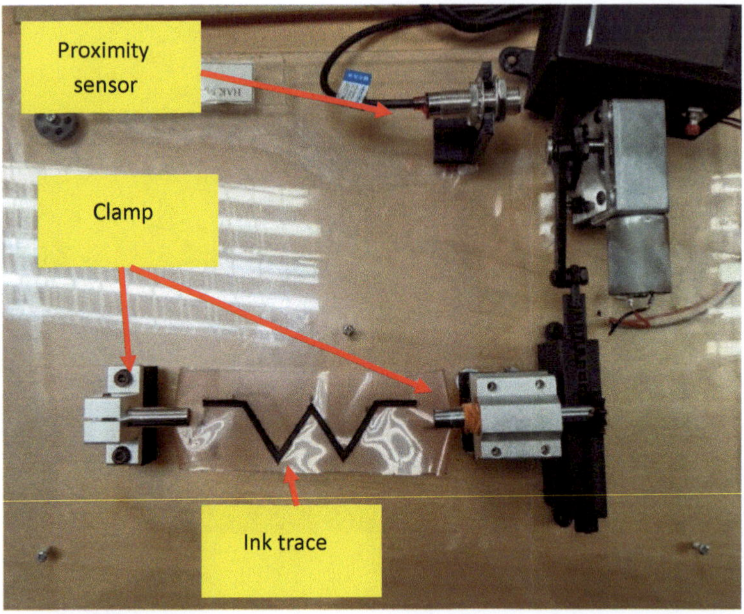

Fig. 4.15 Test-rig model for cyclic torsion

the changes in electrical and mechanical properties over repeated twisting allows for comprehensive evaluation of the material's torsional fatigue strength, fatigue life, and cyclic deformation behaviour.

A key consideration in torsional testing is the design of the specimen holder, which must securely grip the sample and transmit the applied torque without slippage. Improper gripping can result in underestimating the true torsional strain or cause premature failure unrelated to the material's intrinsic properties. Furthermore, motor control systems regulate torque, angular displacement, and cycle count while simultaneously recording data such as torque response, angular displacement, and resistance variation.

4.7 Cyclic Bending Test

In the flexible and SE, conductive films serve as critical components in devices such as wearable sensors, rollable displays, and implantable medical systems. These applications impose stringent requirements on the mechanical resilience and electrical reliability of the conductive films, particularly under cyclic mechanical stress. Among the most common mechanical loads encountered in real-usage is repetitive bending, which can gradually degrade material performance over time. Therefore, systematic evaluation of the film's endurance under such conditions is indispensable for determining its suitability for long-term and high-strain applications.

By addressing this need, a mechanical bend tester was custom-designed and fabricated in-house. The system was developed with the primary objective of simulating cyclic bending conditions while enabling precise, repeatable, and scalable testing. The design integrates key components including a direct current (DC) motor, rotating arms, fixed and mobile clamps, a wheel mechanism, and a proximity sensor to track cyclic motion. The overall mechanical configuration is devised to induce concentrated strain in the central portion of the substrate, which is most representative of actual device operating conditions.

As illustrated in Fig. 4.16 and demonstrated in the operational setup in Fig. 4.17, the mechanical bend tester operates using a dynamic arm-actuation mechanism. The test sample typically a 90 mm long conductive film or substrate is secured at both ends between a stationary holder and a motor-driven mobile holder. Upon activation, the DC motor initiates a rotational motion that drives the arm to displace the movable holder horizontally. This, in turn, induces a consistent flexural motion in the test specimen.

During operation, the substrate is cyclically bent to a defined curvature, producing a bending radius of approximately 12.5 mm at the maximum deflection point in each cycle. The motor is calibrated to complete one full bending cycle every four seconds, allowing for high-throughput fatigue testing while maintaining control over deformation parameters. The incorporation of a proximity sensor enables accurate cycle counting, data logging, and triggering for other instrumentation such as electrical resistance measurement devices or strain gauges, facilitating concurrent mechanical–electrical analysis.

The mechanical bend tester was employed in conjunction with the ASTM D7774-17 [2] standard, which provides comprehensive guidelines for evaluating the fatigue resistance of plastic materials under flexural loading. Specifically, this standard is

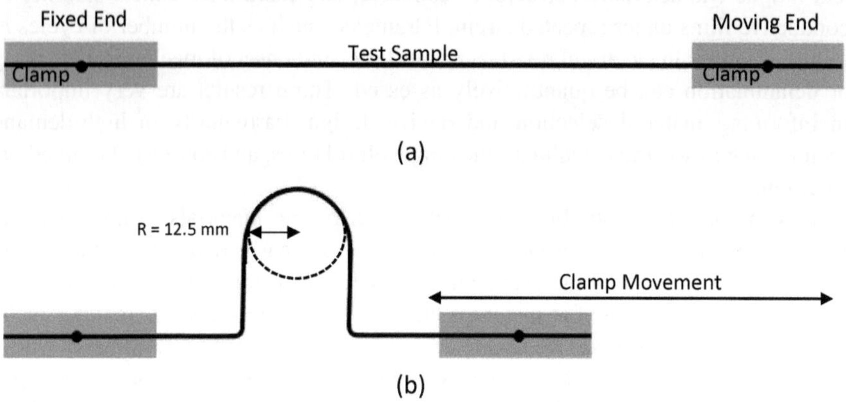

Fig. 4.16 Schematic of the test component in a cyclic bending test rig with a constant bending radius **a** in a flat position and **b** in a bent position

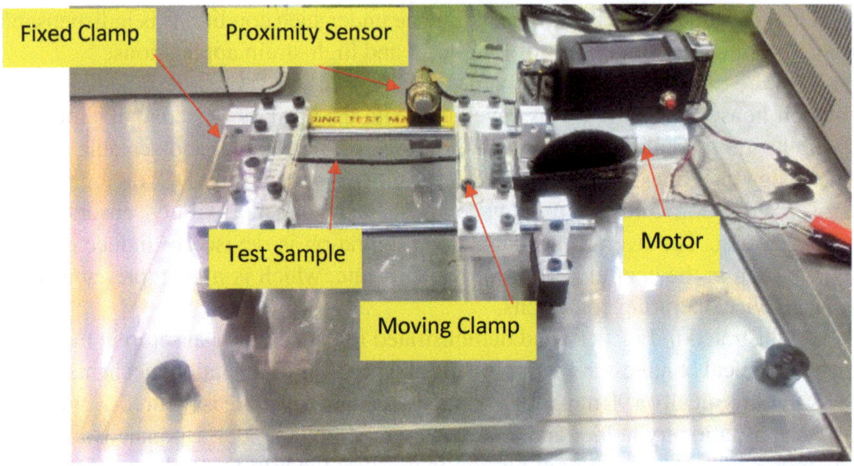

Fig. 4.17 Cyclic bending test rig

designed to assess the mechanical behaviour of rigid and semi-rigid polymer-based substrates or encapsulants when subjected to repeated, reversed-direction flexural deformation. According to ASTM D7774-17 [2], the cyclic test maintains stress and strain levels below the proportional limit of the material, meaning the mechanical loading is confined within the elastic regime, where deformation is fully recoverable upon unloading. Operating within this range ensures that the test measures fatigue performance without introducing permanent plastic deformation or fracture under short-term cyclic loading. Such conditions are crucial for simulating long-term operational environments where the material must undergo many thousands of bending cycles without functional degradation.

The data generated from this test provide essential information on the mechanical fatigue characteristics, structural resilience, and electromechanical stability of conductive films under repeated strain. Parameters such as the number of cycles to failure, changes in electrical resistance over time, and onset of mechanical cracking or delamination can be quantitatively assessed. These results are very important in informing material selection and device design, particularly in high-demand sectors such as wearable health monitoring, soft robotics, and conformal biomedical electronics.

It is important to note, however, that not all plastic materials exhibit identical fatigue behaviour. The presence or absence of a clearly defined elastic region, in which strain is reversible plays an essential role in how materials respond to repeated flexural loading. For materials that lack a pronounced elastic regime, cyclic loading can lead to progressive plastic deformation, even if the initial stress is within presumed safe limits. Such behaviour can alter the programmed displacement amplitude, introduce strain localisation, and compromise test repeatability. Therefore, careful characterisation of each material's stress–strain response is imperative before fatigue testing.

Moreover, external factors such as surface condition, fabrication defects, and internal microstructural features including voids, agglomerates, or phase interfaces, can significantly influence fatigue life. As such, pre-test inspection and standardised sample preparation are essential for obtaining reliable and reproducible results. The mechanical bend tester developed in this work demonstrates not only the feasibility of low-cost, lab-scale fatigue testing but also provides a robust platform for comprehensive mechanical–electrical characterisation of flexible substrates. Its ability to replicate practical bending conditions in a controlled, instrumented environment makes it a valuable tool in the ongoing advancement of flexible and printed electronic technologies.

4.8 Cyclic Tensile Test

The mechanical performance of conductive films under tensile loading is critical for the development of reliable flexible and SE devices, respectively. One of the most fundamental tests for evaluating a material's response to mechanical deformation is the uniaxial tensile test. In this exploration, a preliminary uniaxial tensile test was conducted to establish the failure point of the conductive film when subjected to high strain levels. This initial assessment is essential for understanding the mechanical robustness, ductility, and failure mechanisms of the conductive composite under linear stretching forces.

The uniaxial tensile test involves applying a steadily increasing tensile load along a single axis until the specimen fractures or fails. This test is commonly employed in material science and engineering due to its simplicity and ability to reveal a wide range of material properties. For conductive films used in printed and flexible electronics, this test helps to determine parameters such as tensile strength, elastic modulus, elongation at break, yield point, elastic limit, and toughness. These parameters are critical for assessing the film's ability to maintain electrical functionality under mechanical deformation.

It is important to distinguish between uniaxial tensile testing and cyclic tensile testing. The former is a one-time destructive test used to characterise material behaviour under monotonic loading, while the latter evaluates the material's ability to endure repeated tensile deformation over time. Cyclic tensile testing is more representative of real-applications where flexible devices are often subjected to continuous or intermittent stretching, such as in wearable electronics, bio-integrated sensors, or soft robotics.

During tensile loading, damage to the conductive ink layer typically progresses through a multi-stage mechanism. In the first stage, microscopic cracks begin to emerge within the ink layer as it is stretched. These micro-cracks significantly reduce the stiffness of the film due to the disruption of the conductive pathways. In the second stage, the rate of stiffness degradation decelerates, but the accumulated damage causes interfacial delamination, in which the ink begins to detach from the substrate. Finally, in the third stage, the interface becomes severely compromised, leading to macroscopic failure and total loss of mechanical and electrical integrity. Understanding these failure stages is crucial for designing durable printed electronic systems and for selecting ink-substrate combinations that can better resist tensile fatigue.

In the present experimental setup, the test sample typically a composite of conductive ink deposited on a flexible polymer substrate was securely mounted between two clamps. The sample was then subjected to a steady, unidirectional tensile force until a visible crack or complete failure was observed. The point at which the first crack formed was recorded as the maximum stretch limit for the subsequent cyclic tensile test. Establishing this limit ensures that the sample is only subjected to sub-critical strain levels during cyclic testing, thereby allowing for multiple cycles of deformation without immediate failure.

By preventing sample slippage during the tensile test, special attention was given to the design and configuration of the gripping mechanism. The clamps were adjusted to exert sufficient frictional force on the sample, and silicone padding was incorporated into the grips to enhance the holding capacity while avoiding localised damage to the substrate. Proper sample clamping is essential for reliable test results which as inadequate gripping may lead to premature failure or inconsistent data due to slippage or stress concentrations at the clamped edges. Figures 4.18 and 4.19 illustrate the tensile testing procedure and the actual test rig, respectively. Together, these figures provide visual clarity on the working principles and experimental methodology employed in this section of the work.

The tensile testing system was developed in-house and built to comply with the ASTM D3479/D3479M-19 standard. This international standard outlines procedure for assessing the tensile properties of polymer matrix composite materials, which include conductive films comprised of polymer binders filled with conductive nanomaterials such as silver nanoparticles, carbon black (CB), or GNPs. The ASTM guideline ensures the use of consistent specimen geometries, strain rates, and test environments, enabling data reliability and cross-study comparability.

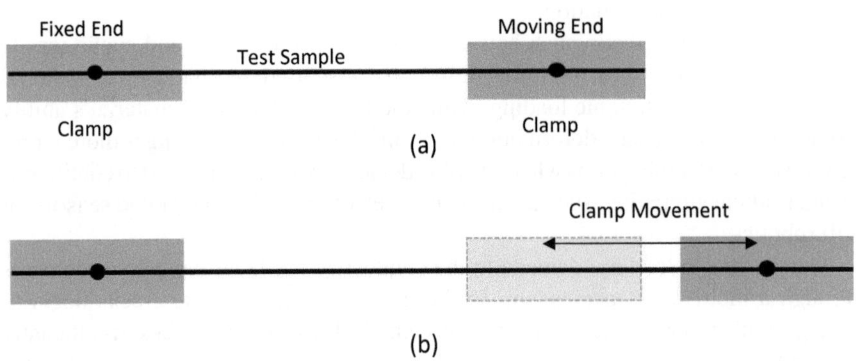

Fig. 4.18 Schematic of the test component in a cyclic tensile test rig **a** in an un-stretch position and **b** in a stretched position

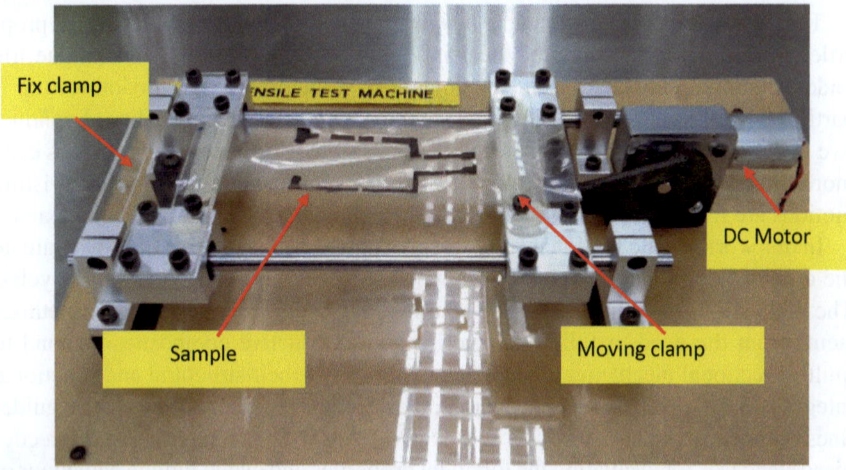

Fig. 4.19 Cyclic tensile test rig

The tensile testing apparatus incorporates a DC motor-driven actuation mechanism similar to that employed in the cyclic bending test rig. The motor operates continuously, displacing the sample at a rate that completes one full stretch–relax cycle every four seconds. This configuration provides controlled and repeatable deformation, making it suitable for extended cyclic testing protocols where hundreds or thousands of cycles may be performed to assess mechanical fatigue resistance and electromechanical stability.

Ultimately, the tensile test not only quantifies the mechanical resilience of the conductive films but also serves as a precursor for more complex mechanical assessments. By integrating mechanical testing with concurrent electrical measurements, such as in situ resistance monitoring, explorers can obtain a comprehensive understanding of the electromechanical coupling behaviour of printed electronic materials. These insights are invaluable for designing next-generation flexible devices that are both mechanically compliant and electrically robust under tensile loading conditions.

4.9 Cyclic Torsion Test

Torsion testing, often referred to as shear testing, is a critical method for evaluating the shear and fatigue performance of materials subjected to rotational or twisting loads. Unlike axial tensile or bending tests, which apply stress along a single axis or in a flexural direction, torsion tests subject the sample to rotational deformation, generating a state of pure shear stress across the cross-section of the material. This form of mechanical loading more accurately replicates the complex real-conditions encountered in applications involving rotational movements, such as in flexible electronic circuits, wearable devices, bioelectronics, and rotating machinery components.

The torsion test provides valuable insights into several fundamental material properties, including shear strength, shear modulus, torsional rigidity, and fatigue life under rotational deformation. In the context of printed and SEs, torsion testing is particularly relevant for assessing the mechanical and electrical stability of conductive traces when embedded or patterned onto flexible substrates. This becomes even more crucial for devices that are integrated into dynamic environments where twisting motions are frequent, such as smart textiles, implantable sensors, or robotic skins.

In this work, cyclic torsion testing was performed on conductive films to evaluate the durability of conductive ink traces when subjected to repeated twisting cycles. The motivation for including torsion testing alongside bending and tensile methods stems from the need to fully characterise how conductive composites respond to multi-directional mechanical stress, thereby ensuring their structural and functional integrity under diverse usage scenarios. The testing protocol followed the guidelines outlined in ASTM A938-18 and ASTM E2207-15, two internationally recognised standards for evaluating the torsional properties and cyclic fatigue behaviour of materials under shear loading. The ASTM A938-18 provides detailed procedures for assessing torsional strength and ductility, while ASTM E2207-15 focuses on cyclic torsion testing to determine the number of cycles a specimen can endure before failure under controlled shear stress amplitudes. These standards ensure that the results obtained are both reproducible and comparable across different studies and material systems.

A key differentiator between cyclic torsion and other mechanical tests is the nature of the stress distribution. In torsion, stress is applied tangentially, inducing shear along planes perpendicular to the axis of rotation. This stress distribution is ideal for evaluating delamination resistance, ink adhesion, and mechanical fatigue at the interface between the conductive ink and its substrate. Materials that perform well under bending or tensile stress may still exhibit premature failure under torsional loading due to different failure modes such as shear-induced cracking, ink trace separation, or interface delamination.

The cyclic torsion test apparatus used in this work was designed to simulate such mechanical stress conditions. It consists of a fixed holder and a movable holder, securely gripping both ends of the test specimen. A DC motor, connected through a pinion gear, actuates the movable holder, thereby twisting the sample. The gear system precisely controls the twisting angle, which in this case ranged from 0° to 90°, and operated at a constant speed of 100 revolutions per minute (RPM). This setup allowed for repeated and consistent application of torsional stress over numerous cycles, simulating realistic service conditions.

As the sample is twisted, the conductive ink layer experiences shear deformation, which may cause crack propagation, interfacial debonding, or permanent distortion depending on the material's mechanical resilience. The ability of the conductive film to maintain electrical conductivity throughout these cycles is a critical parameter for performance evaluation. By monitoring the number of cycles to failure, its can estimate the fatigue life and durability of the ink-substrate composite under torsional loads.

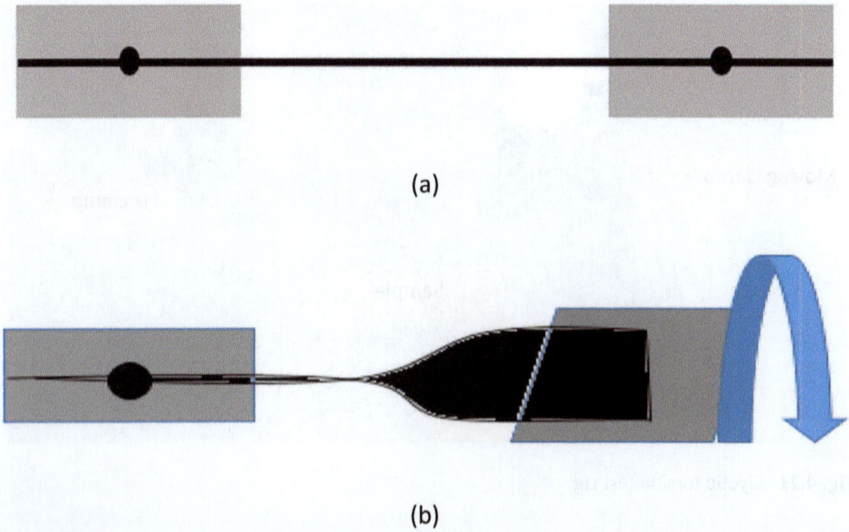

Fig. 4.20 Schematic of the test component in a cyclic torsion test rig **a** in a flat position and **b** in a twisted position

This method is especially valuable for evaluating ductile versus brittle behaviour. Ductile materials can accommodate larger torsional strains by undergoing plastic deformation, which may delay crack formation and extend fatigue life. Conversely, brittle materials typically fail after minimal torsional displacement, exhibiting abrupt fracture behaviour with little warning. In wearable electronics, where devices are frequently bent, twisted, and stretched, such differentiation is an essential for material selection and design optimisation. Figures 4.20 and 4.21 provide visual representation of the cyclic torsion test.

Therefore, cyclic torsion testing complements bending and tensile assessments by offering a unique perspective on how conductive materials respond to rotational fatigue loading. This comprehensive mechanical evaluation framework helps ensure the long-term reliability of printed electronic components, particularly when deployed in environments where dynamic multi-axial loading is prevalent. The insights gained from torsional fatigue data are instrumental in the design, fabrication, and qualification of next-generation of SE and flexible electronic systems.

4.10 Summary

The development and integration of printed electronic components such as flexible circuits, sensors, and conductive interconnects, require a thorough understanding of their electrical and mechanical performance, particularly when subjected to various

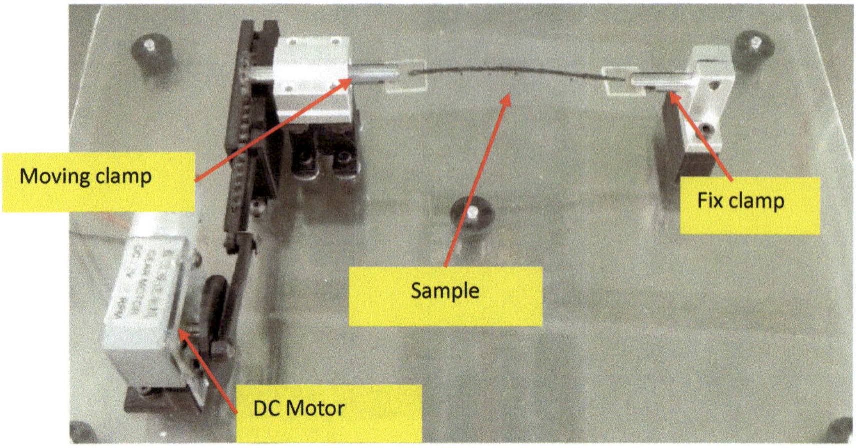

Fig. 4.21 Cyclic torsion test rig

forms of mechanical stress. As these materials are increasingly deployed in applications involving dynamic deformation, such as wearable electronics, bio-integrated devices, and flexible displays, ensuring their structural reliability and functional stability under cyclic loading becomes essential. This chapter provides a detailed overview of the key methodologies and principles used to characterise the electrical and mechanical behaviour of conductive inks and films. Special focus is placed on cyclic bending, tensile, and torsional testing, including the design and fabrication of custom-built test rigs and adherence to internationally recognised standards such as IPC 9204 [1], ASTM D7774-17 [2], ASTM D3479/D3479M-19, and ASTM A938-18/E2207-15.

Conductive films are typically composed of a polymeric or elastomeric matrix embedded with conductive fillers such as silver nanoparticles, CB, or GNPs. While their initial electrical conductivity may be optimised through careful formulation, their long-term performance under mechanical strain must also be verified. As these devices are bent, stretched, or twisted repeatedly during use, the percolation network of conductive particles may be disrupted, leading to increases in resistance, micro-cracking, delamination, and ultimately electrical failure.

Mechanical characterisation, therefore, serves not only to validate durability but also to correlate mechanical stress with electrical degradation pathways, offering insights into how microstructural changes such as crack formation, filler displacement, and ink-substrate debonding, affect conductivity over time. The cyclic bending test is one of the most fundamental assessments used to determine the fatigue performance of conductive films on flexible substrates. The objective of this test is to evaluate how well a material maintains its mechanical integrity and electrical functionality when subjected to repeated flexural stress. In the work presented, a custom mechanical bend tester was developed in-house. The rig consists of a DC motor, proximity sensor, wheel, and both fixed and movable clamps. The motor initiates a continuous bending motion by rotating the arms, which horizontally displace the sample, generating a bending radius of approximately 12.5 mm at its extreme position. The test samples, measuring 90 mm in length, were subjected to cycles lasting 4 s each, simulating long-term mechanical fatigue.

The test followed the ASTM D7774-17 [2] standard, which outlines procedures for assessing the fatigue resistance of rigid and semi-rigid plastics under flexural stress. This standard ensures that the applied stress remains within the elastic region of the material, allowing for meaningful comparison between materials based on their reversible strain behaviour. However, some polymers, particularly those used in flexible electronics, may exhibit nonlinear deformation, including plasticity and viscoelastic effects. Therefore, visual inspection and in situ electrical resistance monitoring are often employed in tandem to detect crack initiation, substrate wrinkling, or trace delamination.

The cyclic tensile test complements the bending test by evaluating the conductive film's response to uniaxial elongation. This type of mechanical loading is representative of strain conditions encountered in SEs, such as those worn on skin or integrated into garments. A preliminary simple uniaxial tensile test was conducted to determine the stretch limit, the point at which the first visible crack appears or resistance changes drastically. This information was critical for defining safe strain levels in the subsequent cyclic tests. The test device was self-fabricated and designed to operate continuously using a DC motor, stretching the sample repeatedly up to 5 mm elongation without surpassing the necking threshold. This limitation ensured that the conductive trace could undergo multiple cycles without catastrophic failure in the first iteration.

The tensile tests adhered to ASTM D3479/D3479M-19, a standard procedure for characterising the tensile fatigue of polymer matrix composites. The gripping mechanism was enhanced using silicone inserts to prevent slippage and ensure even load distribution. Over successive cycles, the degradation of the conductive trace followed a 3-phase failure process. First, initial cracking, where multiple microscopic cracks form, leading to a noticeable reduction in stiffness. Delamination, where the interface between the conductive ink and substrate weakens, causing partial detachment. Complete failure, where separation and mechanical fracture of the ink-substrate interface occur. Each phase was marked by corresponding increases in electrical resistance, indicating a decline in conductive performance. Understanding this progressive degradation provides critical insight for designing more durable ink formulations and substrates. While bending and tensile tests assess flexural and axial mechanical stability, torsion testing introduces a rotational shear stress, a condition frequently overlooked but essential in many real-applications. Wearable electronics and soft robotics, for example, often undergo twisting motions that can lead to early fatigue if not properly accounted for in material design.

In this work, a cyclic torsion test rig was custom-built using a rack-and-pinion gear system, DC motor, and specimen holders. The setup allows the sample to be twisted from 0° to 90° at a rate of 100 RPM, continuously cycling between angular extremes. The test conforms to both ASTM A938-18 and ASTM E2207-15, which are specifically designed to measure material behaviour under torsional loads and to determine the fatigue life based on the number of cycles to failure. The shear stress introduced during torsion can result in interfacial delamination, micro-crack propagation, and plastic deformation especially in ductile materials. Brittle materials, on the other hand, tend to exhibit low torsional tolerance, failing catastrophically after minimal displacement. The cyclic torsion test is thus invaluable for discriminating between materials based on their shear fatigue resistance, helping scientists and engineers identify formulations and composite architectures that can endure rotational stress in real-applications.

The test data, combined with post-mortem microscopy or resistance measurements, provides a complete picture of how mechanical shear influences the conductivity and structural stability of printed electronics over time. The characterisation of conductive films under mechanical stress is essential for developing robust, reliable, and long-lasting printed electronic devices. By integrating cyclic bending, tensile, and torsion testing into a unified evaluation framework, this work provides a comprehensive method for understanding the interplay between mechanical deformation and electrical performance. These insights are particularly valuable in fields such as wearable technology, flexible medical sensors, and stretchable displays, where devices are routinely subjected to complex, multi-directional stress patterns. As the field of printed electronics continues to expand, such characterisation protocols will play a crucial role in guiding material innovation, device design, and performance validation.

References

1. IPC: PC-9204—Standard Only: Guideline on Flexibility and Stretchability Testing for Printed Electronics. https://www.ipc.org/ipc-9204/ipc-9204-standard-only (2017). Accessed 23 July 2022
2. ASTM International: Standard Test Method for Flexural Fatigue Properties of Plastics. Standard ASTM D7774-17 (2017)

Chapter 5
Nanostructured Carbon Materials

5.1 Introduction

Carbon-based conductive inks have garnered substantial interest in recent years, particularly in the domain of flexible and stretchable electronics (SE), due to their inherent advantages in electrical performance, mechanical resilience, and material adaptability. Among the diverse array of conductive fillers available, carbon-based materials such as graphene, graphene nanoplatelets (GNPs), carbon black (CB), and carbon nanotubes (CNTs) stand out for their exceptional physicochemical properties. These include high electrical conductivity, robust mechanical strength, thermal stability, and superior flexibility. This characteristic that are critically important for applications involving frequent deformation, dynamic movement, or integration with non-flat surfaces [1].

The remarkable performance of carbon-based inks is fundamentally rooted in the unique electronic and structural properties of carbon nanomaterials. Graphene and its derivatives, for instance, exhibit a delocalised electron cloud derived from the sp^2 hybridised carbon atoms arranged in a two-dimensional honeycomb lattice. This structure supports exceptionally high charge carrier mobility, which enables rapid and efficient electron transport even under low-voltage conditions. Additionally, graphene's high tensile modulus and excellent mechanical resilience allow it to maintain structural integrity and conductivity under continuous mechanical stress. These properties make GNPs ideal as functional fillers in the formulation of conductive inks.

However, despite their outstanding theoretical advantages, early formulations of screen-printed carbon-based conductive inks were limited by inadequate stretchability and poor reliability under strain. These deficiencies posed significant challenges for applications such as wearable electronics, soft robotics, and flexible sensors, where mechanical compliance must coexist with reliable electrical performance. A central issue in the development of GNPs conductive inks is the formulation of an optimised ink system that can be processed using techniques like doctor

M. A. Salim et al., *Nanotechnology in Conductive Inks*, Engineering Materials,
https://doi.org/10.1007/978-981-95-5708-0_5

blading and direct writing [2]. Achieving this requires precise control over the ink's composition, including its viscosity, dispersion stability, evaporation kinetics, and particle-substrate interactions.

The ink formulation significantly influences the performance of the printed film. It governs not only the electrical conductivity but also the adhesion to different substrates, the uniformity of the printed layers, and the mechanical durability of the final structure. The choice of polymer binders, solvents, surfactants, and rheological modifiers must be carefully considered to ensure compatibility and performance across various flexible substrates such as polyethylene terephthalate (PET), polydimethylsiloxane (PDMS), polyimide (PI), and even paper. Additionally, the dispersion of GNPs must be homogeneous to prevent agglomeration, which could disrupt the continuity of the conductive network.

From a materials science and engineering perspective, the electrical behaviour of these inks is influenced by factors such as filler loading, network connectivity, and interfacial resistance between filler particles. The formation of an interconnected conductive network is crucial, that once a critical concentration of conductive fillers is reached, the material transitions from an insulative to a conductive state. This phenomenon is often discussed in terms of percolation, though without entering into mathematical formulation, it can be described as the point at which sufficient physical contact between conductive particles allows electrons to move efficiently across the material. Furthermore, the ink's dynamic behaviour under operational stresses, such as mechanical bending, stretching, environmental exposure to moisture or temperature fluctuations, plays a key role in determining its practical reliability. Mechanical fatigue, micro-crack formation, and delamination are common failure modes that can compromise both conductivity and structural performance. Therefore, comprehensive mechanical and environmental testing, including cyclic bending and durability assessment under accelerated ageing conditions, is critical in evaluating ink performance.

A carbon-based conductive inks represent a compelling solution for the development of next-generation flexible electronic systems. However, their successful deployment requires a multidisciplinary approach that combines nanomaterials science, applied physics, surface chemistry, and advanced manufacturing. Understanding the complex interplay between material formulation, structural integrity, and device performance is essential for advancing this technology toward practical, commercial applications.

5.2 Influence of Filler Loading and Dispersion Techniques on Sheet Resistance

According to a prior investigation by Mokhlis (2019), a GNPs formulation incorporating epoxy resin as a polymer matrix and a glass slide as the substrate was developed. The ink was prepared through manual mixing of the filler and polymer binder and subsequently thermally cured at 150 °C for 30 min. Filler loading concentrations were varied up to 35 wt.%, and the corresponding sheet resistance values were calculated to evaluate the ink's electrical performance. This work served as a foundational reference, or baseline, for the enhanced ink formulations discussed in the current exploration. Table 5.1 shows the filler concentration loading up to 35 wt.%.

Mokhlis observed that formulations with 10 and 15 wt.% GNPs exhibited no measurable sheet resistance. This lack of electrical conductivity was attributed to insufficient filler particle connectivity, which in turn was a result of particle agglomeration and inadequate percolation. At low filler content, conductive particles are dispersed too sparsely throughout the polymer matrix, making it difficult to form continuous conductive pathways. Agglomeration at these loadings isolates particles within the matrix, thereby inhibiting electron transport. These findings are consistent with those of [3], who reported similar challenges in low-concentration conductive nanocomposites.

As the concentration of GNPs increased from 20 to 35 wt.%, sheet resistance began to decrease significantly, signalling the formation of conductive pathways across the polymer matrix. This improvement in electrical performance is governed by the fundamental principles of percolation theory, which describes the critical point at which isolated conductive particles begin to form a connected network that allows for efficient charge transport. The concept is deeply rooted in statistical physics and is often used to explain transitions in composite systems from insulating to conductive behaviour based on filler content and spatial distribution.

In materials science, the formation of a three-dimensional conductive network is paramount. At low filler content, the charge carriers encounter long, winding pathways with limited conduction bridges, resulting in high resistance or complete electrical failure. As the loading increases and approaches a critical concentration threshold, filler particles bridge gaps more efficiently, establishing inter-particle contact through overlapping, tunnelling, or physical contact. This network reduces both constrictive resistance, which is related to the narrowing of current paths, and tunnelling resistance, where electrons move across small gaps by quantum mechanical tunnelling.

Table 5.1 Value of sheet resistance of the baseline ink and enhance GNPs and CB with different filler loading percentages

wt.% of filler	Sheet resistance (Ω/sq)		
	Mokhlis (2019)	GNP conductive ink	CB conductive ink
10	0	55.52×10^6	12.51×10^6
15	0	1.14×10^6	5.21×10^6
20	10.55×10^6	4.18×10^3	1.03×10^6
25	2.53×10^6	3.98×10^3	3.2×10^3
30	0.27×10^6	2.17×10^3	0.4×10^3
35	0.12×10^6	2.00×10^3	0.3×10^3

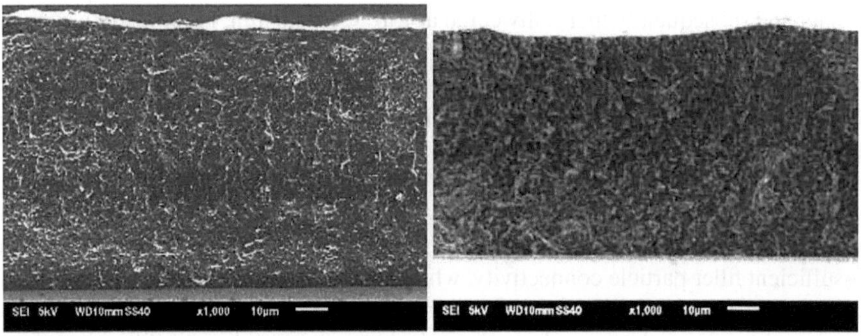

Fig. 5.1 The cross-sectional surface of the baseline ink with 25 wt.% (left) and 30 wt.% (right)

Microscopic analysis of the ink's cross-sectional surface at various GNPs concentrations revealed a strong correlation between microstructural morphology and electrical performance. The baseline ink with 25 wt.% GNPs appeared smooth, suggesting a more continuous polymer phase with limited filler protrusion. This morphology implies weaker resistance to crack initiation and insufficient mechanical anchoring between particles. In contrast, the 30 wt.% GNPs ink demonstrated a much rougher microstructure with intricate, indirect surface features such as river-like fissures, hackles, and ribbons, and indicative of higher filler dispersion, particle interaction, and structural complexity. This increase in microstructural heterogeneity aligns with the formation of robust conductive networks and thus better electrical performance. Figure 5.1 shows the morphology analysis of baseline ink in this work.

Once the GNPs loading exceeded 30 wt.%, the sheet resistance plain, indicating that further increases in filler concentration produced diminishing returns in conductivity. In this tenure, the conductive network is well-established, and additional filler serves primarily to reinforce mechanical properties or marginally reduce path resistance, rather than initiating new conductive pathways. The elevation also implies that the system may be reaching a saturation point where the conductive framework is no longer significantly influenced by small variations in composition or environmental conditions.

A comparative assessment was carried out between the baseline formulation and a new formulation developed in this work, in which a planetary centrifugal mixer was employed instead of manual mixing. This high-shear, high-speed technique drastically improved the uniformity of filler dispersion, minimised agglomeration, and eliminated entrapped air within the ink matrix. These improvements yielded markedly better electrical properties. With the same 150 °C curing temperature and 30 min duration, the enhanced formulation achieved measurable sheet resistance values even at 10 wt.% filler loading, previously unattainable with hand-mixed inks.

The new formulation exhibited sheet resistance values of 55.52×10^6 Ω/sq for GNPs and 12.51×10^6 Ω/sq for CB at 10 wt.%. This demonstrates that both fillers were capable of forming initial conductive pathways at relatively low concentrations. As the filler concentration increased to between 15 and 25 wt.%, a dramatic drop

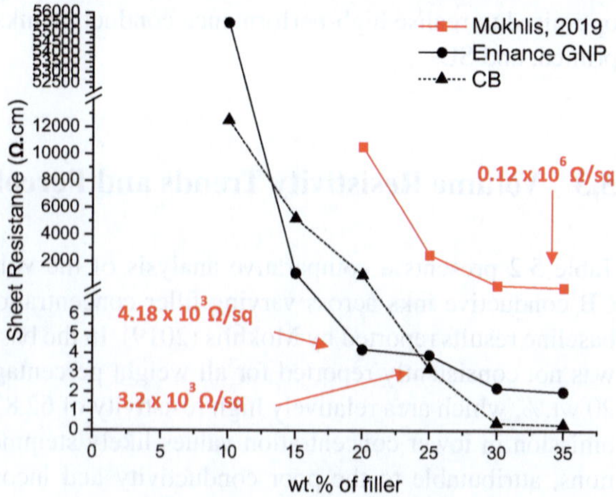

Fig. 5.2 Values of sheet resistance for different filler loading of enhance formulation

in sheet resistance was recorded for both types of ink, reflecting the establishment of well-connected conductive networks. Importantly, the GNPs achieved this percolation threshold at slightly lower filler loading than CB, suggesting more efficient network formation, likely due to the platelet morphology and stable dispersion characteristics of GNPs. Figure 5.2 shows the values of sheet resistance for different filler loading.

The CB inks showed superior electrical performance at higher loadings, achieving a lowest sheet resistance of approximately 0.3 kΩ/sq, compared to 2.0 kΩ/sq for GNPs. This is attributed to the nanoscale particle size and high specific surface area of CB, which provide more active sites for electron transport. The ink particles, with their quasi-spherical (fullerene-like) structure, possess greater curvature and reactivity, allowing for enhanced interfacial contact with the surrounding matrix and neighbouring particles. However, the tendency of ink to agglomerate at higher concentrations necessitates precise mixing and stabilisation strategies to maintain its conductive potential.

Despite its slightly higher resistivity, GNPs-based ink formulations demonstrated an advantage in terms of lower required filler loading to achieve optimal conductivity. The flake-like morphology of GNPs encourages planar overlap and long-range conductive bridges, which facilitate rapid electron movement. The GNPs also exhibit lower surface activity than CB, which reduces agglomeration and improves their dispersion stability in polymer matrices. Consequently, while CB inks may exhibit lower resistivity, GNPs ink achieve comparable performance with less material, thereby offering advantages in cost, viscosity control, and mechanical flexibility. The overall effectiveness of the conductive ink, whether CB or GNPs is ultimately governed by the inherent electronic properties of the filler, such as its intrinsic conductivity, aspect ratio, and compatibility with the polymer matrix. Moreover, the final electrical performance is shaped by processing parameters including mixing technique, curing conditions, and substrate interaction. These factors must be precisely optimised to realise high-performance conductive inks that are suitable for flexible, printed, and SEs.

5.3 Volume Resistivity Trends and Percolation Behaviour

Table 5.2 presents a comparative analysis of the volume resistivity of GNPs and CB conductive inks across varying filler concentrations, benchmarked against the baseline results reported by Mokhlis (2019). In the baseline work, volume resistivity was not consistently reported for all weight percentages of filler, only the result at 20 wt.%, which area relatively high resistivity of 62,831.7 Ω cm was recorded. This omission in lower concentration values likely stemmed from measurement limitations, attributable to the poor conductivity and incomplete network formation in under-filled systems.

A consistent trend is observed wherein the volume resistivity of the inks generally decreases with increasing filler content, culminating in the lowest value of 706.6 Ω cm at 35 wt.% filler loading. This behaviour aligns with the principles of percolation theory, which provides a theoretical framework to describe the electrical transition in composite materials from an insulating to a conductive state. The theory considers that at low concentrations, conductive fillers are dispersed too sparsely within the insulating polymer matrix, preventing the formation of continuous electron pathways. As the filler content increases, the likelihood of inter-particle contact rises, eventually reaching a critical threshold or percolation point. Beyond this point, the material transitions to a conductive state with a rapidly decreasing resistivity profile.

In both GNPs and CB conductive inks, a similar descending trend in volume resistivity is observed as the filler concentration increases. However, the performance of these materials surpasses that of the baseline across all comparable concentrations. This demonstrates the importance of improved dispersion techniques and material optimisation in the ink formulation. At low filler loadings, particularly 10 wt.%, both inks exhibit excessively high and unstable resistivity values. This is attributed to the lack of a continuous conductive network, a condition where filler particles remain

Table 5.2 Value of sheet resistance of the baseline ink and enhance GNP and CB with different filler loading percentages

wt.% of filler	Volume resistivity (Ω cm)		
	Mokhlis (2019)	GNP conductive ink	CB conductive ink
10	0	0.33×10^6	74.55×10^3
15	0	6.81×10^3	31.07×10^3
20	62.83×10^3	0.025×10^3	6.14×10^3
25	15.08×10^3	0.023×10^3	0.019×10^3
30	1.62×10^3	0.013×10^3	2.4
35	0.71×10^3	0.012×10^3	1.8

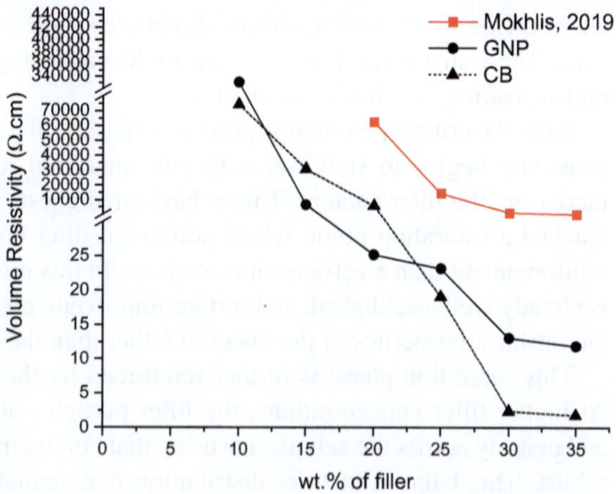

Fig. 5.3 Values of volume resistivity for different filler loading of enhance formulation

isolated or are separated by insulating gaps, preventing efficient charge migration. Such unstable behaviour renders the 10 wt.% formulations unsuitable for reliable electronic applications, and as noted by Li et al. (2011), low filler concentrations frequently result in erratic conduction and are often excluded from practical consideration. Figure 5.3 shows the values of volume resistivity for the different filler loading.

Interestingly, the performance divergence between both inks becomes pronounced at 15 wt.% filler concentration. At this level, GNPs conductive ink demonstrates significantly lower volume resistivity compared to its CB counterpart. This early improvement in conductivity suggests that GNPs, due to their planar morphology and high aspect ratio, are more effective in establishing conductive pathways at lower concentrations. Their flake-like structure allows overlapping and bridging between particles over longer distances, enhancing the probability of network formation even at modest loadings.

This trend continues across the higher concentration range, with GNPs conductive ink consistently exhibiting superior conductivity performance, that is, lower volume resistivity than CB conductive ink at corresponding filler percentages. The underlying reason lies in the differences in intrinsic material properties. The GNPs possess a larger lateral dimension and a more stable dispersion in polymer matrices, which facilitates better alignment and network connectivity. Additionally, their two-dimensional nature supports electron delocalisation across extended areas, reducing interfacial resistance and enhancing charge mobility.

At 20 wt.% loading, the GNPs conductive ink exhibits a sharp drop in volume resistivity, signifying that the critical percolation threshold has been reached. At this point, the conductive network formed by the particles becomes significantly interconnected, enabling efficient electron transport through the composite. The increased filler density allows for more frequent and stable contact between particles, thereby reducing the overall resistance of the material. In contrast, CB conductive ink achieves this substantial conductivity transition slightly later, between 20 and 25 wt.%, owing to its smaller, spherical particle shape and higher tendency to agglomerate. These structural characteristics hinder early-stage connectivity and necessitate higher loading to achieve percolation.

After the critical percolation point is surpassed, the rate of improvement in volume resistivity begins to stabilise, with only marginal reductions observed in further increasing the filler content. This behaviour suggests that the composite system has reached a saturation point, where additional filler contributes more to mechanical reinforcement than electrical enhancement. In this moment, the conductive network is already well established, and further improvements in conductivity are limited by the intrinsic properties of the material rather than the concentration of filler.

This saturation phase is further reinforced by the morphology of the ink layer. At higher filler concentrations, the filler particles are distributed more uniformly and densely across the sample, ensuring that conductive paths are evenly spread and robust. This balance in filler distribution is essential to maintain uniform current flow and reduce localised resistance peaks that can arise from poor dispersion or agglomeration. Thus, the volume resistivity plateau observed at the upper end of the filler range is indicative of optimal filler-matrix integration, where the conductive network percolates uniformly throughout the sample. Fundamentally, the electrical properties of such nanocomposite inks are governed by multiple factors such as the intrinsic conductivity of the filler material, the degree of dispersion and alignment, the interface between filler and matrix, and the overall architecture of the composite. For instance, in nanomaterial-based inks, the efficiency of electron transport depends not only on how well the particles are connected, but also on how easily electrons can hop, tunnel, or percolate across the network. The physics of these transport mechanisms is deeply rooted in the principles of quantum conduction, contact resistance, and barrier potential, all of which are influenced by nanoscale structure and processing methods.

From an engineering perspective, understanding these mechanisms is critical for designing inks modified to specific electronic functions, whether for resistive traces, sensors, or flexible interconnects. The use of advanced processing tools such as high-shear mixers, ultrasonic homogenisers, or solvent-assisted dispersion, can significantly alter the internal morphology of the ink, thereby affecting both macroscopic conductivity and mechanical stability. This interplay between nanostructure and macro-performance underscores the importance of a multidisciplinary approach, integrating materials science, physics, and process engineering to achieve high-performance printed electronics.

Therefore, the observed trends in volume resistivity across GNPs and CB conductive inks illustrate the complex yet predictable behaviour of nanomaterial-based conductive systems. The careful selection of filler type, concentration, and dispersion technique plays a decisive role in determining electrical performance, particularly in applications requiring both flexibility and reliability. These findings not only highlight the superiority of GNPs-based systems at lower filler concentrations but also reinforce the broader relevance of percolation theory and nanoscale morphology control in the design of next-generation conductive inks.

5.4 Mechanical Properties Using Nanoindentation

Mechanical properties are of equal importance to electrical characteristics when evaluating the performance, durability, and application potential of functional materials. In the context of flexible and printed electronics, understanding the mechanical robustness of conductive inks is essential, particularly for applications involving bending, stretching, or repeated mechanical stress. This section details the nanoindentation-based mechanical characterisation of GNPs and CB conductive inks, exploring their hardness and elastic modulus across varying filler loadings.

Nanoindentation, a high-resolution technique ideal for small-scale or thin film materials, was employed to assess these mechanical parameters. This method is particularly valuable in composite and nanomaterial research due to its sensitivity to microstructural features and the minimal sample volume required. All indentation experiments were conducted in accordance with ISO 14577-1 (Annex A) standards. Figure 5.4 illustrates the load–displacement behaviour of samples with varying filler loadings, highlighting how filler concentration directly influences the deformation response of the nanocomposites. There is no load–displacement profiles for the baseline ink, only its final hardness and modulus data are available.

The indentation protocol followed a standardised sequence such as surface approach, loading to peak force, a 5 s hold at maximum load to capture time-dependent deformation, followed by controlled unloading. These stages allow for differentiation between elastic and plastic behaviour, crucial in assessing the viscoelastic nature of polymeric composites. When plotted, the load–displacement curves exhibit a characteristic hysteresis loop between the loading and unloading paths, indicating that energy is dissipated as heat during deformation. This energy loss, known as hysteresis loss, confirms that the ink samples exhibit viscoelastic properties, which is typical of polymer matrix nanocomposites [4]. The area between the loading and unloading curves is proportional to the mechanical energy dissipated during the indentation cycle.

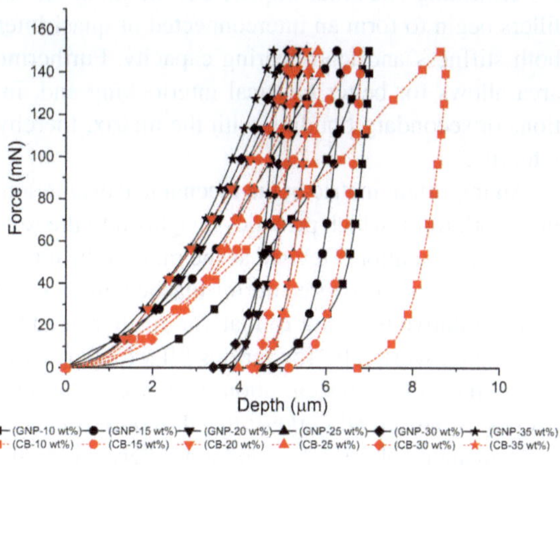

Fig. 5.4 The loading curve of different filler loading percentage of enhance GNP and CB formulation

The mechanical properties such as hardness and elastic modulus were extracted directly from the load–displacement data, without the need to image the residual indent. This was accomplished using the methodology developed by Oliver and Pharr, which has become the gold standard in interpreting nanoindentation data. Their approach relies on the estimation of the projected contact area based on penetration depth, a method particularly effective in heterogeneous nanocomposite systems [5]. A consistent indentation load of 150 mN was applied across all samples, with filler concentrations ranging from 10 to 35 in 5 wt.% increments. These samples correspond to those previously evaluated for electrical properties, ensuring a holistic assessment of electro-mechanical behaviour.

It reveals that both GNPs and CB inks with 10 wt.% filler content demonstrate the deepest penetration depths, indicating the lowest hardness and elastic modulus. This softness is a consequence of the low filler concentration, which results in a polymer-rich matrix with minimal reinforcement. As filler content increases, the indentation depth decreases, reflecting a marked improvement in mechanical resistance. For GNPs-based inks, the depth decreases by approximately 11.8, 11.3, 3.6, 4.6, and 5.8%, respectively with each incremental increase from 10 to 35 wt.%. A similar trend is observed for CB inks, with reductions of 24.8, 14.4, 5.9, 4.0, and 2.3% across the same range. These reductions clearly illustrate the reinforcing effect of conductive fillers within the polymer matrix.

The more substantial reduction in indentation depth observed in the GNPs-based samples suggests that it offers superior mechanical reinforcement compared to CB. This can be attributed to the intrinsic mechanical strength of graphene, which possesses one of the highest Young's modulus among known materials, as well as its high aspect ratio and large surface area, which promote strong filler–matrix interaction and efficient stress transfer. In contrast, CB particles, being more spherical and possessing a higher tendency to agglomerate, present less effective load transfer and a more localised stress distribution. The marked improvement in hardness as the filler content increases can be explained by fundamental nanocomposite mechanics. The incorporation of rigid nanofillers within a ductile matrix introduces stress-concentrating sites that impede deformation. At sufficient volume fractions, these fillers begin to form an interconnected or quasi-interconnected skeleton, enhancing both stiffness and load-bearing capacity. Furthermore, the increased filler surface area allows for better physical interlocking and, in some cases, chemical interactions or secondary bonding with the matrix, thereby further improving mechanical integrity.

An important metric in the mechanical evaluation of these systems is the indentation work ratio, which provides insights into the relative contributions of elastic and plastic deformation. Defined as the ratio of the total work performed during indentation to the work recovered during elastic unloading, the indentation work ratio serves as a qualitative indicator of material resilience and recoverability. The values of the indentation work ratio for various filler loadings are presented in Table 5.3.

As the filler concentration increases, the indentation work ratio generally decreases. This trend suggests that the composite becomes less elastic and more prone to permanent (plastic) deformation, likely due to the rigid nature of the conductive

Table 5.3 Value of indentation work ratio for enhance GNP and CB conductive ink with different loading

wt.% of filler	Indentation work ratio (%)	
	GNP conductive ink	CB conductive ink
10	23.514	22.512
15	17.770	19.542
20	22.135	17.845
25	16.004	21.654
30	11.653	14.894
35	8.598	12.135

network and the reduction in matrix flexibility. While greater filler content improves hardness, it may simultaneously reduce the ability of the composite to recover after mechanical stress, which is a key consideration in applications requiring repeated mechanical flexing.

Particular interest are the 20 and 25 wt.% filler concentrations, which show the highest average indentation depth recovery and relatively high of the indentation work ratio values. This implies that these formulations achieve a desirable balance between stiffness and elastic recovery. Despite not having the highest absolute hardness, they maintain sufficient plastic endurance while still recovering elastically, making them well-suited for mechanically demanding applications such as flexible electronics, wearable devices, and sensors. In contrast, at 35 wt.% filler loading, although the indentation depth is shallow indicating a hard surface, the indentation work ratio is at its lowest for both GNPs and CB inks. This suggests that the material behaves in a predominantly plastic manner, with limited elastic recovery. This loss of elasticity can be attributed to the over-saturation of filler within the matrix, which not only constrains the polymer's ability to flex but also leads to microstructural rigidity and brittleness. Furthermore, excess filler can result in agglomeration, poor dispersion, and void formation, all of which compromise mechanical reliability.

The observations from nanoindentation thus reinforce a crucial principle in nanomaterial composite design, there exists an optimal filler concentration range that maximises mechanical reinforcement while preserving elasticity and durability. Beyond this range, further increases in filler content yield diminishing returns or even adverse effects on mechanical behaviour. The nanoindentation characterisation provides valuable insights into the structure–property relationship of GNPs and CB conductive inks. The improved mechanical performance of GNPs-based inks is strongly tied to the fundamental material properties of graphene, including its stiffness, morphology, and interfacial behaviour with the polymer binder. These findings, when combined with the earlier electrical characterisation results, underscore the importance of comprehensive electro-mechanical analysis for the development of next-generation conductive inks. The ability to adapt mechanical properties through precise control of filler content and dispersion opens pathways for high-performance, durable, and flexible electronic systems.

5.5 Evaluation of Hardness and Elastic Modulus

The mechanical characterisation of conductive ink samples, particularly their hardness and elastic modulus, was carried out using nanoindentation instrumentation, which is a powerful technique that enables precise assessment of material properties at the nanoscale. These parameters are critical indicators of a material's ability to withstand deformation under mechanical load, and it play an essential role in determining the structural performance and durability of materials designed for flexible electronic systems. Table 5.4 summarises the nanoindentation findings obtained for conductive inks containing GNPs and CB at varying filler loadings of 10, 15, 20, 25, 30, and 35%, along with baseline samples devoid of filler. The values presented represent averaged measurements to ensure statistical reliability and reduce the impact of localised material heterogeneity.

In nanoindentation analysis, hardness is defined as the material's resistance to permanent plastic deformation under an applied load. According to Maganty [5], this property can be deduced from the relationship between the applied force and the depth of the resulting indentation. During the indentation process, the indenter, typically a diamond tip with a calibrated geometry is driven into the surface of the sample. As the load increases, the material initially deforms elastically, followed by plastic flow once the yield stress is exceeded. The load–displacement curve generated from this process captures the entirety of the elastic–plastic response, from which hardness and modulus values can be extrapolated. The depth of penetration is inversely related to material hardness: the shallower the depth under a given load, the harder the material.

In the present work, both GNPs and CB conductive inks exhibited an increase in hardness with higher filler concentrations. Specifically, the GNPs-based ink showed a rise in hardness from 0.113 GPa at 10 wt.% to 0.208 GPa at 35 wt.%, while the CB ink increased from 0.076 GPa to the same terminal value. These results corroborate earlier findings by Mokhlis (2019), who reported that the hardness of graphene-based inks rose from 0.109 to 0.277 GPa with increasing graphene content. This incremental

Table 5.4 Value of hardness and elastic modulus of baseline ink, and different enhance GNP and CB loading percentage

wt.% of filler	Baseline conductive ink (Mokhlis 2019)		GNP conductive ink		CB conductive ink	
	Hardness (GPa)	Elastic modulus (GPa)	Hardness (GPa)	Elastic modulus (GPa)	Hardness (GPa)	Elastic modulus (GPa)
10	0.128	8.531	0.113	7.680	0.076	5.440
15	0.156	8.396	0.154	9.521	0.105	7.500
20	0.223	7.995	0.191	10.092	0.125	9.771
25	0.294	6.315	0.187	11.670	0.146	10.300
30	0.315	5.210	0.185	12.491	0.116	8.759
35	0.367	5.112	0.208	12.876	0.208	10.210

improvement in hardness is attributed to the reinforcing effect of the nanofillers and shows in Fig. 5.5. When nanomaterials such as GNPs or CB are integrated into a polymer matrix, they restrict the mobility of polymer chains, effectively enhancing the composite's load-bearing capacity and resistance to indentation. The intrinsic rigidity of these carbon-based fillers, especially graphene with its exceptional in-plane strength due to the sp^2 carbon network, contributes significantly to the improved mechanical properties of the ink.

The GNPs-based inks consistently exhibited higher hardness than their CB counterparts across all tested concentrations. This can be attributed to several material characteristics. It possesses a large aspect ratio and high surface area, which allow for greater mechanical interlocking within the matrix and more efficient stress distribution. Their planar geometry provides superior reinforcement in comparison to the near-spherical morphology of CB, which is less effective at bridging and constraining polymer chains. Furthermore, graphene's two-dimensional crystalline lattice offers higher Young's modulus and tensile strength, leading to a stiffer and more robust composite structure.

However, some deviations from the expected trend were observed. For instance, the GNPs ink at 20 wt.% exhibited a slightly higher hardness than at 25 wt.%, and similarly, a CB ink at 25 wt.% exceeded the hardness of the 30 wt.% formulation. These anomalies may be linked to filler aggregation, a phenomenon that often occurs at intermediate filler concentrations. When nanofillers are not well-dispersed, they tend to cluster into agglomerates that behave differently under mechanical stress. Aggregates can create localised zones of densification, which may temporarily enhance resistance to indentation, thereby producing higher hardness values. However, these zones are also more susceptible to brittle fracture and mechanical failure, as the internal cohesion of agglomerates is typically weak and prone to rupture under shear or compressive stress. Besides, [6, 7] have both discussed the dualistic role of aggregation in polymer nanocomposites, while it may initially

Fig. 5.5 Graph of hardness for different GNPs and CB loading percentage of enhance formulation

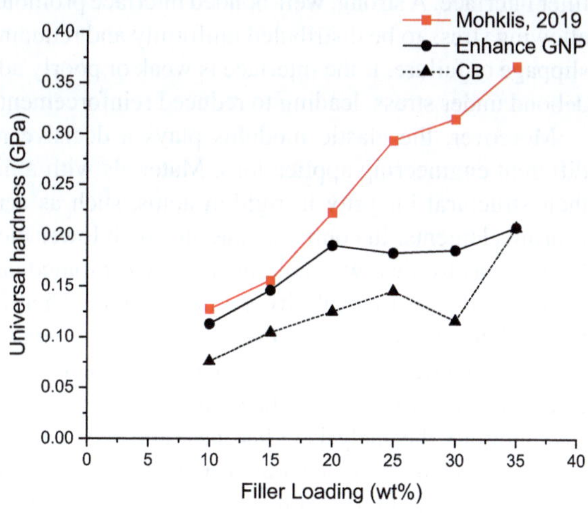

act as a reinforcement mechanism, it also introduces stress concentration sites that degrade long-term mechanical performance.

The elastic modulus, another key parameter derived from nanoindentation testing, measures the stiffness of the material and its ability to resist elastic deformation. It is fundamentally linked to the interatomic bond strength and reflects how much a material deforms reversibly under load. A high modulus indicates a stiffer material, which returns to its original shape more readily upon unloading. This property is particularly important in applications requiring dimensional stability and mechanical endurance, such as structural supports or rigid electronic substrates. In the context of this work, the elastic modulus of the baseline ink without fillers, decreased from 8.531 to 5.112 GPa, which may be indicative of inconsistencies in the sample preparation process. For instance, incomplete mixing or localised filler segregation can result in regions of differing mechanical character, especially if the indentation occurs on a filler-deficient zone. Such variability underlines the importance of homogeneity in the fabrication of nanocomposites, as even minor inconsistencies can produce significant differences in local mechanical responses.

Conversely, the elastic modulus of the experimental GNPs and CB inks increased with higher filler loading. The GNPs-based ink rose from 7.680 to 12.876 GPa, while the CB ink increased from 5.440 to 10.210 GPa across the loading range. These improvements reflect the ability of nanofillers to stiffen the matrix by constraining molecular motion and creating a more continuous stress transfer pathway. The effectiveness of graphene as a reinforcing agent is again apparent, as its higher surface area and mechanical properties offer superior stiffness enhancement compared to CB.

Fundamentally, this stiffness arises from the nanofillers' influence on the local polymer structure. When stress is applied to the composite, a portion of the load is transferred from the softer polymer chains to the rigid filler network. The efficiency of this load transfer depends on several factors, including the aspect ratio, surface functionality, and orientation of the fillers, as well as the nature of the polymer–filler interface. A strong, well-bonded interface promotes better mechanical coupling, allowing stress to be distributed uniformly and reducing the likelihood of interfacial slippage or failure. If the interface is weak or poorly adhered, the filler particles may debond under stress, leading to reduced reinforcement efficiency.

Moreover, the elastic modulus plays a decisive role in material selection for different engineering applications. Materials with a high modulus are favoured for their structural integrity in rigid systems, such as aerospace components or load-bearing elements. In contrast, materials with lower modulus values are essential for flexible electronics, wearable devices, and biomedical sensors, where mechanical compliance and stretchability are outstanding. Therefore, by tuning the concentration and dispersion of nanofillers, it is possible to engineer materials with following by the mechanical properties that match the specific functional demands of emerging electronic and structural applications.

Therefore, the analysis of hardness and elastic modulus provides critical insights into the mechanical behaviour of nanocomposite conductive inks. The incorporation of GNPs and CB significantly enhances these properties, although the extent

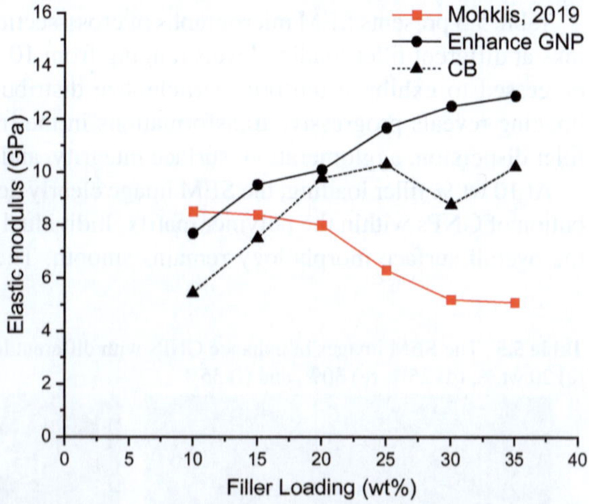

Fig. 5.6 Graph of elastic modulus for different GNP loading percentage of enhance formulation

of improvement is governed by filler type, concentration, and dispersion quality. These findings underscore the complex interrelationship between nanostructure, interface chemistry, and bulk mechanical performance, an understanding that is essential for the rational design of high-performance, multifunctional materials for next-generation technologies. Figure 5.6 shows the graph of elastic modulus for different loading of GNPs.

5.6 Mechanisms of Void Formation, Agglomeration, and Network Formation

The microstructure of conductive inks plays a decisive role in determining both their electrical and mechanical performance. In advanced materials science, the internal arrangement, size, morphology, and spatial distribution of filler particles directly influence macroscopic properties such as conductivity, stiffness, toughness, and durability. In particular, the microstructural evolution of the ink with varying filler loading provides deep insight into the interaction between the conductive phase and the polymer matrix. This interplay governs charge transport, mechanical reinforcement, and long-term stability.

Conductive inks formulated with nanomaterials such as GNPs and CB exhibit microstructural behaviours that are highly sensitive to filler concentration and dispersion quality. Both GNPs and CB represent carbonaceous nanostructures with distinct physical characteristics which are GNPs are two-dimensional, plate-like structures with high aspect ratios, while CB consists of quasi-spherical particles often forming fractal aggregates. These morphological differences lead to divergent behaviour in ink formulations under otherwise identical processing conditions.

Table 5.5 presents SEM micrographs of cross-sections of GNPs-based conductive inks at different filler loading levels ranging from 10 to 35 wt.%, all of which were processed to exhibit a uniform particle size distribution of 5 microns. The SEM imaging reveals progressive transformations in the ink's microstructure, including filler dispersion, agglomeration, surface integrity, and void formation.

At 10 wt.% filler loading, the SEM image clearly depicts a relatively sparse distribution of GNPs within the polymer matrix. Individual platelets are well-spaced, and the overall surface morphology remains smooth. The dispersion appears uniform,

Table 5.5 The SEM images of enhance GNPs with different loading of (a) 10 wt.%, (b)15 wt.%, (c) 20 wt.%, (d) 25%, (e) 30%, and (f) 35%

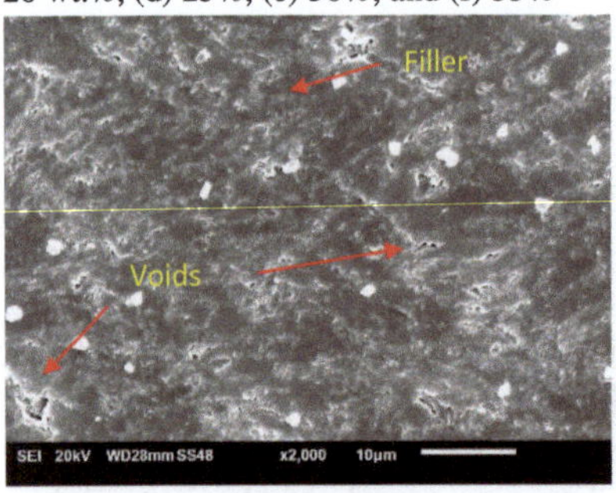

(a)10 wt.%

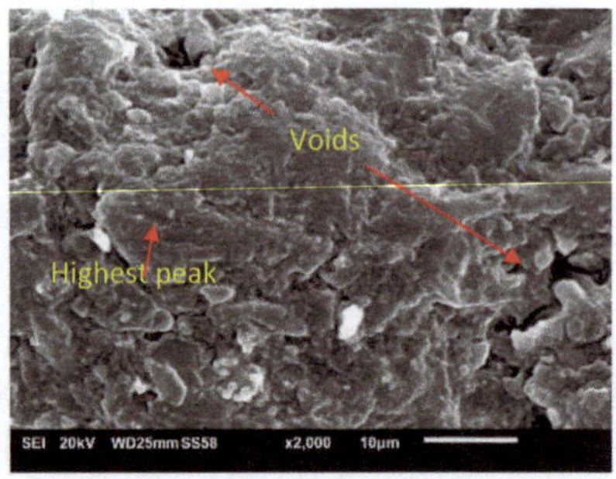

(b)15 wt.%

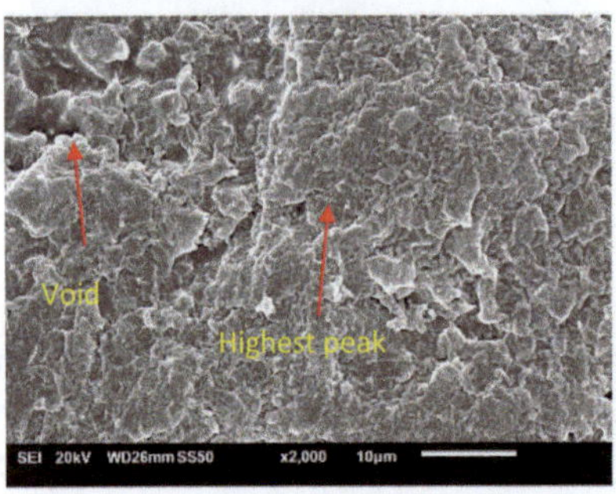

(c)20 wt.%

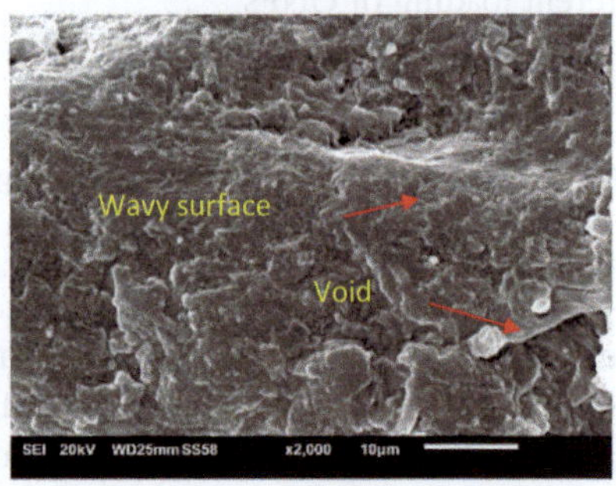

(d)25 wt.%

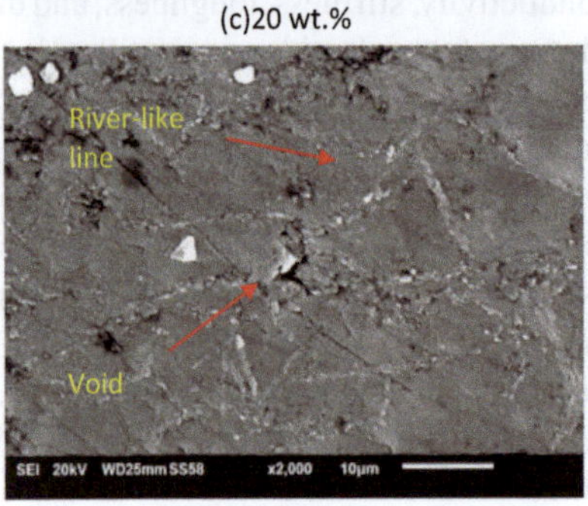

(e)30 wt.%

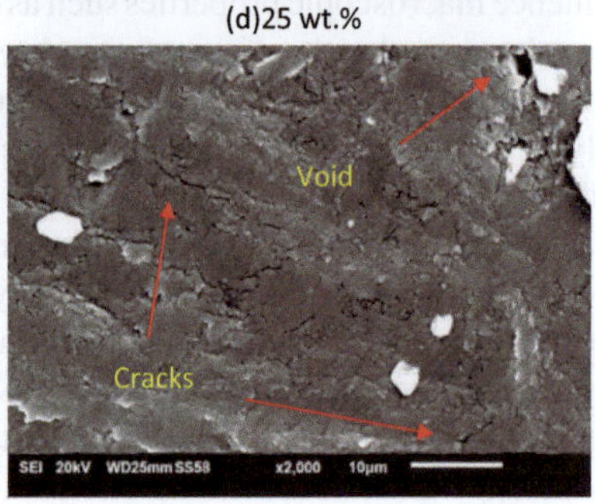

(f)35 wt.%

with minimal agglomeration. However, due to the insufficient volume fraction of conductive filler, inter-particle connectivity remains poor. The lack of percolating pathways restricts both mechanical reinforcement and electron transport, thus limiting the ink's overall performance.

When the filler loading is increased to 15 wt.%, microstructural irregularities begin to emerge. The SEM observations reveal the presence of agglomerated GNPs clusters, which form due to van der Waals interactions and inadequate stabilisation in the polymeric medium. These clusters introduce surface waviness and result in a heterogeneous distribution of the conductive phase. Particularly, increased void formation is observed at this stage. The presence of voids in polymer nanocomposites is detrimental, as they serve as stress concentrators under mechanical loading and act as scattering centres for charge carriers, thereby deteriorating both mechanical toughness and electrical conductivity. Besides, [8] have noted that voids can drastically affect the wear resistance and plastic deformation behaviour of composite materials. In the case of the 15 wt.% sample, such void-induced stress localisation may contribute to premature mechanical failure and sub-optimal functional performance.

By increasing the filler concentration to 20 and 25 wt.%, the microstructure evolves significantly. The SEM images show a much smoother and denser cross-sectional morphology. The GNPs form a more interconnected and continuous network, which is critical in the context of percolation theory, a conceptual framework used to describe the onset of electrical conductivity in disordered systems. Above a critical volume fraction, conductive fillers form continuous pathways that allow for efficient charge transport. At these concentrations, the improved particle-particle contact enhances both electron mobility and load transfer capability within the composite. Furthermore, the reduction in voids and uniform distribution of platelets suggest a more efficient filler–matrix interaction, which is fundamental to stress transfer and crack inhibition mechanisms.

The GNP's high aspect ratio and intrinsic stiffness, coupled with its large surface area, enable superior reinforcement of the surrounding matrix when adequately dispersed. As filler–matrix interfacial adhesion improves, stress is more effectively transferred from the softer polymer to the rigid GNPs. This contributes to an increase in composite modulus and hardness, which is corroborated by the mechanical characterisation data presented earlier. Moreover, the decreased number of voids at this loading results in higher density and mechanical robustness of the ink.

At 30 and 35 wt.% filler loadings, the SEM micrographs begin to show signs of microstructural degradation. The ink appears dry and brittle, with evident surface cracking and a loss of continuity in the matrix. This phenomenon is commonly attributed to insufficient binder availability relative to the high volume of nanofillers. In the absence of sufficient polymer, the ink loses its cohesive integrity, leading to micro-cracking and embrittlement. This condition undermines the ink's suitability for applications involving repeated mechanical deformation, such as flexible circuits or wearable sensors. Although high filler loading enhances electrical conductivity and mechanical stiffness, it often comes at the expense of ductility and processability. A balanced formulation is therefore required to achieve a synergy between conductivity, toughness, and flexibility.

In parallel, Table 5.6 presents SEM micrographs of CB conductive inks across the same filler loading range. The morphology of CB inks evolves differently due to the inherent properties of the particles. At lower filler concentrations (10 and 15 wt.%), SEM images depict a smoother surface, but with the presence of numerous voids. The sparse distribution of ink particles and insufficient interconnectivity result in a less cohesive structure, adversely impacting mechanical and electrical performance. As the CB filler loading increases to 20, 25, 30, and 35 wt.%, a gradual transformation in microstructure is observed. The surface becomes more textured and wrinkled, and the void density decreases markedly. This suggests that the increasing number of ink particles fills the previously empty interstitial spaces, contributing to a denser and more continuous microstructure. Although CB has a lower aspect ratio compared to GNPs, its smaller particle size and greater number of contact points at a given volume fraction facilitate the formation of a percolated network at moderate loading levels. These interconnected clusters form a conductive backbone that enhances electron transport throughout the composite.

Interestingly, even at high filler loadings, CB inks do not show significant signs of micro-cracking, as observed in the GNPs-based formulations. This could be attributed to the intrinsic morphology of it aggregates, which form more compliant and deformable networks due to their quasi-spherical structure. The ability of ink particles to deform and accommodate strain contributes to improved toughness and resistance to crack propagation. Furthermore, the cohesive interaction between CB particles and the polymer binder appears to be sufficient even at elevated loadings, allowing the ink to maintain structural integrity under stress.

The reduction in voids and enhanced filler dispersion in CB inks at high loadings also lead to better charge conduction pathways and improved mechanical properties. A denser microstructure reduces pathways for crack initiation and propagation, while also enhancing thermal and electrical transport due to improved inter-particle connectivity. These observations are consistent with established theories in composite mechanics and percolation, where an optimised network of conductive fillers significantly boosts overall composite functionality.

Consequently, the SEM characterisation of both GNPs and CB conductive inks illustrates the intricate relationship between filler concentration, microstructure, and resulting material properties. Understanding the morphological evolution of these inks is essential for accommodate their performance for specific applications. The insights gained from microstructural analysis not only inform ink formulation strategies but also contribute to the broader field of functional nanocomposites, where the convergence of nanoscience and materials engineering enables the design of next-generation flexible, stretchable, and high-performance electronic systems.

Table 5.6 The SEM images of CB with different loading of (a) 10 wt.%, (b)15 wt.%, (c) 20 wt.%, (d) 25%, (e) 30%, and (f) 35%

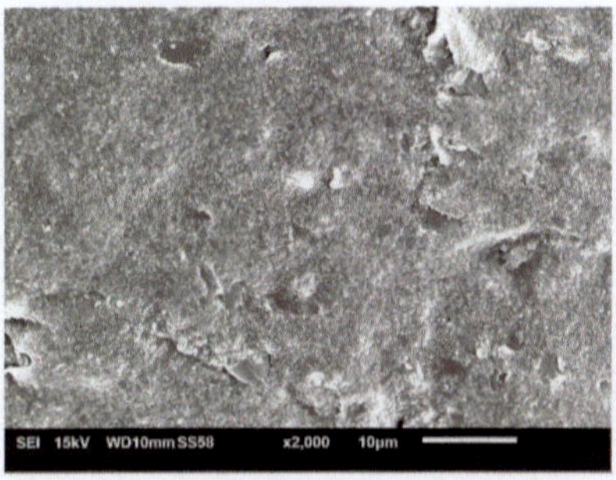

(a)10 wt.%

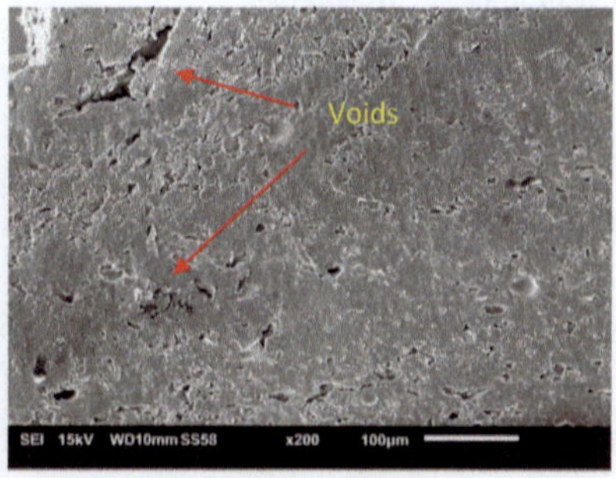

(b)15 wt.%

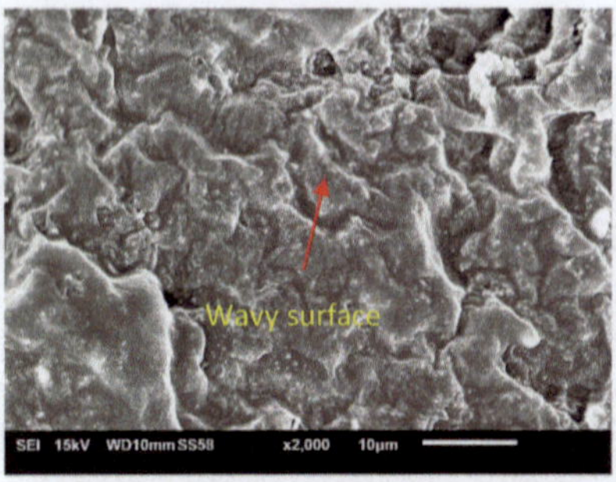

(c)20 wt.%

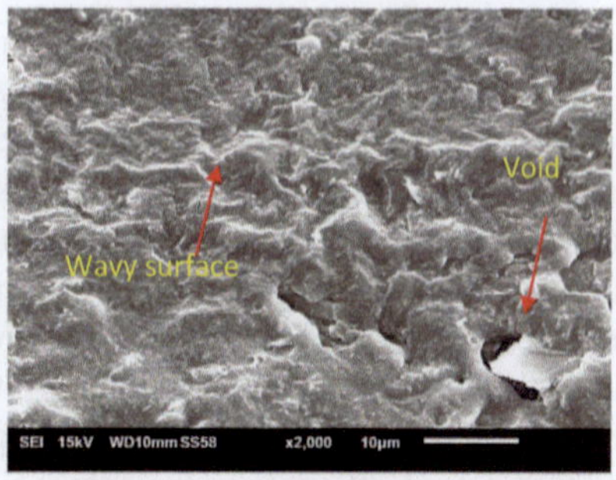

(d)25 wt.%

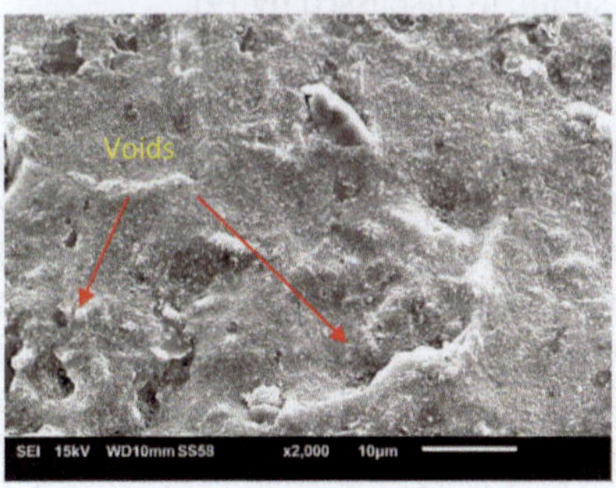

(e)30 wt.%

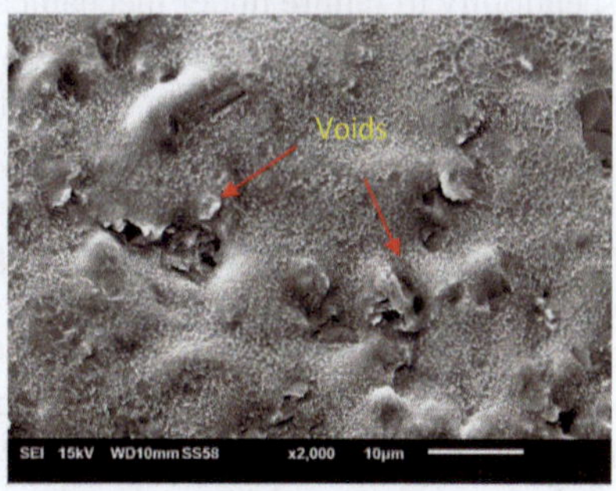

(f)35 wt.%

5.7 Elemental Composition Using EDX Spectroscopy

The EDX analysis results, as presented in Table 5.7, revealed great variations in the elemental weight percentages, particularly carbon and oxygen, across conductive inks formulated with different filler loading concentrations. These variations are instrumental in understanding the material's composition, chemical behaviour, and eventual performance characteristics in electronic applications.

Table 5.7 Filler loading elements' weights percentage

wt.% of filler	Weight (%)					
	GNPs conductive ink			CB conductive ink		
	Carbon, C	Oxygen, O	Total	Carbon, C	Oxygen, O	Total
10	62.47	37.53	100%	62.47	37.53	100%
15	65.33	34.67		65.33	34.67	
20	67.95	32.05		67.95	32.05	
25	69.09	30.91		69.09	30.91	
30	72.52	27.48		72.52	27.48	
35	73.54	26.46		73.54	26.46	

The EDX spectra consistently demonstrated the presence of carbon and oxygen in all samples, with carbon comprising the dominant element. This outcome is anticipated, given the carbon-rich nature of both GNPs and CB fillers. For GNPs-based inks, the lowest carbon content observed was approximately 62.47 at 10 wt.% graphene loading. As the graphene content increased, so did the carbon signal, corroborating the hypothesis that filler concentration is directly related to elemental carbon presence. However, while a higher filler loading contributes positively to conductivity due to the formation of a percolating conductive network, it also introduces challenges related to mechanical integrity. Specifically, increased graphene content can lead to greater brittleness in the composite, resulting in micro-cracks and increased susceptibility to failure under mechanical strain, as discussed by [9].

The predominance of carbon over oxygen in all ink formulations further confirms the conductive nature of these materials, as carbon serves as the primary carrier of electric charge in these composites. The oxygen detected in the inks originates from both the graphene structure itself and the epoxy matrix used as the binder. Graphene oxide and partially reduced graphene typically retain residual oxygen-containing functional groups such as hydroxyl, carboxyl, and epoxy groups. These oxygen functionalities play a dual role. On one hand, they enable better interaction and bonding with the polymeric binder through hydrogen bonding and van der Waals forces. On the other hand, excessive oxygen can impair the conductive pathways by introducing defects and disrupting the sp2 hybridised carbon network essential for delocalised electron transport.

According to Li et al. (2014), the oxygen present on the graphene surface can enhance molecular contact with the epoxy resin by disrupting the resin's hydrogen bonding network, thereby facilitating improved filler dispersion and adhesion. However, this comes at the cost of slightly reduced electronic performance. Moreover, [10] emphasise that residual oxygen can significantly reduce the intrinsic quality of graphene, leading to post-deposition etching and the initiation of multilayer graphene domains. These defects are not only detrimental to electrical conductivity but also promote localised degradation mechanisms. Furthermore, [11] highlight that during the thermal curing phase, oxygen within the ink matrix is ideally replaced by an inert atmosphere, such as nitrogen, to promote reduction and crystallisation of graphene sheets. However, in practical processing conditions, complete elimination of oxygen is often unachievable. The remnant oxygen can contribute to the formation of oxide species, interfere with nucleation kinetics, and reduce the density of high-quality graphene domains. Such imperfections underscore the delicate balance between functionalisation for dispersion and structural purity for conductivity.

Tables 5.8 and 5.9 provide further insights through elemental mapping of optimised GNPs (20 wt%) and CB (25 wt%) conductive inks, respectively. These figures visually depict the spatial distribution of carbon and oxygen across the ink's cross-sectional surface. The carbon distribution appears uniform, forming interconnected networks essential for electron mobility. In contrast, the oxygen distribution exhibits a tendency to concentrate in regions of structural imperfection, such as micro-cracks or corroded domains. This observation aligns with the studies of [12, 13], who reported that oxygen atoms preferentially adsorb or migrate to defect-rich zones, forming oxide layers that compromise the ink's mechanical integrity and electrochemical stability.

The oxygen penetration into the GNPs matrix suggests the existence of microstructural pathways, either inherent or formed during curing through which oxidative

Table 5.8 The EDX mapping images for 20 wt.% of GNP ink

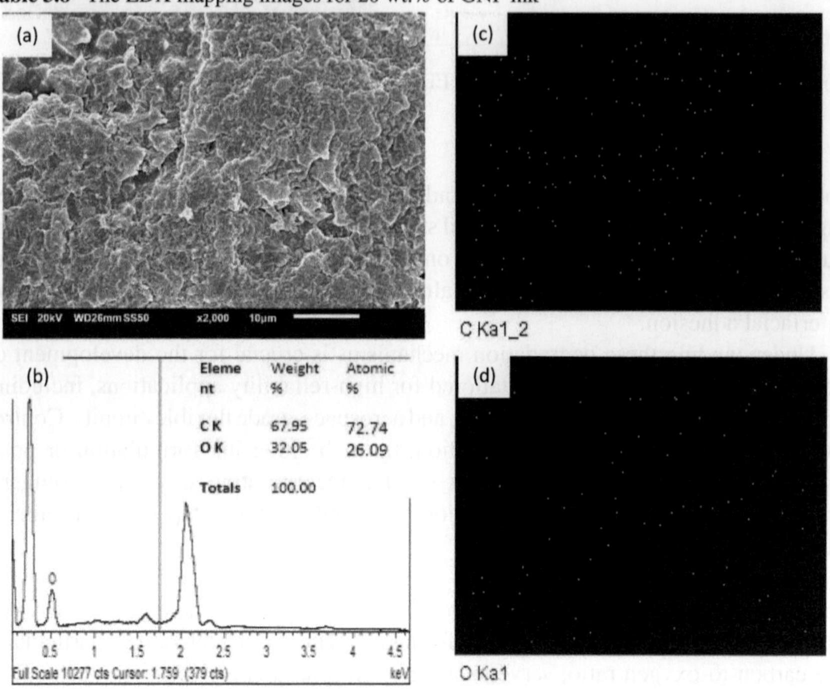

(a) SEM image, (b) EDX element graph, (c) EDX mapping image of Carbon (GNPs), and (d) EDX mapping image of oxygen

Table 5.9 The EDX mapping images for 25 wt.% of CB ink

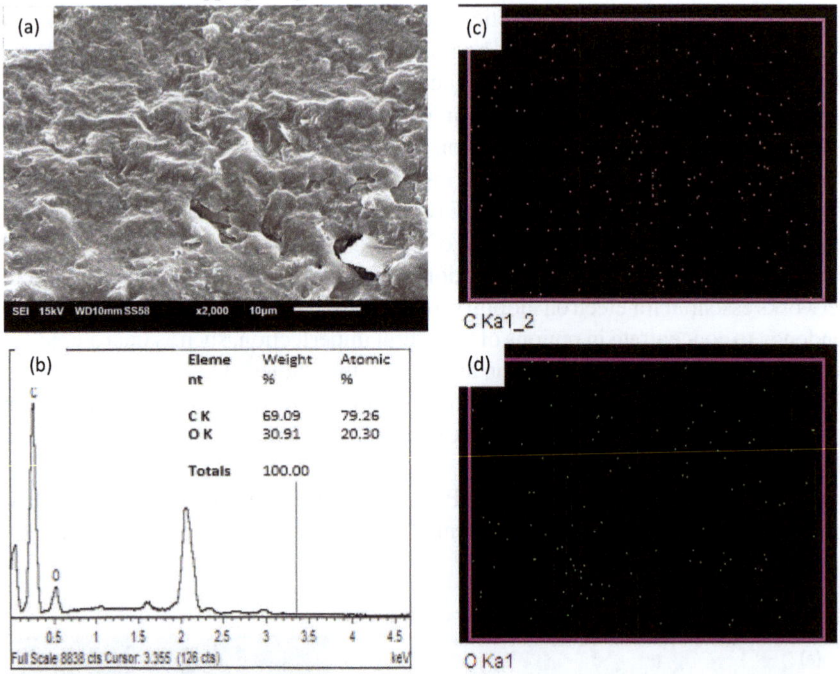

(a) SEM image, (b) EDX element graph, (c) EDX mapping image of carbon (CB), and (d) EDX mapping image of oxygen

species may diffuse. These cracks or voids serve as channels for atmospheric oxygen ingress, especially under environmental stresses such as elevated temperature or high humidity. The resultant oxidation not only weakens the mechanical robustness but also degrades electrical properties by altering surface energy, carrier mobility, and interfacial adhesion.

Understanding these degradation mechanisms is crucial for the development of next-generation conductive inks tailored for high-reliability applications, including printed electronics, stretchable sensors, and aerospace-grade flexible circuits. Control over the oxygen content and distribution, through either ink formulation or post-processing, is paramount. This may involve incorporating oxygen scavengers, enhancing the degree of filler reduction, or adopting multi-step curing protocols under controlled atmospheres.

Finally, the EDX analysis not only verifies the elemental composition of GNPs and CB conductive inks but also provides critical insight into the interplay between material structure and functional performance. The elemental ratios, particularly the carbon-to-oxygen ratio, serve as a proxy for predicting electrical, mechanical, and thermal behaviour of the ink. Future developments in conductive ink technology will benefit greatly from a multidisciplinary approach, by integrating materials science, surface physics, and nanotechnology to optimise ink formulation and ensure durability in demanding operational environments.

5.8 Optimisation of Filler Loading

Building upon prior work that served as a foundational benchmark, this exploration developed a novel conductive ink formulation utilising GNPs and CB as conductive fillers, embedded within an epoxy-based polymer resin binder. This composite ink was stencil-printed onto glass substrates for electrical characterisation. The investigation focused on two critical variables which are filler loading concentration and the employed mixing techniques, both of which are well known to influence filler dispersion and conductive network formation within polymer matrices. The primary goal of this stage was to identify the optimal formulation parameters that enhance the electrical conductivity of the ink system. Electrical conductivity was inferred from sheet resistance values whereby lower values indicate more efficient electron transport. The baseline conductive ink, which incorporated GNPs fillers without process enhancements, was used as a point of reference. The improvements sought in this work revolved around achieving a more interconnected conductive network, reduced defect density, and superior mechanical integrity.

In addition to electrical characterisation, a comprehensive analysis of the inks' mechanical properties and morphological structure was conducted. This multi-faceted approach provided insight into the interrelationships between microstructure, filler content, and performance outcomes, which are central themes in the field of advanced functional nanocomposites. The optimal filler loading concentration for the enhanced GNPs-based conductive ink was determined to be 20 wt.%, yielding a sheet resistance of 4.18×10^3 Ω/sq. For the CB conductive ink, optimal performance was achieved at 25 wt.%, with a lower sheet resistance of 3.20×10^3 Ω/sq. In contrast, the baseline GNPs ink required a much higher filler loading of 35 wt.% to reach a comparatively inferior sheet resistance of 0.12×10^6 Ω/sq. These findings clearly demonstrate that efficient dispersion and network formation can achieve superior conductivity at significantly lower filler concentrations.

The comparative characterisation of the baseline GNPs ink, enhanced GNPs ink, and CB ink, covering electrical, mechanical, and morphological parameters is presented in Table 5.10. This comparison reveals critical insights into the structure–property relationships of each formulation. From the finding, it is evident that both the enhanced GNPs and CB conductive inks exhibit substantially lower volume resistivity compared to the baseline. This improvement can be attributed to the optimised dispersion of the nanofillers, which facilitates the formation of a more continuous and percolative conductive pathway. Such conductive paths are essential for reducing electron scattering and minimising the distance between conductive domains, both of which are crucial for efficient charge transport in polymer-based systems.

Table 5.10 Summary of electrical and mechanical characterisation for enhance GNPs and CB

Conductive ink	Volume resistivity, Ω.cm	Hardness, GPa	Elastic modulus, GPa
Baseline GNP	706.6	0.367	5.112
Enhance GNP	25.5	0.190	10.093
CB	19.1	0.145	10.300

Moreover, these optimised formulations display a moderate decrease in hardness, suggesting a balance between mechanical robustness and flexibility which an important consideration in printed electronics, where devices are frequently subject to bending or stretching. More importantly, both inks exhibit an increase in elastic modulus relative to the baseline, indicating a stiffer and more resilient matrix. A higher elastic modulus correlates with improved resistance to elastic deformation, which is particularly beneficial in applications requiring mechanical integrity over prolonged operational cycles, such as flexible sensors and wearable devices.

The mechanical reinforcement observed is grounded in nanomaterial theory, wherein the intrinsic strength of carbon nanostructures (such as GNPs and CB) enhances the mechanical properties of polymer composites through stress transfer mechanisms. These mechanisms rely heavily on the aspect ratio, surface area, and degree of interfacial bonding between the nanofillers and polymer chains. When these interactions are optimally tuned with appropriate mixing and dispersion. The resulting composite exhibits enhanced stiffness, while maintaining sufficient flexibility and stretchability. Furthermore, the morphological characterisation confirms that the refined formulations possess a more homogeneous filler distribution, reduced void density, and stronger interfacial adhesion. These microstructural improvements contribute not only to better electrical conductivity but also to mechanical durability, as voids and poor dispersion often serve as initiation points for mechanical failure or electrical discontinuities.

Therefore, the optimisation of filler loading and processing conditions has yielded conductive inks with superior multifunctional performance. With these parameters defined, the improved formulations are now suitable for advanced applications. This will enable a deeper understanding of their behaviour under dynamic mechanical conditions and their viability in next-generation flexible electronics.

References

1. Kedambaimoole, V., Shirhatti, V., Neella, N., Ray, P., Rajanna, K., Nayak, M.M.: PDMS encapsulated graphene-nickel composite film as flexible tactile sensor. In 2017 IEEE Sensors (pp. 1–3). IEEE (2017, October)
2. Ghediya, P. R., Chaudhuri, T.K.: Microwave-processed copper zinc tin sulphide (CZTS) inks for coatings in solar cells. In Advanced Ceramic and Metallic Coating and Thin Film Materials for Energy and Environmental Applications (pp. 121–174). Springer International Publishing, Cham (2017)

3. Lee, S., Koo, J., Kang, S.K., Park, G., Lee, Y.J., Chen, Y.Y., Rogers, J.A.: Metal microparticle – Polymer composites as printable, bio/ecoresorbable conductive inks. Mater. Today **21**(3), 207–215 (2018)
4. Poon, P.C.B.: A critical appraisal of nanoindentation with application to elastic-plastic solids and soft materials. Dissertation (Ph.D.), California Institute of Technology (2009)
5. Maganty, S.: Microstructures and their relationship to nanomechanical behavior of electronic packaging materials. State University of New York at Binghamton (2012)
6. Zare, Y.: Study of nanoparticles aggregation/agglomeration in polymer particulate nanocomposites by mechanical properties. Composites: Part A **84**, 158–164 (2016)
7. Dorigato, A., Dzenis, Y., Pegoretti, A.: Filler aggregation as a reinforcement mechanism in polymer nanocomposites. Mech. Mater. **61**, 79–90 (2013)
8. Elbishari, H., Silikas, N., Satterthwaite, J.: The effect of filler size on the presence of voids within resin composite. J. Int. Dent. Med. Res. **11**(2) (2018)
9. Mohammed, A.A.: Development of a new stretchable and screen printable conductive ink (Doctoral dissertation, University of Maryland, College Park) (2017)
10. Burton, O.J., Babenko, V., Veigang-Radulescu, V.P., Brennan, B., Pollard, A.J., Hofmann, S.: The role and control of residual bulk oxygen in the catalytic growth of 2D materials. J. Physic. Chem. C **123**(26), 16257–16267 (2019)
11. Ding, D., Solís-Fernández, P., Yunus, R.M., Hibino, H., Ago, H.: Behavior and role of superficial oxygen in Cu for the growth of large single-crystalline graphene. Appl. Surf. Sci. **408**, 142–149 (2017)
12. Wahid, A.K.A., Lim, Y.T., Cheong, J.Y., Lim, S.K.: Driving monitoring system application with stretchable conductive inks: a review. Int. J. Nanoelectron. Mater. **13**(Special Issue), 327–346 (2020)
13. Hikku, G.S., Jeyasubramanian, K., Venugopal, A., Ghosh, R.: Corrosion resistance behaviour of graphene/polyvinyl alcohol nanocomposite coating for aluminium-2219 alloy. J. Alloy. Compd. **716**, 259–269 (2017)

Chapter 6
Pattern Geometry Effects on Conductive Ink Performance

6.1 Introduction

Electrical characterisation is a fundamental aspect in the evaluation and optimisation of conductive inks, providing critical insight into their performance in practical applications such as printed circuits, sensors, and stretchable electronics (SE). At the core of this characterisation lies the precise measurement of electrical resistance, which serves as a gateway to understanding a range of properties, including electrical conductivity, charge mobility, carrier continuity, and the quality of the ink-substrate interface. Several methods are routinely employed for resistance measurement in printed nanocomposite systems. These include sheet resistance, bulk resistance, and volume resistivity measurements, each offering different perspectives into the ink's structural and electrical behaviour.

Sheet resistance measurement is a particularly valuable metric in planar or thin-film configurations where the conductive ink is patterned in two-dimensional geometries. It reflects the resistance across a square unit of the printed trace and is highly sensitive to variations in ink thickness, filler dispersion, and surface uniformity. As such, it serves as an effective tool for evaluating the uniformity of the ink deposition process. Any heterogeneity in filler distribution such as local agglomerations or voids and it can manifest as spatial variation in sheet resistance across the patterned area. This is especially critical in nanoscale or microscale circuits, where even small deviations can impair signal integrity or sensor responsiveness. Sheet resistance measurements also act as a proxy for determining the continuity and percolation threshold of the conductive filler network, particularly in formulations involving nanomaterials like graphene nanoplatelets (GNPs) or carbon black (CB).

Bulk resistance measurements, on the other hand, offer insight into the total resistance across a defined volume of the ink, incorporating the thickness of the film and its internal structure. This metric provides a macroscopic view of the ink's conductive network and is especially useful in assessing whether the conductive paths remain intact and continuous throughout the ink body. Bulk resistance is often influenced

M. A. Salim et al., *Nanotechnology in Conductive Inks*, Engineering Materials,
https://doi.org/10.1007/978-981-95-5708-0_6

by filler loading, inter-particle contact resistance, and the degree of polymer cross-linking. A poorly dispersed filler phase or inadequate binder-filler interaction will typically result in higher bulk resistance due to limited charge transport pathways.

Volume resistivity offers the most intrinsic measurement of a material's electrical performance, representing the inherent resistance of a unit cube of the ink to current flow. It is a function of both bulk composition and internal microstructure, including factors such as filler morphology, interfacial conductivity, and the presence of dielectric gaps. In nanocomposite inks, volume resistivity is highly sensitive to the quality of filler percolation and the physical connectivity of conductive particles. For instance, well-exfoliated GNPs with strong π–π interactions form effective conductive bridges when optimally loaded within a polymer matrix, resulting in lower volume resistivity. Conversely, oxygenated graphene or oxidised CB can introduce barriers to electron mobility, thereby elevating resistivity.

The electrical behaviour of the patterned area is of particular interest, especially in modern flexible and wearable electronics where mechanical reliability and electrical stability must coexist. The patterned ink traces, often created through screen printing, inkjet printing, or stencil techniques, can experience stress concentrations at junctions, edges, and corners due to geometric discontinuities. These localised stress fields can induce micro-cracks or delamination in the ink layer, disrupting the conductive pathway and leading to resistance drift or failure under cyclic loading.

In materials science perspective, such degradation mechanisms are intimately tied to the ink's microstructure. Inks formulated with nanomaterials rely on a well-connected network of conductive fillers within a polymer binder. When subjected to mechanical deformation, these networks can undergo fragmentation or slippage at the nanoscale, particularly if filler-filler or filler-matrix interactions are weak. The degree to which the ink can withstand such stress without electrical degradation is linked to its percolation robustness, interfacial adhesion, and ductility of the matrix phase. Furthermore, temperature, humidity, and environmental exposure can influence resistance measurements by affecting the ink's microstructural integrity. For instance, oxidation of the filler or matrix can alter the charge transport mechanism, especially in systems containing residual oxygen functionalities or hydrophilic polymer chains.

By addressing these challenges, the investigators often employ in-situ resistance monitoring during bending or tensile tests, allowing the mapping of electrical performance against mechanical deformation. This approach provides a dynamic profile of how patterned conductive inks behave under realistic operational conditions. Optimisation strategies may include enhancing the ink formulation with hybrid fillers (for example combining GNPs and CB), adjust the surface energy of the substrate to improve adhesion, or modifying the pattern geometry, such as serpentine or fractal designs to distribute stress more uniformly.

An electrical characterisation of conductive inks especially in patterned configurations is a multidimensional task requiring integration of electrical, mechanical, and materials-based perspectives. By coupling precise resistance measurements with an understanding of nanomaterial behaviour, interface science, and stress distribution, it is possible to develop inks that are not only highly conductive but also robust under operational demands. This understanding is essential for the advancement of printed, stretchable, and flexible electronics where reliability and performance are inseparable.

6.2 Bulk Resistance Analysis of Patterned Conductive Traces

The bulk resistance measurement of conductive ink traces plays a crucial role in evaluating their electrical performance under both static and cyclic mechanical conditions. This is particularly important in applications such as flexible electronics, stretchable sensors, and wearable devices, where the ink is deposited in complex patterns and subjected to dynamic deformation. The measured electrical resistance, in combination with the geometry of the printed trace, provides the basis for determining volume resistivity which a key property that reflects the intrinsic conductivity of the material across various filler compositions, particularly GNPs and CB.

In this section, resistance values were obtained for various geometric patterns including straight lines, curves, squares, and zigzags by using a Two-Point probe method, which involves direct contact at the ends of each trace. This technique is suitable for comparative analysis and is commonly applied in in-situ measurements during functional testing. Such in-situ resistance monitoring is an important for verifying the ongoing conductivity of the ink during experimental conditions, especially when samples are subjected to repeated mechanical loading or thermal cycling.

Figure 6.1 illustrates the trends in bulk resistance across different patterns and filler types. Particularly, the resistance decreased as the width of the trace increased. This behaviour is consistent across all geometries and can be explained by considering the effective conducting cross-sectional area. Wider traces provide a greater area for electron transport, thus reducing the overall resistance. The analogy with fluid dynamics is often used to conceptualise this effect, wherein a wider pipe allows more fluid to pass with less pressure, similarly, a wider conductor facilitates a greater flow of electrons with lower resistance.

From a materials physics perspective, the geometry of the printed trace plays a non-negligible role in current distribution, especially in patterned structures. In simple linear geometries, such as straight lines, the current experiences minimal obstruction and flows uniformly across the trace. However, in complex patterns such as curves or zigzags, the current is forced to navigate around corners and through turns, introducing non-uniform current density, known as current crowding. These regions of concentrated current can increase localised resistive heating, induce electromagnetic interference, and contribute to mechanical fatigue, especially under repeated loading.

The sharp corners and tight boundary in these complex patterns create stress concentration points at the microstructural level. In such areas, the conductive pathway becomes less efficient, and the overall resistance increases. This phenomenon is not merely a geometric inconvenience but is also fundamentally

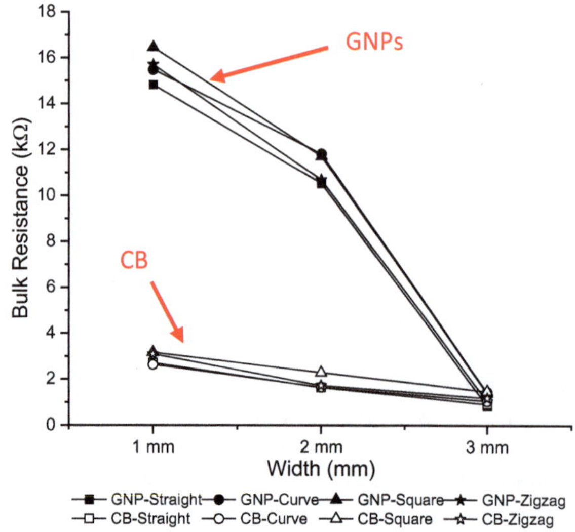

Fig. 6.1 Graph of bulk resistance for patterned GNP and CB conductive ink trace without load

rooted in electronic transport physics, where electron scattering becomes more pronounced in disrupted or indirect pathways. Moreover, the interaction between the conductive filler particles and the polymer matrix at these high-strain zones may further deteriorate due to interfacial slippage or micro-crack formation.

Besides, the morphology and size distribution of the fillers play an essential role in determining resistance. The GNPs fillers used in this work, with particle sizes of approximately 5 μm and surface areas exceeding 500 m^2/g, act as effective micro-fillers capable of forming extended, planar conductive networks. Their high aspect ratio facilitates long-range electron transport with fewer particle-particle junctions, thereby reducing contact resistance. Moreover, their relatively large and stable particle size helps mitigate aggregation issues, resulting in consistent electrical performance across trace widths.

On the other hand, some CB particles, which range from 15 to 350 nm in size, contribute to conductivity primarily through their high surface activity and ability to form percolating networks at low loadings. However, due to their nanoscale dimensions and tendency to aggregate, the resultant network structure can vary significantly with the quantity of ink and the geometry of the printed pattern. As the trace width increases, more ink is deposited, allowing carbon black particles to form denser and more interconnected conductive networks, thereby reducing resistance more noticeably.

Empirical results further support these theoretical considerations. For GNPs-based inks, the straight-line pattern yielded the lowest resistance values due to its uninterrupted pathway. As the width increased from 1 to 3 mm, resistance dropped sharply by up to 178%, and highlighting the positive correlation between width and conductive efficiency. Similar trends were observed in curved and zigzag patterns, albeit with higher absolute resistance values due to geometric complexity.

For CB inks, while the initial resistance values were lower than those of GNPs due to the high surface reactivity of CB, the drop in resistance with increasing trace width followed the same general trend. This reinforces the understanding that trace width and filler interaction together dictate the final conductive performance. Importantly, the observations underscore that the straight-line configuration remains the most efficient in minimising electrical resistance, offering a baseline for comparison. As Novak (2017) pointed out, in direct current systems, electrons prefer the path of least resistance. In meandering geometries, electrons are compelled to reroute around inner corners, increasing the effective resistance due to angular deviation and potential self-heating.

Therefore, this analysis confirms that both trace geometry and filler properties must be carefully optimised to ensure reliable conductivity in patterned conductive inks. This is especially an essential in applications requiring consistent electrical performance under strain, deformation, or cyclic use, hallmarks of next-generation flexible electronics. A deeper understanding of how resistance behaves in relation to pattern design will facilitate more efficient circuit layout and improve material selection in future device manufacturing.

6.3 Influence of Pattern Geometry on Sheet Resistivity

Sheet resistivity is a critical electrical property widely used to characterise thin films and conductive coatings, particularly in semiconductor and printed electronics applications. It is especially valuable for engineers and circuit designers, as it provides a means to standardise and simplify resistance calculations across various material configurations without requiring precise thickness measurements, an advantage when dealing with layered, flexible, or miniaturised systems. In the field of printed electronics, where conductive traces are deposited via additive manufacturing techniques such as inkjet printing or stencil printing, sheet resistivity becomes a foundational parameter for evaluating the quality, uniformity, and functionality of the printed lines. Conductive inks formulated from nanomaterials such as GNPs and CB exhibit behaviours that are sensitive to not only material properties but also geometric layout and processing conditions.

The work of [1] is particularly insightful in demonstrating the electrical implications of pattern geometry. His study on serpentine patterns, an arrangement of interconnected squares, has been highlighted that such meandering layouts can nearly double the resistance compared to an equivalent straight-line path. This increase arises from the longer current path and the introduction of multiple sharp turns, which impede the natural flow of electrons and introduce areas of current crowding, particularly at corners or bends.

In the present work, Figure 6.2 compares the sheet resistivity of four distinct trace patterns, which are straight, curved, square, and zigzag, and each printed using two types of conductive inks such as one based on GNPs and the other on. Across all patterns and both ink types, a consistent trend emerges such as the width of the

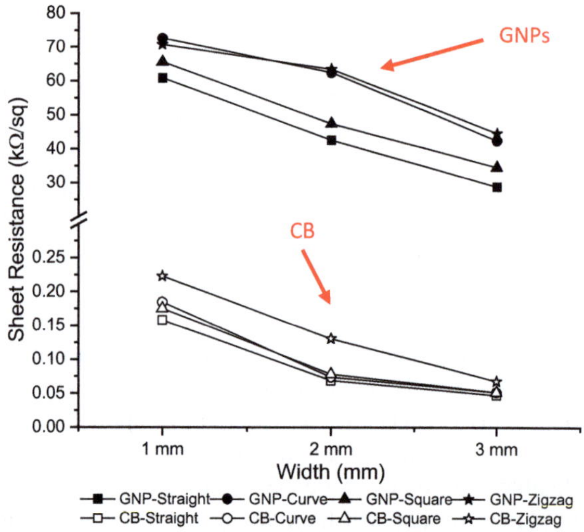

Fig. 6.2 Graph of sheet resistance for patterned GNPs and CB conductive ink trace without load

trace increases, the sheet resistivity decreases. This behaviour can be attributed to the increased availability of conductive filler particles within the broader traces. Wider traces accommodate a larger volume of ink, thereby distributing the conductive fillers more effectively. This leads to the formation of more continuous and overlapping percolation pathways, facilitating better electron transport.

This observation is in line with findings by [2], who reported that the electrical conductivity of printed patterns improved with wider traces due to enhanced filler dispersion and inter-particle contact. In materials science, this phenomenon is explained by the percolation theory, which describes how the electrical connectivity of a composite system increases when the concentration of conductive elements exceeds a certain threshold. Wider patterns allow for better spatial accommodation of these elements, increasing the probability of forming a continuous conductive network.

At the microscopic level, GNPs fillers, with their large planar surface areas and high aspect ratios, function as micro-scaled bridges for charge carriers. Their structure promotes directional conductivity across the trace when adequately dispersed. However, in narrower patterns, there is a tendency for these particles to aggregate due to spatial confinement, resulting in reduced effective contact and consequently, increased resistance. This clustering or agglomeration can obstruct current pathways and create discontinuities in the conductive network, especially where deposition is uneven.

The CB, by contrast, consists of spherical to semi-spherical particles with high surface area and fine particle sizes, often in the range of 15 to 350 nm. These particles tend to form extensive, fractal-like conductive networks at relatively lower loading levels. Their smaller size allows them to occupy interstitial spaces and complement larger fillers such as GNPs in hybrid systems. However, in very narrow traces, even

CB particles can exhibit excessive aggregation due to capillary forces during ink drying, which diminishes their electrical performance.

Quantitatively, the GNPs-based ink showed reductions in sheet resistance of 35.28 and 71.36% for the straight-line pattern as the width increased from 1 to 3 mm. Similar reductions were observed across the curved, square, and zigzag patterns, though the magnitude varied based on geometric complexity. The CB ink traces exhibited even more pronounced improvements, with sheet resistance decreasing by over 100% in some cases. This reflects the inherently better network-forming capability of CB in confined geometries, though this is also a function of how well the ink spreads and dries during deposition.

The performance disparity among patterns is particularly well-known in the zigzag and square geometries, which involve abrupt changes in direction and sharp angles. These features introduce disruptions in current flow, elevating local resistance. The meandering paths, especially in the zigzag and serpentine layouts, and also contribute to non-uniform current distribution, which may not only increase resistivity but also lead to thermal hotspots under continuous operation. Merilampi et al. [3] noted that these structures are often susceptible to uneven material deposition, resulting in thinner areas or micro-gaps that act as resistance bottlenecks.

Such challenges underscore the importance of trace geometry in the overall electrical performance of printed electronic components. From a design perspective, understanding how width, pattern complexity, and filler morphology interact can guide the optimisation of circuit layouts in space-constrained applications. For instance, while meandering patterns are often used to increase the effective length of a trace in a limited area, this comes at the cost of increased resistance and potentially reduced reliability under stress or flexing. Ultimately, the present work highlights that sheet resistivity is not solely a material-dependent property, but one that is highly sensitive to pattern geometry and printing fidelity. For advanced applications such as flexible circuits, antennas, and strain sensors, the interplay between pattern design and nanomaterial behaviour must be thoroughly understood. This enables the engineering of ink formulations and printed structures that are both electrically efficient and mechanically robust.

6.4 Pattern Influence in Volume Resistivity of Conductive Traces

The discussion herein explores the electrical resistance of conductive ink traces, framed in terms of volume resistivity which a critical parameter used to quantify how well a material resists electric current throughout its volume. Expressed in ohm-centimetres, volume resistivity offers insights into the intrinsic conductive behaviour of the material itself, independent of its geometry. For thin films and printed electronics, this measurement is particularly informative, as it allows comparison across various material systems and structural configurations. Volume resistivity is influenced by a multitude of factors including filler dispersion, particle interaction, film thickness, and most significantly, the pattern geometry and width of the printed trace. Additionally, imperfections such as micro-voids, grain boundaries, impurity inclusions, and lattice dislocations introduce scattering centres for charge carriers. In the context of conductive inks, these scattering events, governed by the microstructure of the material and it can severely hinder charge transport.

From a materials science side, even homogenised materials can exhibit distorted crystal lattices due to internal stress, agglomeration of particles, and uneven drying kinetics. These effects are pronounced in composite materials such as graphene GNPs and CB conductive inks, where the distribution and morphology of the filler phase dominate electrical performance. Figure 6.3 presents a comparative analysis of volume resistivity derived from both bulk resistance and sheet resistance measurements for GNPs and CB conductive ink traces with varying widths (1, 2, and 3 mm). The data reveals nuanced interactions between material type, trace geometry, and measurement method.

For GNPs conductive ink, the volume resistivity measured using bulk resistance increases as the trace width increases from 1 to 2 mm. This suggests a reduction in conductivity, likely due to less efficient percolation pathways or localised agglomeration effects that emerge during the ink curing or drying process. Interestingly, as the width increases from 2 to 3 mm, the volume resistivity decreases, indicating improved conductivity. This shift implies that the broader trace at 3 mm allows for

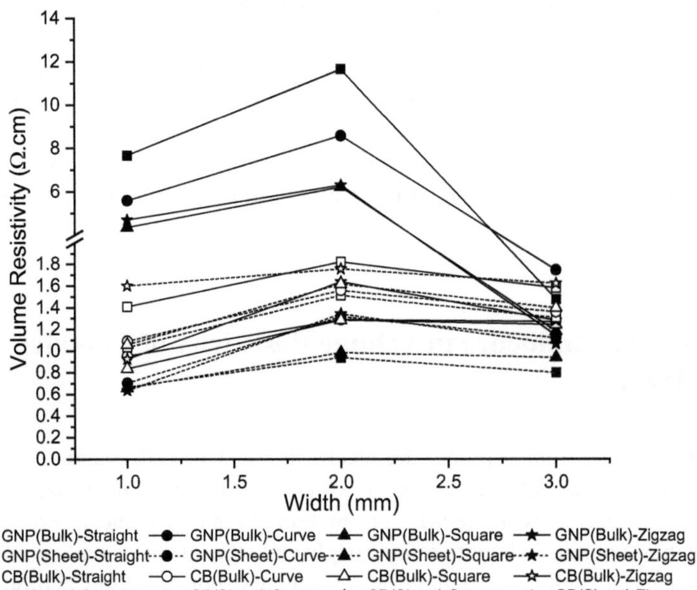

Fig. 6.3 Graph of comparison of volume resistivity for GNP and CB conductive ink trace

better particle distribution and network formation, reducing the number of resistive boundaries.

When volume resistivity is calculated from sheet resistance, a similar but more controlled trend is observed. The values increase moderately from 1 to 2 mm and then decrease at 3 mm width. The relatively lower values in sheet resistance-derived measurements compared to bulk suggest that the Four-Point probe method, typically used for sheet resistance is more sensitive to the conductive layer itself, excluding contact resistance and other extrinsic contributions. This is particularly beneficial for thin films and structured patterns where uniformity and surface conduction dominate.

For CB conductive ink, a similar trend is observed the volume resistivity increases from 1 to 2 mm, then slightly decreases from 2 to 3 mm for both bulk and sheet resistance measurements. This behaviour can be explained by the inherent properties of the ink. Unlike GNPs, CB particles are smaller, more spherical, and form conductive networks through point contacts rather than planar overlap. While it is capable of forming dense percolation networks, it typically exhibits higher intrinsic resistivity compared to graphene, due to its lower aspect ratio and smaller electron mobility. The volume resistivity values derived from sheet resistance measurements for the ink also suggest better conductivity compared to bulk-derived values. This reinforces the assumption that measurement methodology significantly influences the interpretation of a material's performance, especially in patterned geometries where surface contact and ink-substrate interaction play critical roles.

A key insight from the analysis is that pattern geometry and trace width exert a substantial influence on volume resistivity, beyond what would be expected from material properties alone. For instance, more complex geometries such as zigzag, serpentine, or square waveforms often introduce non-uniform current paths, increasing resistivity. In these patterns, current lines are forced to navigate through sharp corners or looped paths, leading to current crowding, localised heating, and longer effective path lengths.

Conversely, straight-line patterns provide direct, unobstructed paths for electrons to travel, reducing both resistive losses and the risk of bottlenecks. As a result, straight-line traces consistently exhibit lower resistivity values in both bulk and sheet measurements, especially when trace width is optimised. The performance difference observed in pattern types underlines the necessity of considering not just material composition but also geometric design in printed electronics. This becomes increasingly critical in applications such as flexible circuits, stretchable sensors, and RFID antennas, where spatial constraints often necessitate creative or compact trace designs.

Another dimension to this finding is the impact of the measurement technique used. Volume resistivity calculated from bulk resistance, often measured using a Two-Point probe, includes the contact resistance between the probe and the sample. This method is susceptible to inaccuracies in thin or narrow traces due to localised resistance variations at the probe contact area. In contrast, sheet resistance measurements typically use a Four-Point probe, which separates the current injection and voltage measurement paths, effectively eliminating contact resistance. As demonstrated by

Li et al. (2012), the Four-point probe method enables precise, reproducible measurements of thin films, especially when high spatial resolution and minimal sample alteration are required.

These distinctions are not merely procedural but they affect design decisions in real-electronic components. For instance, designers opting for narrow traces in wearable sensors or flexible displays must account for increased resistivity and reduced signal integrity unless compensated for through ink formulation or trace redesign. The findings suggest that volume resistivity is not a static material property, but rather a function of trace geometry, ink composition, and measurement approach. While GNPs-based inks offer superior electrical performance due to their large surface area and anisotropic structure, their performance is more sensitive to trace design and width. A CB inks, although more isotropic in nature, offer consistency across patterns but with higher resistivity.

Moreover, the observed discrepancies between bulk and sheet resistance-derived values underscore the importance of adopting the correct measurement protocol for a given application. For precision applications, especially those relying on patterned conductive paths in confined geometries, sheet resistance measurements using the Four-Point probe provide a more accurate representation of true material performance. The selection of a 3 mm wide trace emerges as optimal, balancing ink flow, filler dispersion, and measurement accuracy across all pattern types. These insights provide a solid foundation for future development of high-performance, patterned conductive structures in next-generation electronic systems.

6.5 Mechanical Behaviour of Patterned Conductive Inks

The load-displacement curve obtained from nanoindentation experiments offers profound insights into the mechanical performance of materials, particularly in micro- and nanoscale applications such as printed and flexible electronics. This curve illustrates the relationship between the applied force and the resultant penetration depth of an indenter as it interacts with a material surface. Through careful interpretation of the curve's profile, one can discern key material responses, including elastic recovery, plastic deformation, and viscoelastic effects the all critical to understanding how materials behave under mechanical stress.

For conductive ink traces, especially those embedded with nanostructured fillers, nanoindentation allows for the localised assessment of their surface mechanical properties. These include hardness, elastic modulus, maximum penetration depth, and residual depth after unloading. Such characterisations are crucial not only for ensuring mechanical durability under operational stress but also for guiding ink formulation and pattern design strategies. In nanoindentation, the elastic-plastic transformation observed in the load-displacement curve reflects the dual mechanical response of materials which is an initial elastic behaviour followed by permanent plastic deformation. During the loading phase, the indenter displaces the material both reversibly and irreversibly. Upon unloading, the elastic portion recovers, while the plastic component remains as a permanent impression.

The steepness of the hardness curve, when plotted against penetration depth, provides a measure of how quickly a material resists further deformation. A steeper curve indicates a material that stiffens rapidly under load which a behaviour often observed in nanocomposites reinforced with high-modulus fillers, such as GNPs. In contrast, a flatter slope in the elastic modulus curve suggests a more moderate increase in stiffness, highlighting the nuanced interplay between filler geometry, matrix viscosity, and filler-matrix adhesion at the nanoscale. In the present exploration, nanoindentation results were used to compare the mechanical properties of GNPs and CB conductive inks across various pattern geometries such as straight, curved, square, and zig-zag. Figure 6.4 illustrates that GNPs ink traces consistently exhibited shallower maximum indentation depths compared to their carbon black counterparts, despite being subjected to higher applied forces. This observation underscores the superior load-bearing capability and higher stiffness of GNPs-filled composites.

For instance, indentation depths for GNPs patterns ranged between 5.25 to 8.31 μm, while for CB patterns, depths extended to 15.60 μm. Despite the pattern type, this consistent trend reveals that the GNPs-reinforced matrix is significantly more resistant to plastic deformation, owing to the unique structural and mechanical properties of graphene. The GNPs are composed of two-dimensional sheets of sp^2-bonded carbon atoms, arranged in a hexagonal lattice. This atomic configuration imparts extraordinary in-plane strength and Young's modulus, with values surpassing most conventional filler materials. These properties make graphene an ideal nanofiller for reinforcing soft polymer matrices, providing both mechanical reinforcement and electrical conductivity without requiring large volume fractions. Conversely, CB, although electrically conductive, lacks the same level of structural sophistication. Its spherical morphology and relatively lower surface area lead to less effective stress transfer between the filler and matrix during mechanical loading. As a result, the inks deform more easily, as evidenced by their greater indentation depths under lower maximum loads.

Interestingly, the data suggest that the pattern shape whether straight, curved, square, or zig-zag, has minimal influence on maximum indentation depth for both ink types. This indicates that under concentrated point loads, such as those applied during indentation, the local mechanical response is dominated by the filler-matrix interaction rather than the macro-pattern configuration. However, when considering wider device-level mechanical behaviour, pattern geometry becomes increasingly critical, particularly under distributed stress or cyclic bending conditions.

Nonetheless, subtle effects are still observable. For instance, GNPs traces in zig-zag patterns showed slightly higher maximum indentation depths and variability, possibly due to ink flow anomalies or edge effects during printing and curing. In contrast, straight-line patterns tended to yield the most uniform and predictable mechanical responses, supporting their widespread use in precision applications where stress uniformity is significant. Beyond pattern geometry, the width of the printed trace significantly affects indentation performance. Wider traces generally

Fig. 6.4 Load-penetration depth curves for different pattern GNPs and CB ink trace with **a** 1 mm, **b** 2 mm, and **c** 3 mm widths without load

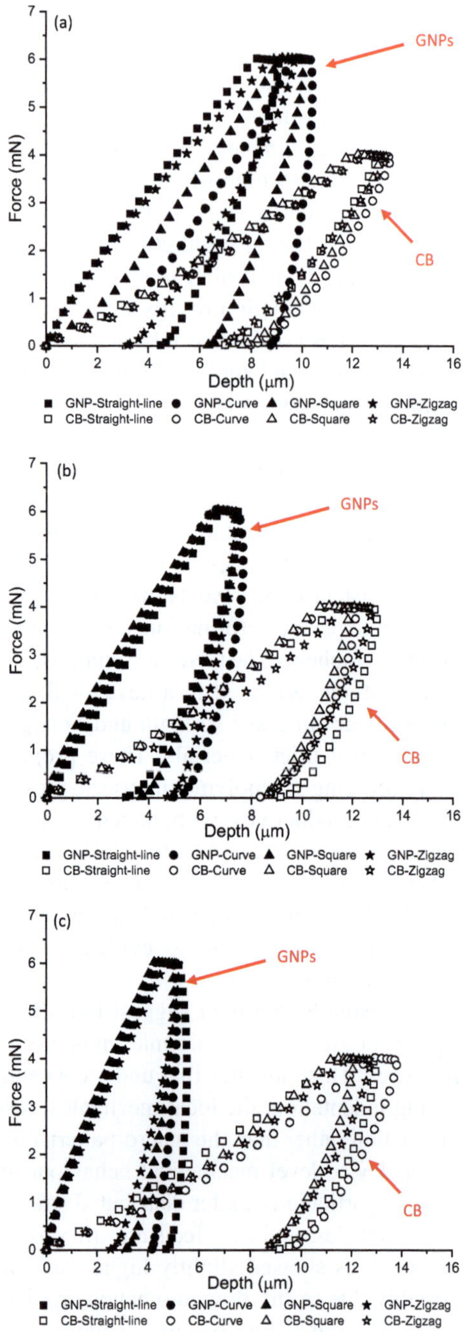

exhibit reduced penetration depths and smaller error margins, indicative of better mechanical stability and enhanced filler packing density. For example, 3 mm-wide GNPs traces resisted indentation more effectively than 1 mm traces, a consequence of improved network percolation and uniform filler dispersion in larger print areas.

The deviation in indentation depth at 1 mm width suggests heterogeneity in filler distribution or possible edge drying effects, which are common in narrow prints due to rapid solvent evaporation and surface tension gradients. This variability highlights the importance of optimising trace width for mechanical reliability, particularly in high-resolution printed electronics where space is at a premium.

The hysteresis loop observed in the load-displacement curve during unloading is a hallmark of viscoelastic and time-dependent plastic recovery. While part of the unloading response is purely elastic, a fraction is governed by reverse plasticity which a phenomenon where internal stresses cause partial recovery beyond typical elastic limits. This behaviour is especially pronounced in polymeric matrices, which exhibit time-dependent strain due to chain rearrangement and relaxation.

The final indentation depth decreases with increasing trace width, supporting the inference that wider traces, rich in filler content, are better able to resist penetration. These traces present a denser and more entangled filler network that obstructs the indenter's progress, promoting energy dissipation and reducing irreversible deformation. The superior indentation response of GNPs traces confirms the critical role of filler selection in defining the mechanical performance of printed conductive features. The inherent anisotropy and high aspect ratio of GNPs allow for efficient stress distribution, minimising localised failure under mechanical loading. Their interaction with the surrounding polymer matrix ensures that even under extreme indentation forces, structural integrity is largely retained.

Moreover, the clear dependency of indentation performance on trace width but not pattern type suggests that material engineers should prioritise optimised line width and filler dispersion over geometrical complexity when designing mechanically robust circuits. As printed electronics evolve toward stretchable, wearable, and flexible platforms, understanding such material-geometry relationships becomes indispensable for ensuring both functional reliability and long-term durability. This comprehensive indentation work basically reveals the intricate interplay between nanofiller type, pattern geometry, and trace width in determining the mechanical behaviour of conductive ink traces. The GNPs-based inks outperformed CB inks in terms of resistance to deformation, attributed to graphene's intrinsic material strength and structural characteristics. While pattern type showed negligible influence on indentation depth, trace width emerged as a critical factor, with wider traces providing improved mechanical stability. The use of nanoindentation as a diagnostic tool offers precise and localised insights into material behaviour, enabling fine-tuning of ink formulation and printed design for next-generation electronic devices. These findings serve as a foundational reference for materials scientists, engineers, and researchers involved in the development of high-performance, mechanically resilient, patterned nanocomposite systems.

6.6 Nanoindentation into Pattern-Dependent

The mechanical properties of nanostructured conductive materials, have gained significant interest in the field of flexible and printed electronics. These properties are typically characterised using nanoindentation techniques, which provide a precise method for determining the hardness and elastic modulus of materials at micro to nanoscale dimensions. Nanoindentation involves the pressing of a sharp indenter into the surface of a material under controlled loading and unloading conditions. The load-displacement curve produced from this process is fundamental to understanding both the elastic and plastic deformation responses of a material. One critical output of this curve is the maximum indentation depth, from which hardness values are derived using methods such as the Oliver and Pharr approach. In this technique, the projected contact area between the indenter and the sample is estimated from the unloading segment of the curve, and hardness is defined as the maximum applied load divided by this contact area.

The GNPs-based conductive inks demonstrated superior mechanical properties compared to CB inks across all tested patterns, with hardness values for GNPs decreasing from approximately 8.86 to 6.30 MPa as the trace width increased from 1 to 3 mm. This trend was consistent across straight-line, curved, square, and zigzag patterns. The decline in hardness with increasing width is attributed to greater dispersion of GNPs flakes within the matrix, leading to a larger effective contact area during indentation. As the flake dispersion increases, the interfacial bonding between filler and matrix weakens slightly, allowing easier penetration by the indenter and thus lowering hardness. In contrast, CB inks exhibited a modest increase in hardness with increasing trace width. This can be attributed to the smaller size and more isotropic morphology of particles, which enables denser packing and greater particle-particle interaction as the pattern area enlarges. The enhanced packing improves load transfer and restricts matrix flow, thereby increasing resistance to indentation.

The geometry of the printed patterns plays a crucial role in determining mechanical behaviour. More complex geometries, such as zigzag or square patterns, introduce non-uniform stress distribution and facilitate greater dispersion of conductive fillers within the resin matrix. This leads to localised differences in modulus and hardness compared to simpler geometries like straight lines. The meandering paths in curved or zigzag designs can also increase the overall surface area of the trace, which contributes to stress delocalisation but reduces the effective stiffness due to filler-matrix dilution.

The elastic modulus, a fundamental measure of a material's stiffness, was derived from the unloading portion of the load-displacement curve. It reflects the material's ability to recover elastically after deformation. For GNPs-based inks, the modulus ranged from 0.23 to 0.29 GPa, with variations across different pattern geometries. The Zigzag patterns typically showed the highest modulus values, while straight-line patterns exhibited the lowest. These findings suggest that the structural complexity inherent in non-linear patterns increases resistance to elastic deformation, likely due to the indirect load paths that inhibit crack propagation and matrix yielding.

In comparison, ca CB traces exhibited significantly lower moduli, generally below 0.08 GPa, but with an increasing trend as the trace width increased. This indicates that while CB fillers contribute to stiffness improvements with increased trace size, they remain inherently less stiff than their graphene counterparts due to their lower aspect ratio and mechanical strength. The observed differences in mechanical behaviour between GNPs and CB inks can be traced back to their nanoscale structure and interactions within the composite matrix. The GNPs possess a two-dimensional lamellar structure, with high aspect ratio and outstanding in-plane mechanical properties, including high tensile strength and Young's modulus. These characteristics enable GNPs to reinforce the matrix effectively, particularly when well-aligned or uniformly dispersed. However, as trace width increases, filler alignment becomes more random and matrix-dominated areas increase, resulting in reduced reinforcement efficiency. Figures 6.5 and 6.6 shows the results of hardness and elastic modulus for GNPs and carbon black, respectively.

The CB particles, on the other hand, consist of nearly spherical aggregates with high surface area but lower stiffness. Their small size allows them to fill voids and interstices effectively, improving packing density and enhancing load-bearing capacity at higher trace widths. However, due to their relatively poor mechanical reinforcement capability, their contribution to modulus improvement is modest compared to GNPs. The elastic recovery during unloading also reveals the viscoelastic nature of the polymer matrix used in ink formulation. Especially in patterns with larger widths or more tortuous paths, residual stresses and reverse plasticity may contribute to variations in final indentation depths and modulus values.

From a design perspective, understanding how pattern geometry and trace width influence mechanical properties is critical for optimising the performance of flexible electronic circuits. Applications such as stretchable sensors, soft robotics, and

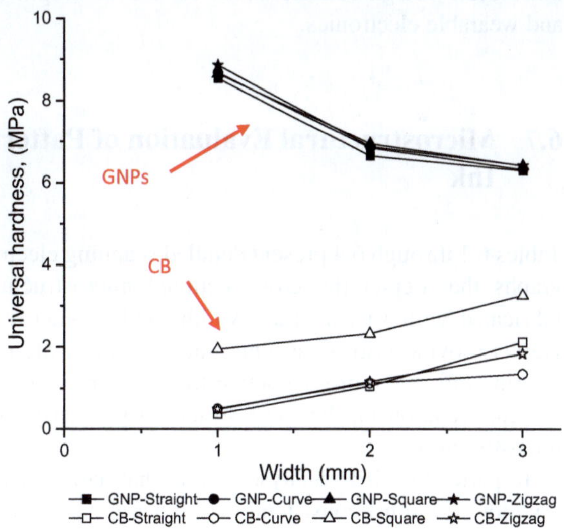

Fig. 6.5 Graph of hardness for GNsP and CB ink trace for different patterns without load

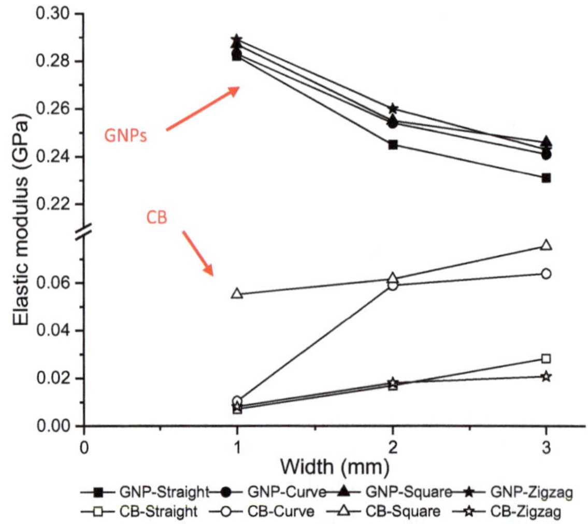

Fig. 6.6 Graph of elastic modulus for GNPs and CB ink trace for different patterns without load

wearable electronics demand materials that can withstand mechanical stress while maintaining structural and functional integrity. As such, the choice of filler (GNPs and CB), pattern design (straight versus meandering), and dimensional parameters (trace width) must be carefully balanced to meet specific application requirements.

The nanoindentation results underscore the superior performance of GNPs-based inks in terms of both hardness and elastic modulus, particularly in complex pattern geometries. The sensitivity of these properties to pattern shape and size provides valuable insights for tailoring mechanical performance in nanostructured conductive inks. This study reinforces the importance of integrating nanoscale material science with macroscale design considerations for the development of next-generation flexible and wearable electronics.

6.7 Microstructural Evaluation of Patterned Conductive Ink

Tables 6.1 through 6.4 present detailed scanning electron microscopy (SEM) micrographs that depict the cross-sectional microstructure of conductive ink traces fabricated with varying trace widths and geometric pattern designs. The micrographs provide critical insights into the nanoscale organisation, distribution, and packing efficiency of conductive fillers in epoxy resin matrices, underscoring the intricate relationship between structural geometry, filler behaviour, and resulting microstructure.

In particular, SEM analyses reveal that narrower traces (specifically those with a width of 1 mm) consistently exhibit a greater incidence of voids and irregular surface topography compared to wider traces of 2 and 3 mm. This microstructural

Table 6.1 Cross sectional images of straight-line pattern of GNPs and CB ink trace microstructure with line width of 1, 2, and 3 mm

GNPs ink trace	CB ink trace
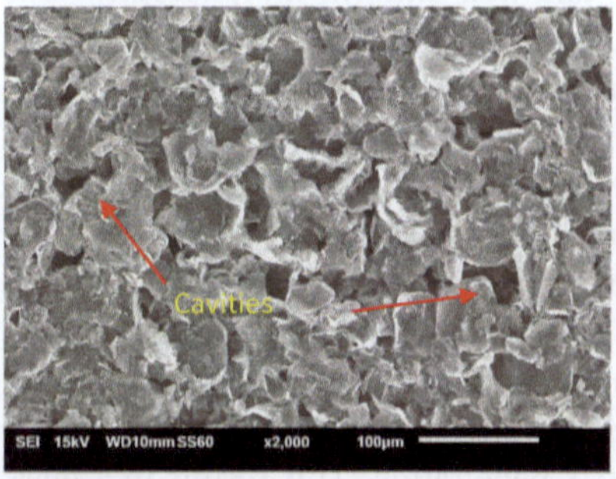	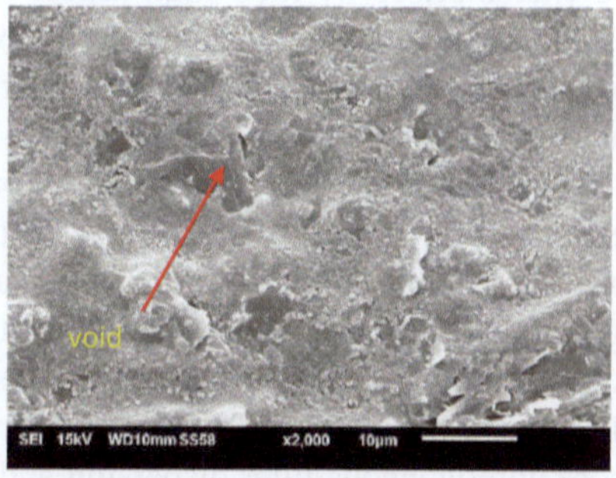
1 mm trace width	
GNPs ink trace	CB ink trace
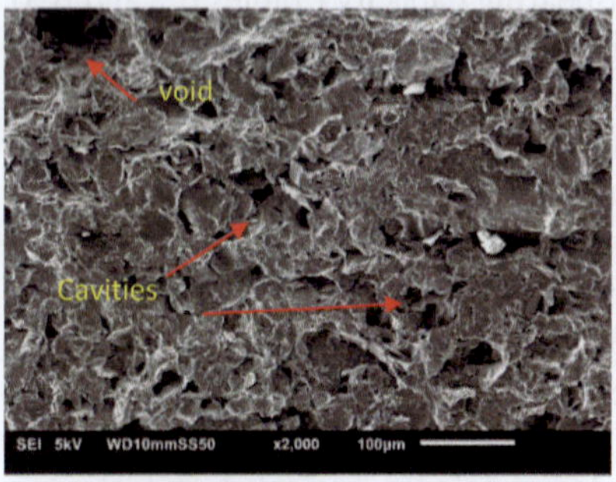	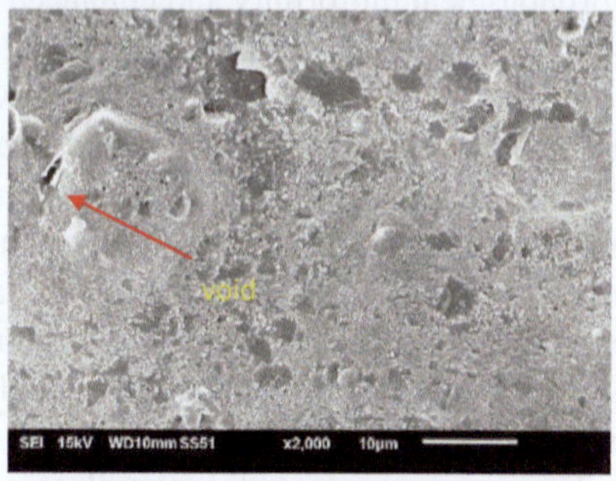
2 mm trace width	
GNPs ink trace	CB ink trace
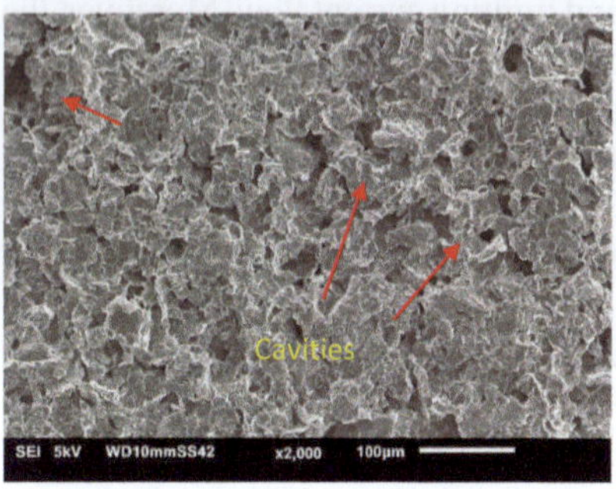	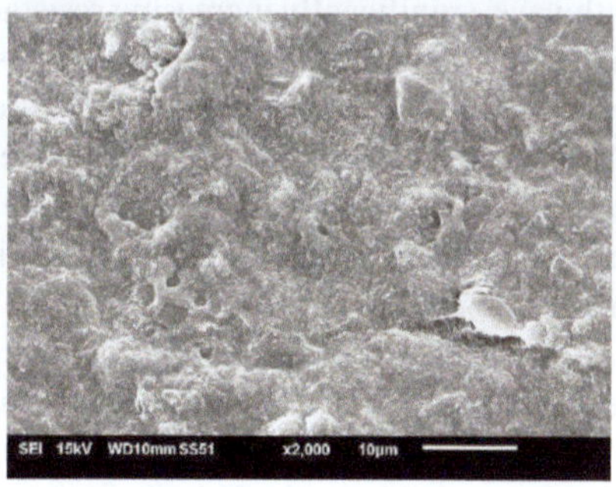
3 mm trace width	

inconsistency appears across all geometric patterns, including straight, curve, square, and zigzag configurations. The presence of voids in the narrower traces is primarily attributed to the incomplete dispersion and poor packing of large GNPs fillers, which are characterised by their high aspect ratios and non-uniform morphologies. The GNPs tend to agglomerate due to strong van der Waals interactions, forming clusters that resist uniform suspension and integration within the polymer matrix during the ink formulation and deposition stages [4].

This phenomenon is especially pronounced in narrow traces, where confinement restricts resin mobility and filler orientation, limiting the ability of the polymer to infiltrate and bond within the compact particle networks. Consequently, these agglomerates create micron-scale interstitial voids that compromise both structural cohesion and functional performance. As the trace width increases, however, spatial allowances for resin infiltration and particle rearrangement improve, allowing for more homogeneous filler dispersion and matrix penetration. This results in a mark reduction in void formation, yielding smoother surfaces and mechanically robust interfaces. Thus, increasing trace width has a direct correlation with improved morphological uniformity and enhanced load-bearing capacity (Tables 6.2 and 6.3).

From a materials science perspective, the influence of trace width on filler dispersion is tied to rheological principles governing flow dynamics and surface tension during ink deposition. Broader traces permit less viscous shear flow and slower solvent evaporation, enabling the conductive nanoparticles to rearrange and form more consistent conductive pathways. The result is a more densely packed composite with fewer flaws and improved electrical and mechanical performance. However, excessive width increases may compromise flexibility, particularly for substrates used in flexible electronics or stretchable devices, due to reduced compliance and increased brittleness of the cured ink.

The presence of voids, regardless of scale, remains a critical defect class in polymer nanocomposites, with wide-reaching implications. From an electrical standpoint, voids act as discontinuities that disrupt electron tunnelling and percolation pathways, significantly increasing resistance and limiting current flow. Moreover, the interfacial tension between void surfaces and conductive fillers facilitates localised oxidation, which can further degrade the electrical integrity of the material over time. Mechanically, voids serve as stress concentrators, particularly under cyclic or impact loading. These stress concentrations can precipitate micro-crack initiation and eventual crack propagation, thereby undermining the fatigue strength and durability of the composite [5].

Further SEM evaluations of CB ink traces show less pronounced microstructural degradation at smaller trace widths compared to GNPs-based inks. In straight-line configurations, a CB ink traces demonstrate smoother and more continuous surfaces with fewer visible voids. This observation is likely due to the smaller particle size and more spherical morphology of the fillers, which promote superior dispersion within the resin matrix. Their lower aspect ratio and more isotropic structure also mitigate the likelihood of agglomeration, enhancing the uniformity of the deposited film. In contrast, meandering and complex geometrical patterns such as curve, square, and zigzag configurations introduce increased surface waviness and asymmetry in the

Table 6.2 Cross sectional images of curve pattern of GNPs and CB ink trace microstructure with line width of 1, 2, and 3 mm

GNPs ink trace

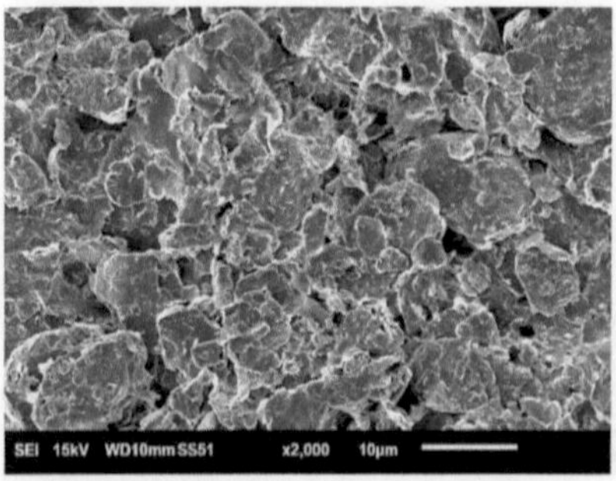

CB ink trace

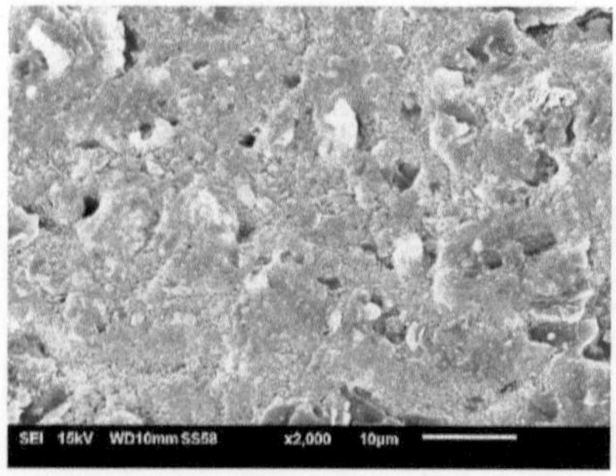

1 mm trace width

GNPs ink trace

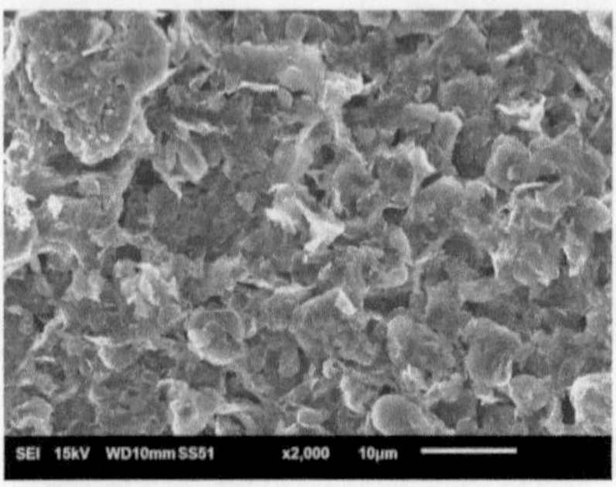

CB ink trace

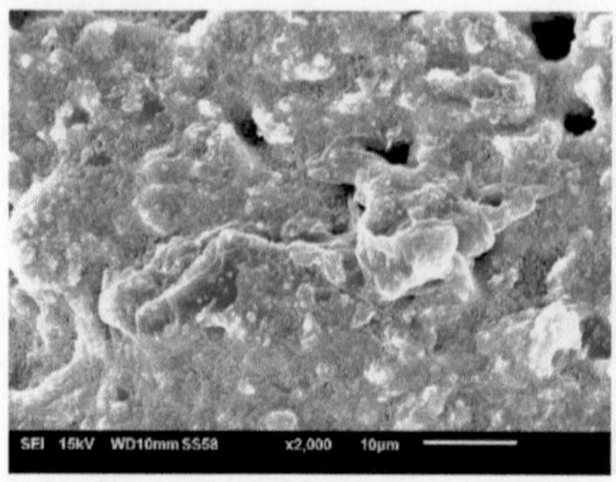

2 mm trace width

GNPs ink trace

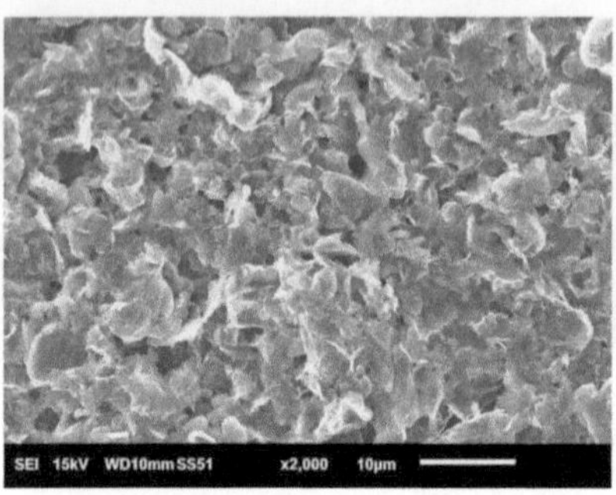

CB ink trace

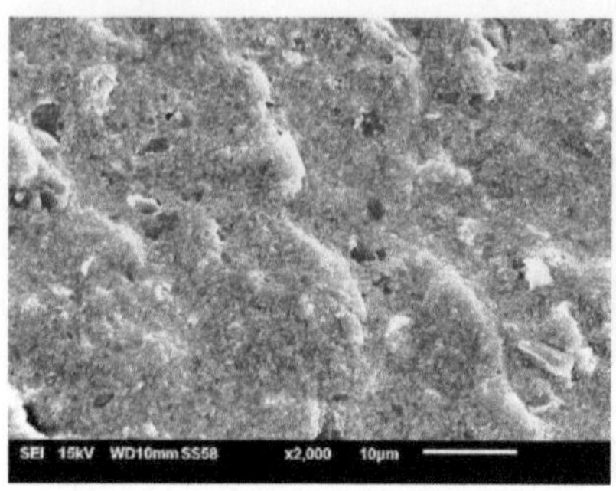

3 mm trace width

Table 6.3 Cross sectional images of square pattern of GNPs and CB ink trace microstructure with line width of 1, 2, and 3 mm

GNPs ink trace	CB ink trace
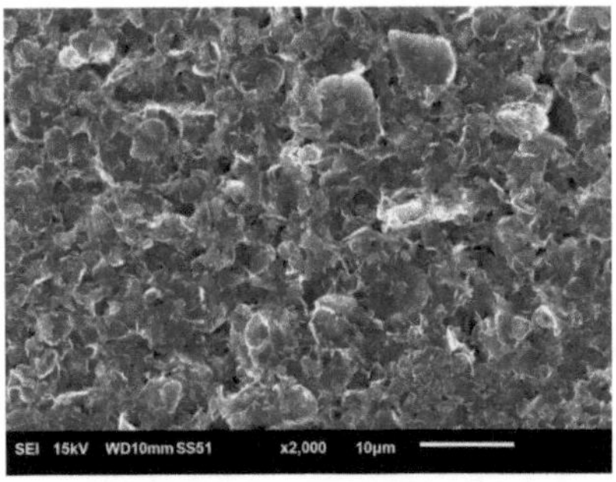	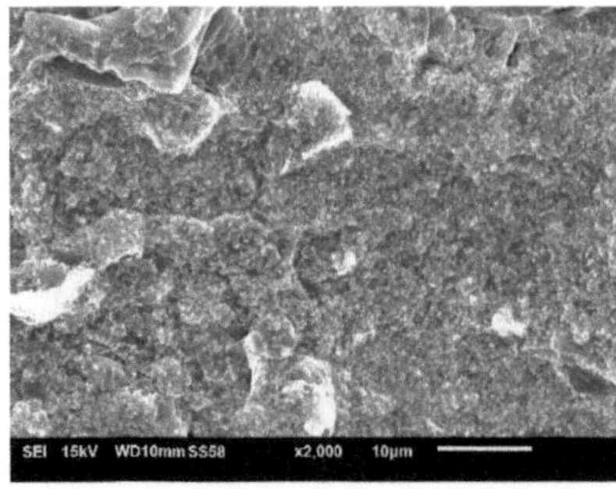
1 mm trace width	
GNPs ink trace	CB ink trace
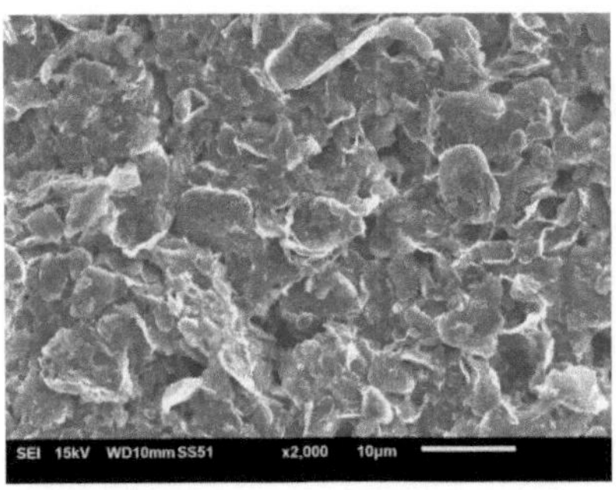	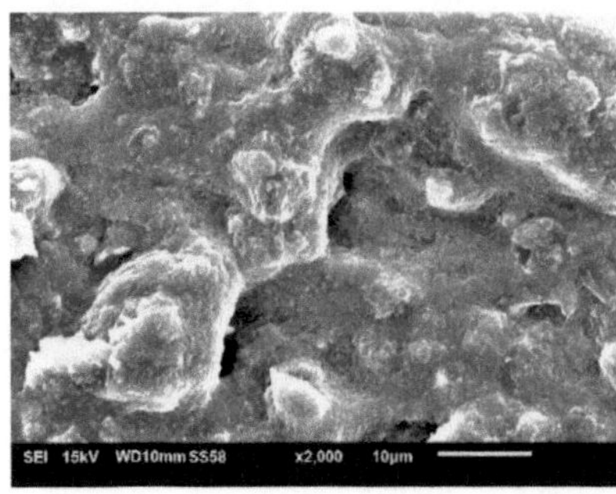
2 mm trace width	
GNPs ink trace	CB ink trace
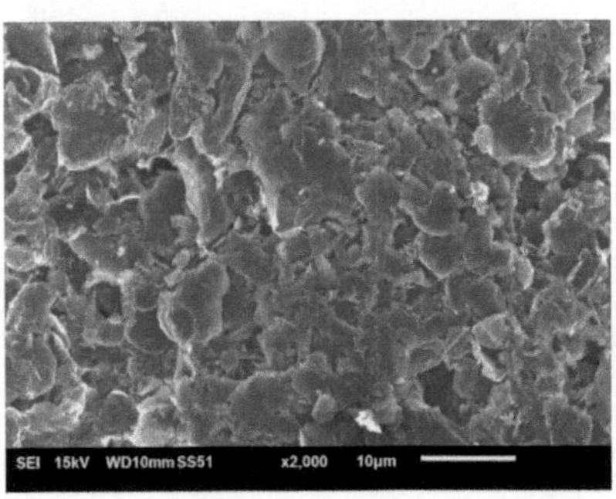	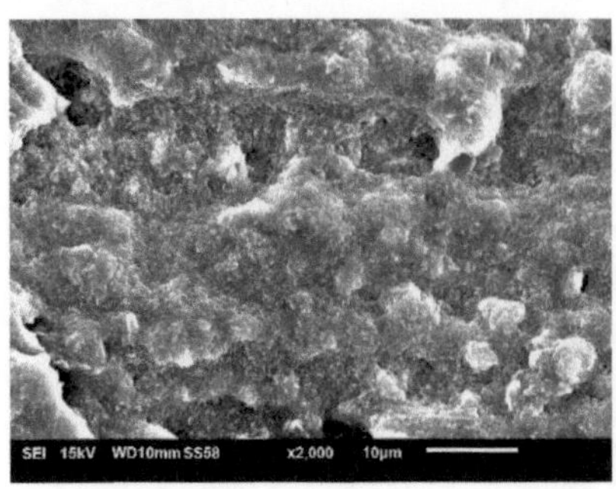
3 mm trace width	

Table 6.4 Cross sectional images of zigzag pattern of GNsP and CB ink trace microstructure with line width of 1, 2, and 3 mm

GNPs ink trace	CB ink trace
	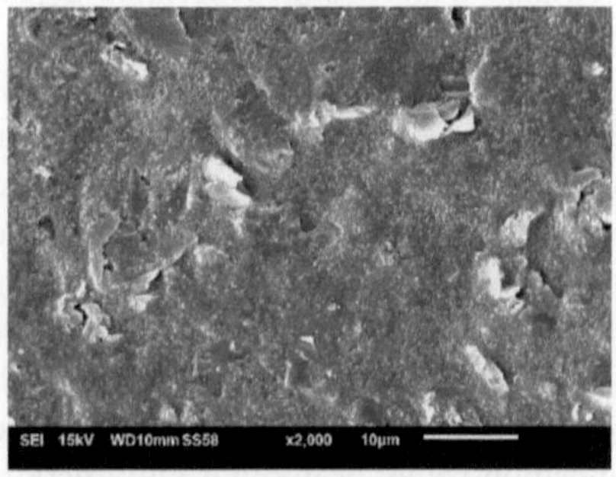
1 mm trace width	
GNPs ink trace	CB ink trace
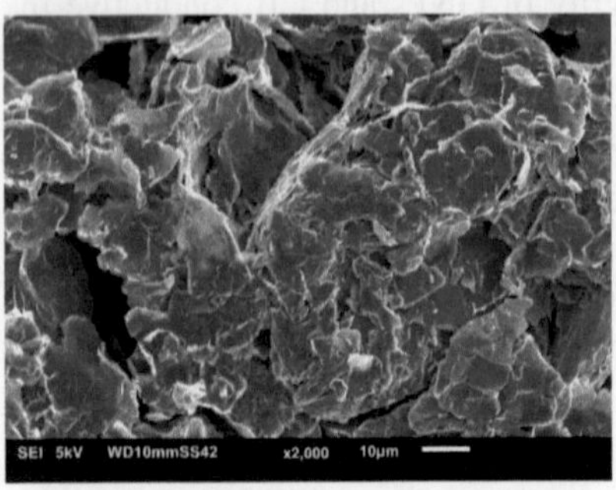	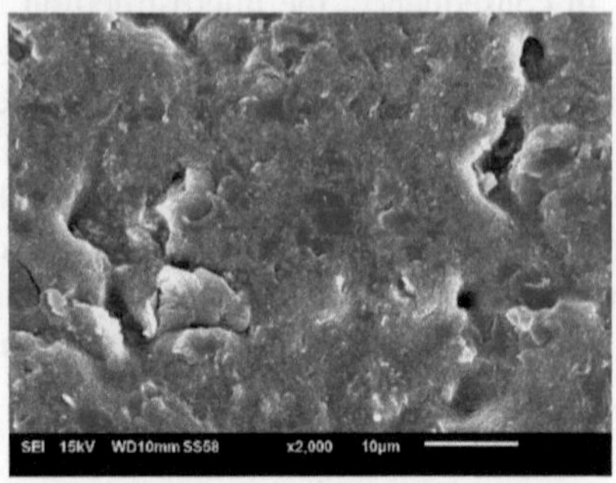
2 mm trace width	
GNPs ink trace	CB ink trace
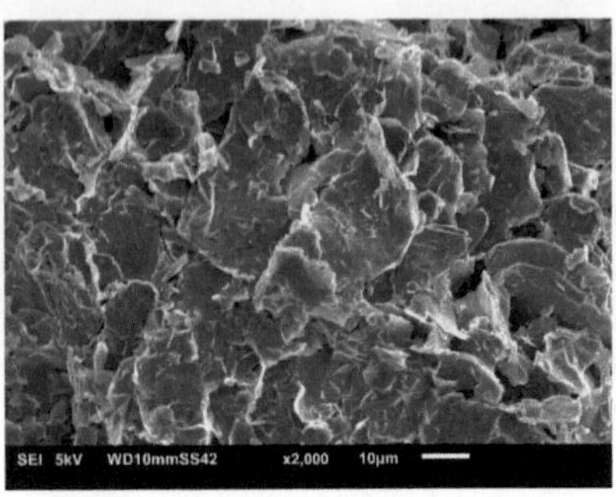	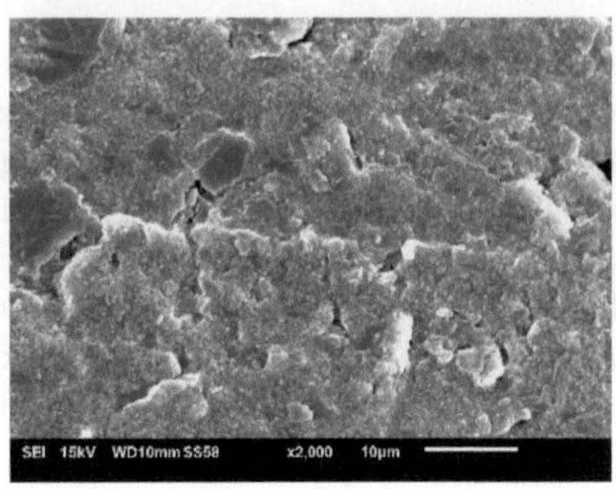
3 mm trace width	

microstructure of the ink traces. These intricate designs inherently complicate ink flow dynamics and promote localised solvent pooling and particle migration, which may hinder complete void filling. As a result, SEM micrographs of these patterns often show a more rugged morphology, interspersed with micro-voids and resin-poor regions.

The impact of geometric patterning on microstructure cannot be overstated. Meandering designs increase the surface area-to-volume ratio, which can be advantageous for strain accommodation in flexible systems, but simultaneously raise challenges in achieving uniform filler distribution. Ink deposition over curved paths requires additional considerations of capillary action, surface tension gradients, and edge effects, all of which can lead to uneven particle settling and resin segregation. Additionally, cross-sectional SEM evidence supports the conclusion that filler-matrix interactions are inherently affected by design geometry. Straight patterns offer a more linear deposition route, promoting unidirectional alignment and more stable microstructures. Conversely, complex geometries introduce multidirectional stresses during curing, which may exacerbate interfacial decohesion and morphological defects.

In consequence, the microstructural integrity of GNPs and CB conductive ink traces is a strong function of both trace width and pattern complexity. Narrower widths and meandering designs correlate with higher void formation and morphological inconsistencies, especially in GNPs composites where agglomeration is more severe. While increased width promotes better dispersion and reduced void content, excessive width may reduce flexibility and hinder certain device applications. Design engineers and material scientists must therefore balance these variables judiciously to optimise electrical performance, mechanical robustness, and structural reliability in conductive ink-based devices.

6.8 Influence of Oxidation and Elemental Distribution

Table 6.5 presents the elemental composition of conductive ink traces, obtained using energy-dispersive X-ray spectroscopy (EDX), specifically focusing on the relative weight percentages of carbon and oxygen. These measurements were conducted across various ink trace widths and pattern geometries to evaluate the extent of oxidation and its potential influence on electrical and mechanical behaviour. The EDX results indicate a higher oxygen content in ink traces with narrower widths, particularly those of 1 mm, as compared to their broader counterparts (2 and 3 mm). This observed trend underscores the influence of geometrical constraints on the degree of oxidation in conductive ink systems. Oxygen incorporation within the carbon matrix often leads to the formation of oxygen-containing functional groups such as carbonyl (C=O), epoxy (C–O–C), hydroxyl (–OH), and carboxyl (–COOH), predominantly localised on the basal planes and at the edges of the carbonaceous fillers. These functionalities disrupt the conjugated sp2-hybridised carbon network, which is responsible for charge delocalisation and high electrical conductivity in materials like graphene and its derivatives.

Table 6.5 Carbon and oxygen weight percentage inside patterned GNPs and CB ink trace for each pattern measured from EDX

Trace pattern	GNPs ink trace						CB ink trace					
	Carbon weight percentage, %			Oxygen weight percentage, %			Carbon weight percentage, %			Oxygen weight percentage, %		
Width, mm	1	2	3	1	2	3	1	2	3	1	2	3
Straight line	70.2	69.1	67.5	29.8	30.9	32.5	56.1	46.6	55.1	43.9	53.4	44.9
Curve	69.8	65.3	67.5	30.2	34.7	32.5	70.2	62.0	64.4	29.8	38.0	35.6
Square	68.8	68.0	65.8	31.2	32.0	34.2	56.8	65.3	59.1	43.2	34.7	40.9
Zigzag	69.5	67.6	68.6	30.5	32.4	31.4	63.9	59.1	65.3	36.1	40.9	34.7

An oxidation alters the electronic structure of carbonaceous materials by introducing defect sites and breaking the π-conjugation, leading to reduced charge carrier mobility. This phenomenon is particularly critical in conductive inks used for printed electronics, where optimal electrical performance is necessary. Moriomoto (2016) highlighted that oxygen functionalities significantly degrade the electrical conductivity of carbon-based materials once a critical threshold is exceeded. It also emphasised the importance of post-treatment or formulation strategies aimed at reducing oxygen functionalities to preserve the electrical integrity of the material.

Furthermore, the microstructural integrity of the conductive traces is compromised in highly oxidised environments. Oxygen-induced defects in the sp2-carbon framework can act as stress concentration sites, fostering crack initiation and propagation under mechanical load. In thin or intricately patterned traces, such as those with meandering geometries, the increased surface area and structural complexity exacerbate this issue by promoting uneven distribution of stress and facilitating oxidation at sharp corners and edges. Tables 6.6 and 6.7 extend this analysis by presenting the spatial distribution of carbon and oxygen across the cross-sections of patterned ink traces with a curve geometry and a width of 3 mm. These figures display elemental mapping results obtained through EDX for both GNPs and CB inks. For the GNPs ink, the distribution comprised 67.5% carbon and 32.5% oxygen, while the CB ink exhibited 64.4% carbon and 35.6% oxygen. These findings suggest a relatively uniform distribution of oxygen throughout the cross-section of the ink traces, which is an encouraging indicator of consistent ink formulation and curing.

The more homogeneous oxygen distribution observed in wider traces and simple patterns may be attributed to improved filler dispersion and curing uniformity. Wider ink traces provide more volume for the migration and settlement of particles, which may reduce the formation of agglomerates and ensure a more isotropic matrix. In contrast, narrower or more complex patterns increase the likelihood of particle entrapment and microstructural heterogeneities, which can act as nucleation sites for oxidation or mechanical failure.

The nature and morphology of the carbon filler also influence the elemental distribution. The GNPs, characterised by their platelet-like structure and high aspect ratio, provide a more extensive surface area and planar geometry, which facilitates greater

Table 6.6 The EDX mapping images for patterned GNPs ink (curve and 3 mm)

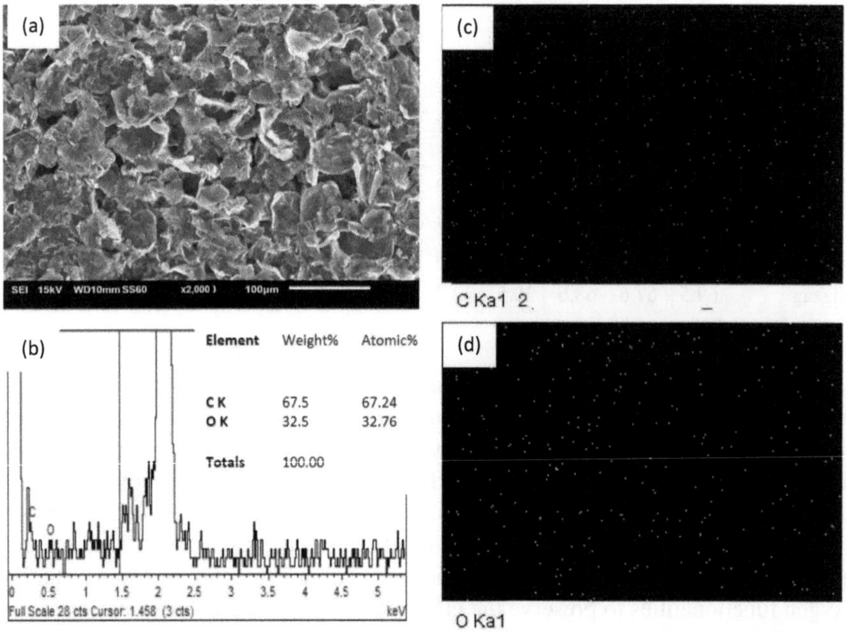

(a) SEM image, (b) EDX element graph, (c) EDX mapping image of Carbon (GNPs), and (d) EDX mapping image of oxygen

surface contact with the binder and enhances dispersion. Their lamellar stacking allows for the formation of interconnected networks that support efficient charge transport and consistent carbon distribution. Conversely, CB particles, being quasi-spherical and typically aggregated due to their high surface energy, exhibit a tendency to cluster. This behaviour often results in poor dispersion, non-uniform distribution of carbon, and increased void formation, which further promotes oxidation.

The oxidation behaviour and elemental composition are intricately linked to the design of the printed trace. Pattern complexity, such as that found in curve or zigzag geometries, increases the surface-to-volume ratio, creating more reactive sites for oxygen interaction. Moreover, edge effects in these patterns lead to localised stresses during curing and application, contributing to microstructural disruptions that facilitate oxidation.

The curing environment significantly influences the performance of conductive inks. Insufficient thermal or UV curing can result in residual solvents and incomplete cross-linking, promoting oxygen ingress and long-term degradation. Additives like antioxidants, dispersants, and crosslinkers help counter these issues by enhancing stability and dispersion. However, their concentrations must be carefully optimised to preserve the ink's rheological behaviour, conductivity, and mechanical integrity, especially in patterned configurations where uniform curing is critical.

Table 6.7 The EDX mapping images for patterned CB ink (curve and 3 mm)

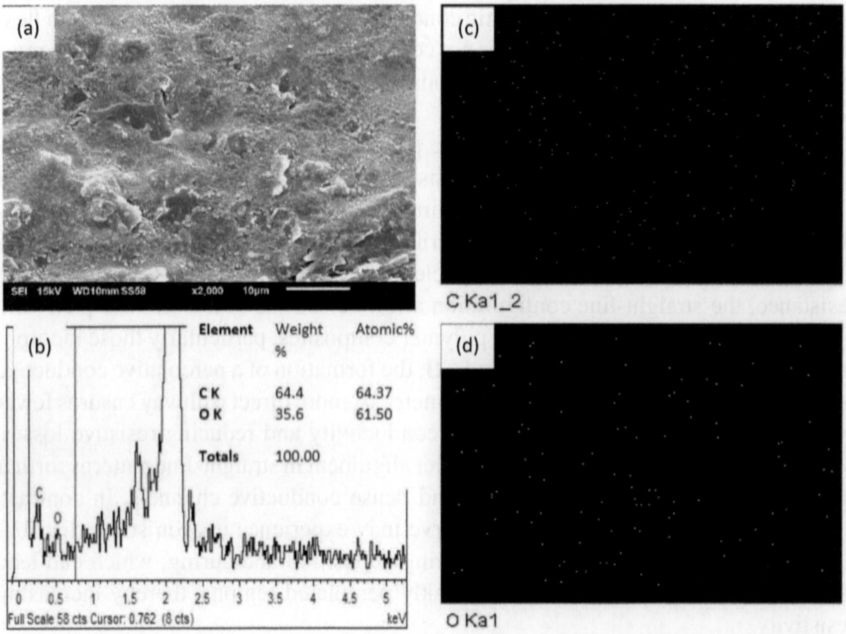

(a) SEM image, (b) EDX element graph, (c) EDX mapping image of Carbon (CB), and (d) EDX mapping image of oxygen

Therefore, the oxidation profile and elemental distribution in GNPs and CB conductive inks are significantly influenced by pattern design, trace width, and filler morphology. Engineers and researchers must consider these interrelated parameters when designing ink formulations and printed patterns for advanced electronic applications. Particularly, minimising oxygen content and ensuring uniform carbon distribution are essential for enhancing the long-term electrical and mechanical reliability of carbon-based inks.

6.9 Effect of Pattern Geometry and Trace Width

In this section, the optimised conductive inks were systematically applied to fabricate ink traces using four geometrically distinct patterns which are straight-line, curve, square, and zigzag. Each pattern was produced in three varying trace widths, such as 1, 2, and 3 mm, to assess the influence of both geometry and dimensional scaling on the performance of the printed features. This approach is critical in advanced printed electronics, where performance is often dictated not only by material composition but also by the structural layout of conductive paths. The investigation centred on evaluating these patterned ink traces under static load conditions, specifically analysing their electrical resistivity, mechanical hardness, and elastic modulus in the absence of external mechanical forces. This static condition reflects many real-scenarios in flexible electronics where deformation is not continuous, but material performance must remain stable during prolonged operational rest periods. The outcomes, presented comprehensively in Table 6.8.

Firstly, ink traces with a straight-line pattern consistently exhibited the lowest volume resistivity across all trace widths and for both GNPs and CB inks. This observation can be attributed to the minimised path an unevenness in the straight-line design. Unlike curved or zigzag patterns, which introduce more junctions, bends, and directional changes that may disrupt electron flow or introduce additional contact resistance, the straight-line configuration allows electrons to traverse the path with minimal interruption. In conductive polymer composites, particularly those incorporating nano-scale fillers like GNPs and CB, the formation of a percolative conductive network is highly sensitive to pattern geometry. A more direct pathway ensures fewer barriers for carrier mobility, enhancing conductivity and reducing resistive losses. Moreover, the relatively homogeneous filler alignment in straight-line patterns further supports the formation of continuous and dense conductive channels. In contrast, different patterns such as zigzag and curve may experience uneven stress distributions and filler dispersion anomalies during deposition and curing, which can lead to microstructural discontinuities or weakly percolated regions, thereby increasing resistivity.

Secondly, the mechanical properties specifically hardness and elastic modulus of GNPs and CB ink traces were found to be largely comparable across all patterned configurations. This similarity suggests that, under static conditions and consistent fabrication parameters (such as ink formulation, curing environment, substrate adhesion, and layer thickness), the macroscale mechanical performance is not significantly influenced by the type of conductive filler, provided both inks are well dispersed within the polymeric matrix. This phenomenon is rooted in the intrinsic properties of the matrix-filler interface and the degree of filler-matrix interlocking, rather than the filler type alone. Although GNPs possesses a two-dimensional platelet morphology with a higher aspect ratio compared to the near-spherical particles of CB, both fillers,

Table 6.8 Summary of electrical and mechanical characterisation of static load test for patterned GNPs and CB (straight-line pattern)

		Width, mm		
		1	2	3
GNPs	Volume resistivity, Ω.cm	0.668	0.938	0.799
	Hardness, MPa	8.540	6.650	6.300
	Elastic modulus, GPa	0.282	0.245	0.231
CB	Volume resistivity, Ω.cm	1.033	1.510	1.305
	Hardness, MPa	0.370	1.054	2.120
	Elastic modulus, GPa	0.007	0.017	0.028

when uniformly embedded within the same matrix can contribute to enhanced stiffness and surface hardness through load transfer mechanisms. The elastic modulus, indicative of a material's ability to resist elastic deformation, is influenced not just by filler stiffness but also by the interfacial adhesion and spatial distribution within the host polymer. In this case, the uniformity in mechanical response suggests effective filler dispersion and robust interfacial bonding, irrespective of filler geometry.

Additionally, the morphological integrity of the ink traces under static conditions was preserved across all patterns and widths, suggesting that neither the geometry nor the size of the pattern significantly compromised structural cohesion. This highlights the importance of carefully tuned rheological properties of the ink and optimised curing strategies, which ensure consistent cross-linking and minimise internal defects like voids or micro-cracks that could otherwise affect mechanical stability. These findings underscore the interdependence between pattern design and functional performance in printed conductive traces. While electrical properties are evidently more sensitive to pattern geometry due to changes in current flow dynamics and filler alignment, mechanical properties remain more resilient, likely due to uniform material processing and filler-matrix interaction. For engineers and researchers working in the field of printed flexible electronics, smart textiles, and stretchable sensors, this insight is critical for designing reliable devices that maintain both conductivity and mechanical integrity under operational conditions.

References

1. Randall, M.S.: The definitive guide to surface mount resistor selection (2015)
2. Merilampi, S.L., Bjorninen, T., Vuorimaki, A., Ukkonen, L., Ruuskanen, P., Sydanheimo, L.: The effect of conductive ink layer thickness on the functioning of printed UHF RFID antennas. Proc. IEEE **98**(9), 1584–1590 (2010)
3. Li, H., Kang, Z., Liu, Y., Lee, S.T.: Carbon nanodots: synthesis, properties and applications. J. Mater. Chem. **22**(46), 24230–24253 (2012)
4. Merilampi, S., Laine-Ma, T., Ruuskanen, P.: The characterization of electrically conductive silver ink patterns on flexible substrates. Microelectron. Reliab. **49**(7), 782–790 (2009)
5. Safdari, M., Al-Haik, M.: Electrical conductivity of synergistically hybridized nanocomposites based on Graphene Nanoplatelets (GNPs) and Carbon Nanotubes (CNTs). Nanotechnology **23**(40), 405202 (2012)
6. Liu, D., Pons, D.J.: Crack propagation mechanisms for creep fatigue: a consolidated explanation of fundamental behaviours from initiation to failure. Metals **8**(8), 623 (2018)

Chapter 7
Cyclic Bending Characteristics of Patterned Carbon-Based Conductive Inks

7.1 Introduction

Fatigue testing plays a critical role in the development and evaluation of advanced electronic systems, particularly in applications such as rollable, flexible, and wearable displays. These technologies are increasingly prevalent in consumer electronics, where end-users may frequently bend, fold, unravel, or re-roll the display during its operational lifetime. As a consequence, the materials used in these devices are subjected to repeated mechanical deformation, which imposes cyclic stress and strain on both the substrate and the functional layers deposited on it. From a mechanical engineering standpoint, such cyclic loading leads to the initiation and propagation of microstructural defects, including dislocations, micro-cracks, and delamination at interfaces, which can severely impair mechanical integrity. Physically, this is a manifestation of material fatigue which a phenomenon governed by principles of fracture mechanics and the accumulation of plastic deformation under oscillatory loading. Over time, this results in the deterioration of both structural and functional performance.

Specifically, in flexible electronics, the substrate and conductive layers must maintain their electrical functionality despite enduring repetitive flexural strain. If the conductive pathways fracture or experience significant resistance drift, the electronic performance degrades, potentially leading to device failure. Therefore, the reliability and operational longevity of such devices are strongly dependent on the electromechanical robustness of the materials employed. Fatigue testing, in this context, serves as an essential methodology to quantify the durability of conductive films under repeated mechanical flexing.

Experimental data gathered from fatigue tests reveal that the evaluated films exhibit promising durability and stability, marking them as feasible candidates for next-generation flexible electronics. The electromechanical efficiency of the conductive materials also can define as the ability to retain electrical conductivity under

M. A. Salim et al., *Nanotechnology in Conductive Inks*, Engineering Materials,
https://doi.org/10.1007/978-981-95-5708-0_7

mechanical deformation, and it was assessed through post-cycling measurements of parameters such as sheet resistance, conductivity retention, and structural integrity.

The insights gained from these tests not only validate the suitability of the materials for practical deployment but also contribute to the fundamental understanding of stress-strain-electrical coupling in deformable conductors. The results underline the importance of optimising film morphology, substrate adhesion, and interfacial toughness to mitigate fatigue-induced degradation. Moreover, from a materials science perspective, properties such as ductility, yield strength, and strain-to-failure must be carefully engineered to align with the mechanical demands of flexible systems. Besides, the fatigue performance of these conductive films underscores their potential for use in robust, mechanically-resilient electronics. Their ability to sustain electrical functionality under repeated mechanical loading positions them as strategic enablers for the advancement of wearable and flexible display technologies. Such findings not only inform the material selection process but also drive innovation in mechanical design, packaging strategies, and overall device architecture.

7.2 Electromechanical Behaviour

The change in electrical resistance of conductive ink samples during continuous cyclic bending was systematically observed and quantified using a custom-designed bending tester rig. This setup enabled real-time resistance monitoring of the cured ink patterns while they were subjected to alternating tensile and compressive stresses at a controlled rate of 20 cycles per minute. The samples were mounted such that the electrical resistance was continuously recorded as they were bent and released cyclically. Cyclic bending introduces repeated mechanical stress and strain, which directly impacts the physical integrity of the conductive pathways. The mechanical deformation causes micro-crack initiation, filler–binder debonding, and progressive interfacial delamination, all of which contribute to an increase in bulk electrical resistance. This phenomenon is well-supported by principles in fatigue mechanics, where cyclic loading leads to material deterioration due to dislocation motion, void coalescence, and phase separation at the microscopic level.

The test protocol maintained a constant bending frequency, as this variable critically affects the rate of damage accumulation. As highlighted by Biga (2008), the rate of cyclic stress application influences the dynamics of crack propagation and viscoelastic deformation in polymeric systems. A constant rate thus ensures experimental consistency and enhances the comparability of results across different ink types and pattern geometries. Figure 7.1 illustrates the resistance evolution of graphene nanoplatelet (GNPs) and carbon black (CB) ink traces under prolonged cyclic bending, plotted as a function of both pattern geometry and trace width.

The work revealed that the geometry and length of the conductive trace pattern exert a substantial influence on the bulk resistance behaviour under mechanical stress. The bulk resistance in this context refers to the total measured resistance, which encompasses both the ink layer and its interaction with the underlying substrate. For

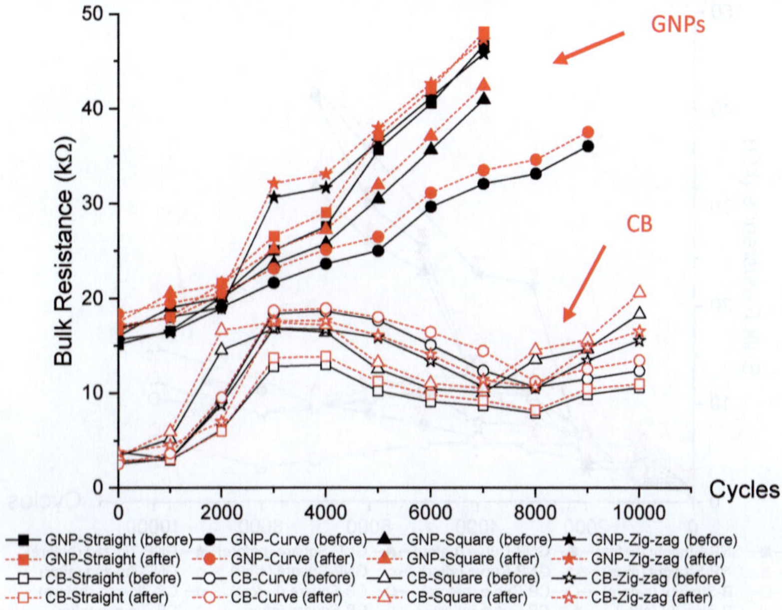

(a)

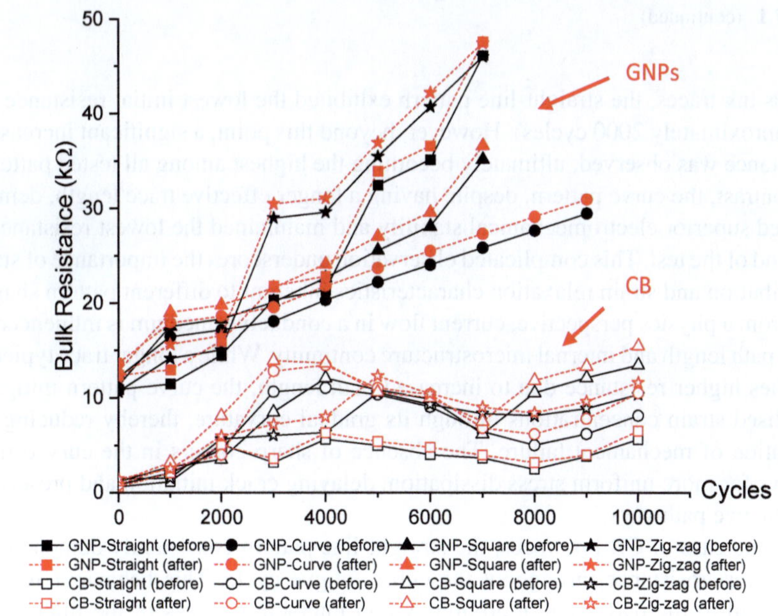

(b)

Fig. 7.1 Graph of resistance value for GNPs and CB ink trace with **a** 1 mm width, **b** 2 mm width, and **c** 3 mm width for 10,000 times of cyclic bending

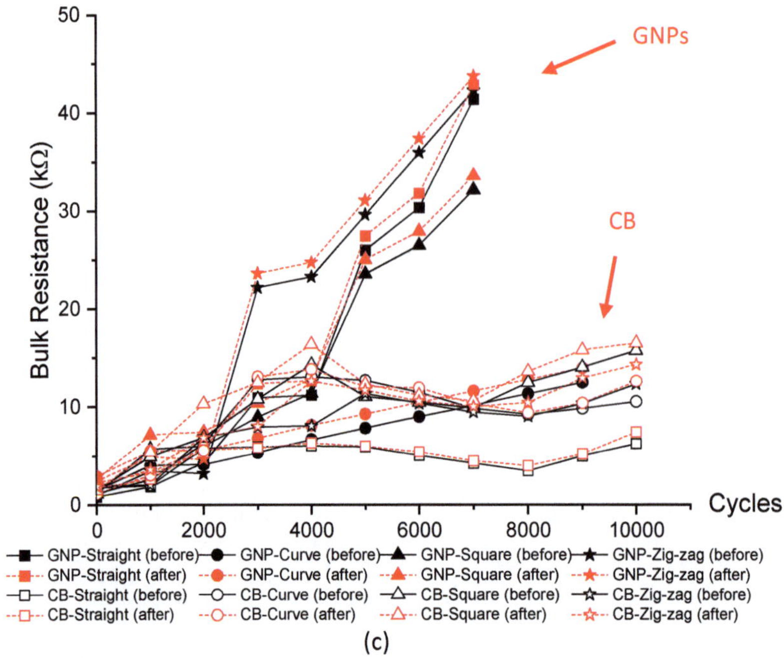

Fig. 7.1 (continued)

GNPs ink traces, the straight-line pattern exhibited the lowest initial resistance (up to approximately 2000 cycles). However, beyond this point, a significant increase in resistance was observed, ultimately becoming the highest among all tested patterns. In contrast, the curve pattern, despite having a longer effective trace length, demonstrated superior electromechanical stability and maintained the lowest resistance at the end of the test. This complicated observation underscores the importance of stress distribution and strain relaxation characteristics inherent to different pattern shapes.

From a physics perspective, current flow in a conductive medium is influenced by both path length and internal microstructure continuity. While a longer trace typically implies higher resistance due to increased path length, the curve pattern mitigates localised strain concentrations through its gradual curvature, thereby reducing the initiation of mechanical failure. The absence of sharp corners in the curved trace allows for more uniform stress dissipation, delaying crack initiation and preserving conductive pathways.

In the case of CB ink traces, the straight-line pattern consistently demonstrated the lowest resistance throughout the entire fatigue cycle. This behaviour may be attributed to the particulate morphology and interaction forces of the fillers. This particle exhibits lower aspect ratios and tend to aggregate due to weaker van der Waals forces, forming conductive clusters that are somewhat resistant to initial mechanical breakdown but susceptible to longer-term agglomeration.

Then, the resistance measurements showed a marked inflection point around 3000 cycles, where a sharp increase in bulk resistance was recorded across most patterns, especially in the GNPs samples. This suggests the onset of microstructural failure such as crack coalescence, filler disconnection, or polymer fatigue softening. These failures are exacerbated under tensile stress where filler particles become physically separated, increasing tunnelling distance and reducing percolation connectivity, thus impeding charge transport. Interestingly, failure thresholds differed across patterns. For GNPs ink: (i) straight-line, square, and zigzag patterns failed around 7000 cycles, (ii) curve pattern endured up to 9000 cycles before failure and (iii) none reached 10,000 cycles without significant resistance degradation.

For CB ink, while resistance initially rose until around 3000 cycles, it then plain or slightly improved up to 8000 cycles before rising again. This abnormal behaviour may result from strain-induced reorganisation of ink aggregates, temporarily enhancing conductive connectivity due to particle rearrangement. Nonetheless, micro-cracking and de-bonding effects resumed dominance as deformation accumulated beyond 8000 cycles.

The differing electromechanical responses between GNPs and CB inks are fundamentally linked to the morphology and physics of the filler particles: (i) the GNPs fillers, with high aspect ratios and planar structures, are prone to mechanical fracture under flexural loading. Their large surface areas promote good electrical conductivity but also create larger stress concentration zones and, (ii) the CB fillers form dense agglomerates due to weaker inter-particle bonding. Upon cyclic bending, these aggregates may undergo compaction and rearrangement, slightly improving contact networks in the short term before failure mechanisms dominate.

The degradation mechanisms in both inks follow a progressive fatigue model, where early-stage performance may remain stable or improve due to rearrangement, followed by a sharp degradation as irreversible damage accumulates. Microstructural weakening increases the inter-particle distance, diminishing electron tunnelling probability, and hence, conductivity. The comparative analysis between different geometrical patterns further reveals that sharp corners and meanders, as seen in square and zigzag patterns, are structurally vulnerable. Such features act as stress concentrators, amplifying localised strain during bending. This is a classic outcome predicted by stress concentration theory in materials science, where discontinuities in geometry lead to non-uniform distribution of mechanical loads. Travelling paths with 90° turns promote higher resistance due to: (i) crack initiation at corners, (ii) current path divergence and (iii) turbulence-like electron scattering in high-resistance zones. These findings are consistent with prior work by Znajdek et al. [1], and Tesson and Barbier (2012), who demonstrated that smoother geometries facilitate stable current density and uniform mechanical response under cyclic loading.

Moreover, data indicate that resistance values were consistently higher in the fully bent state than in the flat state. This is expected, as the bent configuration imposes both maximum tensile strain on the outer radius and compressive stress on the inner radius. These opposing stress modes act simultaneously, amplifying structural distortion, inducing micro-yielding in the polymer matrix, and widening existing micro-cracks. Conversely, in the flat state, the structure is mechanically relaxed, with deformation being primarily elastic or recoverable, except for plastic zones and residual damage. This detailed scientific discussion presents a comprehensive understanding of the mechanical–electrical coupling in printed conductive films and highlights the critical role of material choice, pattern design, and cyclic deformation dynamics in determining device longevity. The insights provided are of significant relevance to researchers and engineers engaged in the design of flexible, stretchable, and wearable electronics, where mechanical reliability is as an essential as electrical performance.

7.3 Sheet Resistance Behaviour

Sheet resistance is a fundamental electrical property employed to characterise the lateral conductivity of thin films, particularly in semiconducting and conductive materials. It is defined as the resistance of a square area of material, independent of the square's size. This property is especially crucial in evaluating printed electronic films, where conductivity must be maintained under mechanical deformation. In this work, the evolution of sheet resistance in GNPs and CB ink traces under cyclic bending conditions is thoroughly investigated to assess durability and structural reliability.

Figure 7.2 illustrates the effect of cyclic mechanical loading on the sheet resistance of GNPs and CB ink traces. As the number of bending cycles increases, a general rise in sheet resistance is observed for all patterns and widths, confirming degradation in the conductive pathways. For GNPs ink traces, four distinct trace patterns (straight, curve, square, and zigzag) with widths of 1, 2, and 3 mm were analysed. It is important to note that the trace geometry significantly influences electrical performance due to variations in path length, stress concentration zones, and strain distribution.

The GNP ink traces displayed a three-stage evolution in sheet resistance under cyclic loading:

Stage I (Initial Stress Accumulation):

- From 0 to approximately 3000 bending cycles, sheet resistance increased moderately (17 to 20%), attributed to minor elastic deformation and early microstructural dislocation within the conductive network. At this stage, reversible damage mechanisms dominate, such as inter-particle slippage and slight debonding between the filler and polymer binder matrix.

Stage II (Crack Initiation and Propagation):

- Between 3000 and 4000 cycles, a sharp increase in sheet resistance (up to 97%) was recorded. This substantial rise is indicative of crack nucleation and propagation within the ink matrix. These cracks disrupt the percolation network of GNPs flakes, impeding the electron tunnelling and hopping mechanisms essential for charge transport. Additionally, due to GNP's two-dimensional flake structure, cyclic strain may cause layer misalignment, increasing inter-flake resistance and altering electron mobility due to band structure distortion.

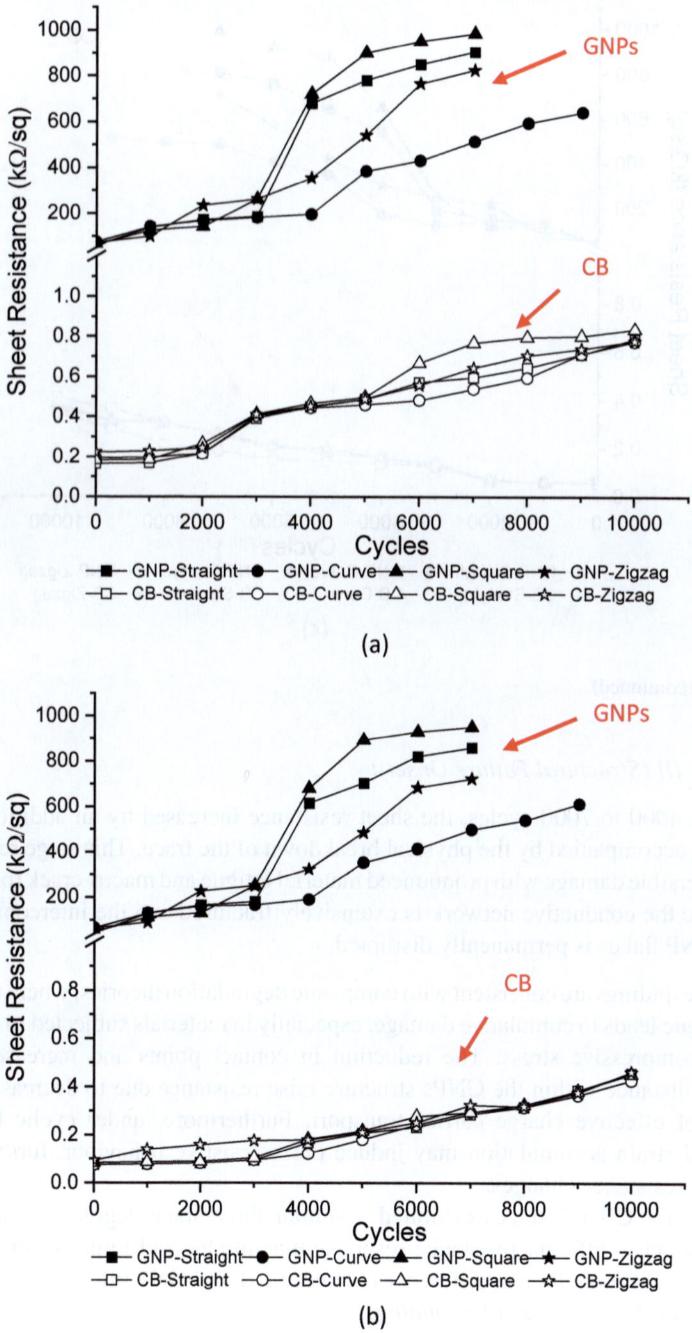

Fig. 7.2 Graph of sheet resistance value for GNP and CB ink trace with **a** 1 mm width, **b** 2 mm width, and **c** 3 mm width for 10,000 times of cyclic bending

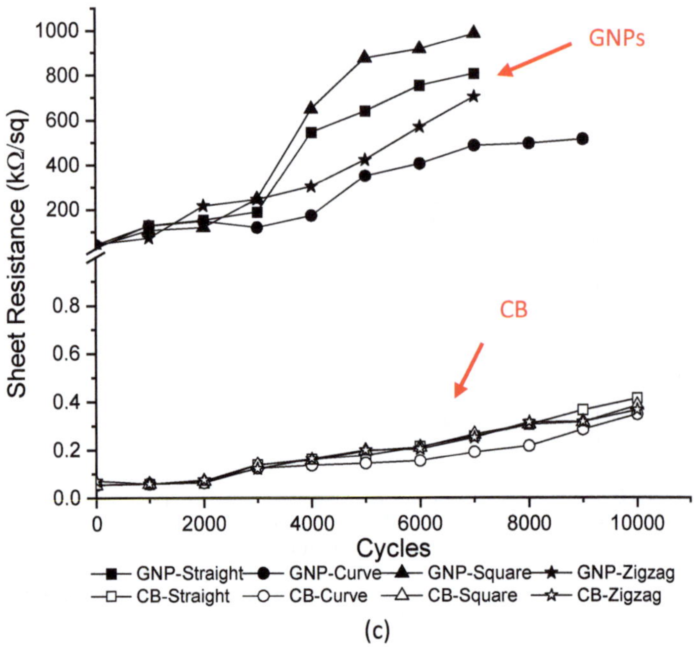

Fig. 7.2 (continued)

Stage III (Structural Failure Onset):

- From 4000 to 7000 cycles, the sheet resistance increased by an additional 6 to 16%, accompanied by the physical breakdown of the trace. This stage represents irreversible damage with pronounced material fatigue and macro-crack formation, where the conductive network is extensively fractured and the interconnectivity of GNP flakes is permanently disrupted.

These findings are consistent with composite degradation theories, where mechanical fatigue leads to cumulative damage, especially in materials subjected to repeated tensile-compressive stress. The reduction in contact points and increased inter-particle distance within the GNPs structure raise resistance due to decreased probability of effective charge carrier transport. Furthermore, under cyclic bending, localised strain accumulation may induce piezo resistive behaviour, further exacerbating resistance changes.

Then, for CB ink traces exhibited a similar three-stage degradation trend but with distinctly different resistance response magnitudes and timings, reflective of the intrinsic morphological differences between CB and GNP fillers:

Stage I (Dispersion and Debonding):

- From 0 to 2000 cycles, the sheet resistance increased gradually (1.95 to 25%), attributed to mechanical dispersion of the particle agglomerates and gradual debonding at the filler-matrix interface. This particle, due to their zero-dimensional (0D) spherical structure, rely heavily on van der Waals interactions and physical contact points to establish conductive pathways.

Stage II (Percolation Path Breakdown):

- Between 2000 and 4000 cycles, the sheet resistance surged by 42 to 75%. This escalation signifies partial collapse of the conductive network, where increased inter-particle distances due to matrix deformation compromise the continuity of electron transport pathways. Unlike GNP's flake-based connections, carbon black pathways are highly susceptible to spatial rearrangements, making them less resilient to tensile and shear forces.

Stage III (Micro-crack Development and Matrix Weakening):

- From 4000 to 10,000 cycles, a more subdued increase in sheet resistance (4 to 12%) was observed, possibly due to saturation in the damage progression. At this stage, micro-cracks become prevalent but fail to fully sever conductive paths. This behaviour highlights carbon black's relative flexibility and the ink's ability to partially accommodate stress without immediate catastrophic failure.

During prolonged cyclic deformation, both conductive inks demonstrate a progressive breakdown in their conductive networks. However, GNP's superior intrinsic conductivity and aspect ratio offer better initial performance but lower mechanical resilience, while CB's particulate nature affords better endurance at the cost of lower initial conductivity.

The geometric design of the ink traces plays an important role in determining their mechanical and electrical response under cyclic loading. Patterns incorporating sharp turns, such as squares and zigzags, experience localised stress concentration at corners. These stress risers initiate early cracking and delamination due to higher strain gradients, leading to elevated sheet resistance. Conversely, curved patterns distribute strain more uniformly, mitigating stress build-up and prolonging structural integrity. Additionally, wider traces typically exhibit lower sheet resistance due to increased cross-sectional area for current flow, but may also accumulate more strain energy, accelerating fatigue in certain geometries. The interplay between trace width and pattern complexity underscores the need for optimised trace design in flexible electronics.

From a physics standpoint, the mechanical loading induces strain energy within the ink trace. In conductive films, electron mobility is sensitive to microstructural defects, and cyclic deformation introduces these defects by disrupting the percolative network. As strain energy increases, bond rupture between filler and binder reduces charge carrier mobility, while crack formation alters the dielectric properties of the medium. The sheet resistance rise thus becomes a direct manifestation of accumulated mechanical and electrical fatigue.

The findings of this investigation are significant for the design of reliable, flexible printed electronics. For applications involving repetitive flexing, such as wearable sensors, flexible circuits, and e-textiles, the long-term electrical stability of the ink under mechanical duress is established. Engineers must consider not only the conductive filler type but also the trace geometry and substrate interactions to ensure performance longevity. Ultimately, the evaluation reveals that GNPs inks are favourable for high-conductivity applications where mechanical strain is limited, while CB inks offer better endurance in high-strain scenarios at the expense of conductivity. Design optimisation should balance electrical performance and mechanical resilience, incorporating considerations such as filler morphology, ink formulation, and trace layout. By deepening the understanding of the underlying material science and physics of deformation-induced resistance changes, this work contributes to the field of advanced materials engineering and helps guide the development of next-generation flexible electronic systems.

7.4 Evaluation of Volume Resistivity

Figure 7.3 presents a comparative graph illustrating the volume resistivity derived from both bulk and sheet resistance measurements for GNPs and CB conductive ink traces. The graph is plotted to assess the influence of two parameters which are trace shape geometry and trace width under progressive cyclic bending stress. This comparison is an essential in understanding the degradation mechanisms affecting the electrical performance of conductive inks under mechanical deformation.

For GNPs-based conductive inks, the evolution of volume resistivity exhibited consistent trends across both measurement approaches which are bulk resistance and sheet resistance, an d it demonstrating a progressive increase with the number of cyclic bending cycles. During the initial phase of the test (up to approximately 5000 cycles), the volume resistivity rose gradually, with increases closely aligned with the trace geometry. This phase signifies the onset of interfacial debonding between the GNPs flakes and the polymer matrix, leading to minor disruptions in the conductive pathways.

The anisotropic morphology of the particles, having a high aspect ratio and layered structure, contributes to a relatively stable conductive network in early deformation stages. However, as the cyclic stress intensifies, particularly beyond 5000 cycles, a marked and more erratic increase in volume resistivity occurs, especially for straight-line patterns. This can be attributed to the propagation of micro-cracks and the breakdown of π-π interactions between adjacent GNPs, causing substantial disruption in the percolating network. The deformation alters the inter-particle distances, which in turn increases tunnelling barriers for electron flow, as governed by the exponential dependence of resistance on inter-flake separation according to quantum tunnelling theory.

In contrast, the CB conductive inks exhibited a distinct response. The volume resistivity determined from bulk resistance initially increased during early cyclic loading, consistent with minor disaggregation of ink agglomerates and microstructural rearrangements. Interestingly, a reduction in volume resistivity was observed during the mid-phase of bending cycles. This counterintuitive behaviour is likely attributed to

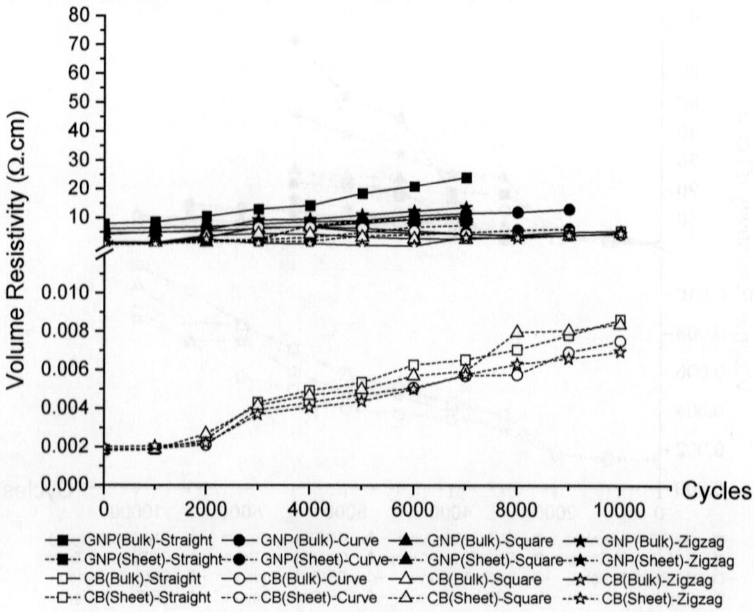

(a)

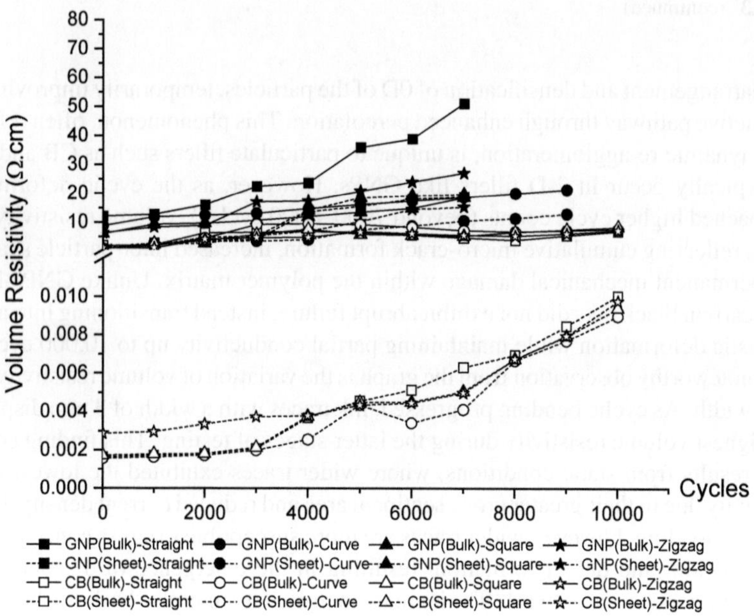

(b)

Fig. 7.3 Graph of volume resistivity for GNPs and CB ink trace with **a** 1 mm width, **b** 2 mm width, and **c** 3 mm width for 10,000 times of cyclic bending

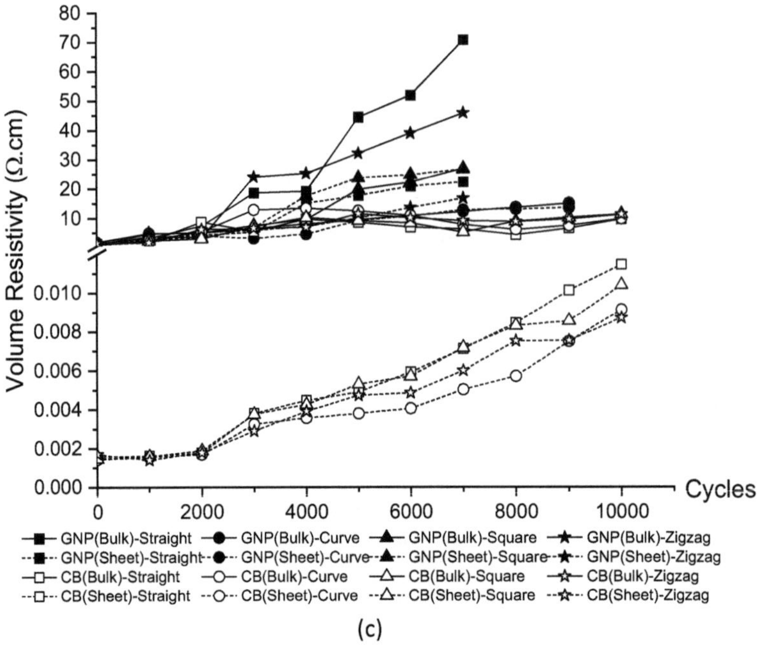

Fig. 7.3 (continued)

the rearrangement and densification of 0D of the particles, temporarily improving the conductive pathway through enhanced percolation. This phenomenon, often referred to as dynamic re-agglomeration, is unique to particulate fillers such as CB and does not typically occur in 2-D fillers like GNPs. However, as the cyclic deformation approached higher cycle counts (beyond 7000–8000 cycles), volume resistivity rose again, reflecting cumulative micro-crack formation, increased inter-particle spacing, and permanent mechanical damage within the polymer matrix. Unlike GNPs-based inks, carbon black inks did not exhibit abrupt failure, instead transitioning into a state of plastic deformation while maintaining partial conductivity up to 10,000 cycles.

A noteworthy observation from the graph is the variation of volume resistivity with trace width. As cyclic bending progressed, ink traces with a width of 3 mm displayed the highest volume resistivity during the latter stages of testing. This finding contradicts results from static conditions, where wider traces exhibited the lowest initial resistivity due to their greater cross-sectional area and reduced current density. Under dynamic loading, however, wider traces are more susceptible to stress concentration, delamination, and defect formation, resulting in the emergence of discontinuities within the conductive network. The increase in resistivity may also be due to heterogeneous ink distribution during printing, as ink rheology can influence film uniformity and particle dispersion across varying widths.

Additionally, considering a cubic centimetre of material, a trace width of 1 mm exhibited a denser distribution of conductive fillers, for both GNPs and CB, compared to wider traces. In GNPs ink traces, the layered flakes remained partially interconnected despite deformation, thereby maintaining a minimal constriction resistance. In CB ink traces, the higher concentration of spherical ink particles formed a relatively dense conductive network, even after substantial mechanical cycling, preserving a degree of electrical continuity. This observation is consistent with percolation theory, which states that below a critical particle volume fraction, the probability of forming a continuous conductive path decreases exponentially.

Importantly, the volume resistivity calculated from sheet resistance values was consistently lower than that derived from bulk resistance, underscoring the accuracy and sensitivity of the Four-Point probe method. This technique mitigates contact resistance issues, particularly significant for thin films and anisotropic materials, thereby offering a more intrinsic measurement of the material's electrical properties. This distinction was more pronounced in CB inks, potentially due to the more pronounced surface heterogeneities and interface resistances inherent to spherical agglomerated particles. Conversely, the difference between the two measurement approaches in GNPs inks was relatively minor, indicating a more homogeneous conduction path due to the planar alignment of flakes. In summary, the analysis of volume resistivity from bulk and sheet resistance measurements provides a deeper insight into the electrical degradation mechanisms of conductive inks under mechanical fatigue. These findings highlight the complex interplay between particle morphology, trace geometry, and mechanical stress, all of which must be considered for the development of robust flexible electronic systems.

7.5 Investigation Using Nanoindentation Method

Figure 7.4 presents the nanoindentation data highlighting the mechanical response of GNPs and CB conductive ink traces under cyclic bending loads, plotted as the maximum applied force versus the corresponding indentation depth. This data provides critical insight into the progressive degradation of mechanical integrity in printed conductive traces when subjected to repeated mechanical deformation, particularly in flexible electronic applications.

A clear trend emerges, with increasing cyclic bending (up to 10,000 cycles), both conductive inks experience a deterioration in mechanical performance, evidenced by the reduction in force required to achieve a greater depth. This inverse relationship indicates a decrease in both hardness and elastic modulus, suggestive of material softening. While, trace width significantly influences the rate and extent of degradation, the trace geometry exerts comparatively negligible influence on the indentation characteristics, implying that mechanical degradation is more sensitive to dimensional factors and internal material composition than to external trace topology.

In the case of GNPs conductive ink, after just 1000 cycles of cyclic bending, a pronounced increase in indentation depth was recorded under a force similar to that applied to the undeformed sample. This early-stage softening suggests an initial disruption of van der Waals interactions and interfacial bonding between stacked

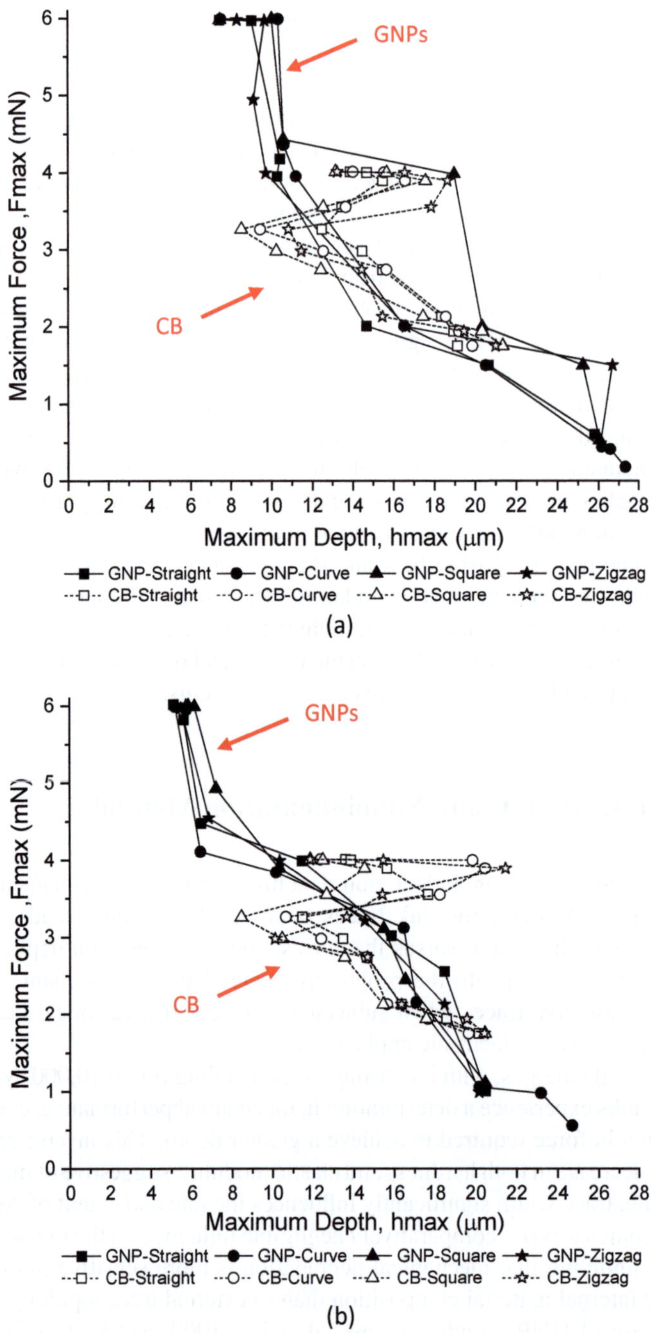

Fig. 7.4 Graph of maximum force versus maximum load of GNPs and CB ink trace with different pattern ink trace with **a** 1 mm, **b** 2 mm, and **c** 3 mm widths of cyclic bending

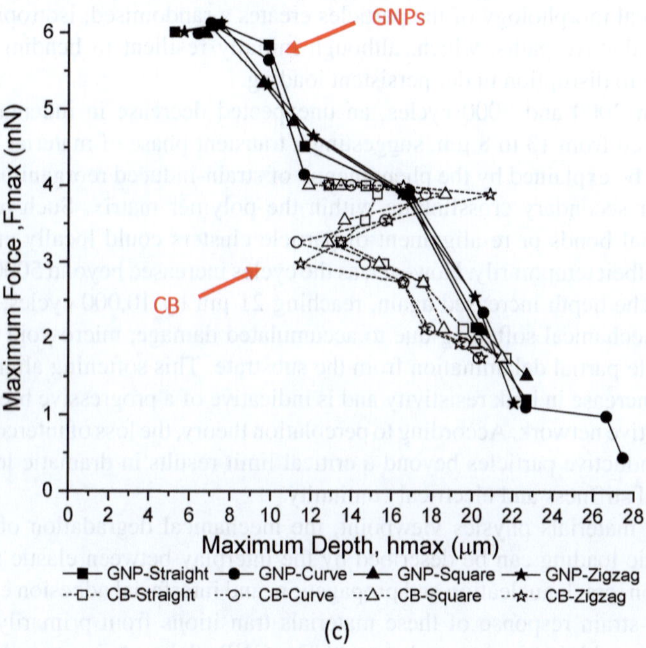

Fig. 7.4 (continued)

GNPs flakes. As the number of cycles progressed to 2000 and beyond, a consistent pattern of increasing depth and decreasing force was observed, indicating progressive loss of mechanical stiffness and cohesion within the conductive matrix.

This behaviour is attributable to multiple coupled mechanisms. Firstly, repetitive bending induces plastic deformation at the microscale, initiating micro-cracks and interfacial debonding among GNPs particles and the polymeric binder. Secondly, these cracks serve as stress concentrators, amplifying local strain fields and accelerating fracture propagation, which is a classical phenomenon described by Griffith's theory of brittle fracture and confirmed by micromechanical failure models. The accumulation of strain energy over cycles further weakens the particle-to-particle conductive pathways, resulting in a material with reduced resistance to indentation. Moreover, as the internal network geometry becomes increasingly disrupted, mechanical anisotropy increases, and the percolation threshold required to maintain a continuous load-bearing network is no longer met, thereby decreasing the effective elastic modulus.

In contrast, CB conductive ink exhibited a comparatively more stable response, though still indicative of mechanical fatigue. Up to approximately 3000 cycles, a slight increase in depth from 15 to 18 μm was recorded, corresponding with an observable increase in bulk electrical resistance. This suggests the onset of weakening due to the untangling or dispersion of carbon black agglomerates within the polymer matrix.

The spherical morphology of the particles creates a randomised, isotropic distribution of conductive paths, which, although initially resilient to bending, becomes susceptible to disruption under persistent loading.

Between 3000 and 5000 cycles, an unexpected decrease in indentation depth was observed from 13 to 8 μm, suggesting a transient phase of material hardening. This could be explained by the phenomenon of strain-induced reorganisation of CB particles or secondary crosslinking within the polymer matrix. Such reformation of interfacial bonds or re-alignment of particle clusters could locally enhance the modulus, albeit temporarily. However, as the cycles increased beyond 5000, the trend reversed. The depth increased again, reaching 21 μm by 10,000 cycles, consistent with the mechanical softening due to accumulated damage, micro-void formation, and possible partial delamination from the substrate. This softening aligns with the observed increase in bulk resistivity and is indicative of a progressive breakdown of the conductive network. According to percolation theory, the loss of interconnectivity among conductive particles beyond a critical limit results in dramatic loss of both mechanical stiffness and electrical continuity.

From a materials physics viewpoint, the mechanical degradation of both inks under cyclic loading can be described by the interplay between elastic and plastic deformation, crack nucleation and propagation, and interfacial adhesion energy loss. The stress-strain response of these materials transitions from primarily elastic to elastoplastic with increasing cycle count. The GNPs flakes, being two-dimensional and plate-like, are more susceptible to shearing and interfacial sliding under bending forces, leading to early onset of damage. Meanwhile, CB particles, as zero-dimensional and spherical, are embedded in a manner that better accommodates stress redistribution, admitting less efficiently at higher cyclic deformations.

Additionally, cyclic bending imposes alternating tensile and compressive stresses that accumulate over time, promoting viscoelastic creep and fatigue phenomena in the polymer binder. This leads to stress relaxation and permanent deformation, further reducing mechanical resistance during indentation testing. The fracture mechanics of composite thin films reveal that once crack density reaches a threshold, the material no longer exhibits recoverable deformation, hence the rapid increase in indentation depth.

The nanoindentation results underscore the critical importance of mechanical stability in printed conductive traces, particularly for devices designed for repeated mechanical interaction such as wearable electronics, rollable displays, or soft robotics. The GNPs ink, while initially more conductive and possessing better percolation networks, suffer from greater mechanical degradation due to their susceptibility to interfacial failure and crack propagation. The CB inks, while slightly more robust mechanically, show complex deformation behaviour that may be advantageous in specific load-bearing applications. Therefore, the nanoindentation profiles provide strong evidence of internal structural degradation in both ink types, driven by cumulative cyclic mechanical loading. The GNPs traces exhibited more significant mechanical fatigue, correlating with more rapid electrical degradation, whereas carbon black traces demonstrated a more enhanced response with intermittent hardening likely due to particle rearrangement. The comparative analysis of depth and force, therefore offers valuable insights for the design of resilient printed electronics where mechanical endurance is a critical parameter.

7.6 Hardness and Elastic Modulus Degradation

Basically, Fig. 7.5 illustrates the progressive decline in the hardness of conductive ink traces as they are subjected to repeated cyclic bending. The hardness values, derived from load–displacement curves during nanoindentation, reveal distinct behavioural trends across different trace widths (1, 2, and 3 mm) and pattern geometries. Generally, hardness decreases as the number of bending cycles increases, indicative of cumulative fatigue damage and microstructural degradation within the ink matrix.

For traces with a width of 1 mm, initial hardness values range from 8.54 to 8.86 MPa. With continued cyclic loading, these values deteriorate significantly, ultimately reaching as low as 0.99 to 1.84 MPa before mechanical failure. Similar trends are observed for 2 and 3 mm widths, where initial hardness begins at 6.84 and 6.46 MPa, respectively, and drops steadily over time. This degradation behaviour is a consequence of the initiation and propagation of micro-cracks, matrix fragmentation, and interfacial debonding within the ink and between the ink and the substrate. During the early stages (up to 1000 cycles), the decrease in hardness is modest (2.2%), with no significant surface damage observed. However, beyond 2000 cycles, a more pronounced drop of approximately 25% in hardness becomes evident. This behaviour is consistent with classical fatigue mechanisms in polymer composites, where the initiation of microscopic flaws leads to material softening. At approximately 5000 cycles, optical and SEM observations confirm the presence of surface cracks, evidence of stress concentration effects and strain localisation in the conductive network.

The hardness degradation is further compounded by changes in material ductility. Under high-cycle fatigue, the ink undergoes local plastic deformation, manifesting as fluid-like stretching. This behaviour arises from the breakdown of the cross-linked polymer matrix and slippage between conductive filler particles and the matrix. According to the principles of continuum mechanics and plasticity theory, when a material undergoes repetitive stress-strain cycles, energy dissipation through internal friction leads to bond breakage and rearrangement of molecular chains, which in turn contributes to ductility and reduced resistance to indentation [2, 3].

Comparatively, CB ink traces demonstrate more complex trends. Although their hardness is generally lower than that of GNPs-based inks, the behaviour under fatigue loading is non-monotonic. Initial reductions in hardness are followed by intermittent increases or fluctuations before a final decline at 10,000 cycles. These fluctuations may be explained by the temporary reorganisation or densification of filler networks due to repeated mechanical agitation which a phenomenon similar to strain-induced crystallisation or rearrangement seen in certain elastomers. The curve pattern consistently shows higher hardness than other geometries across all widths, possibly due to more even stress distribution and reduced localised deformation. The pattern shape

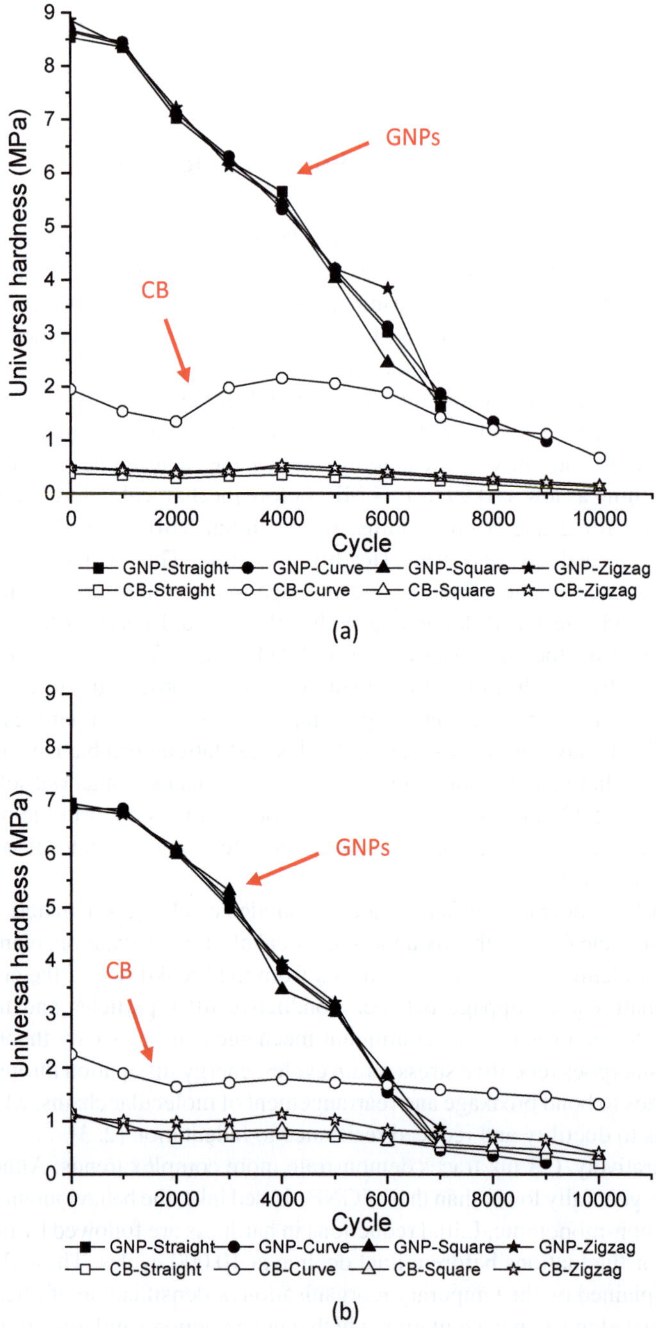

Fig. 7.5 Graph of universal hardness of GNPs and CB ink trace with different pattern ink with **a** 1 mm, **b** 2 mm, and **c** 3 mm widths of cyclic bending

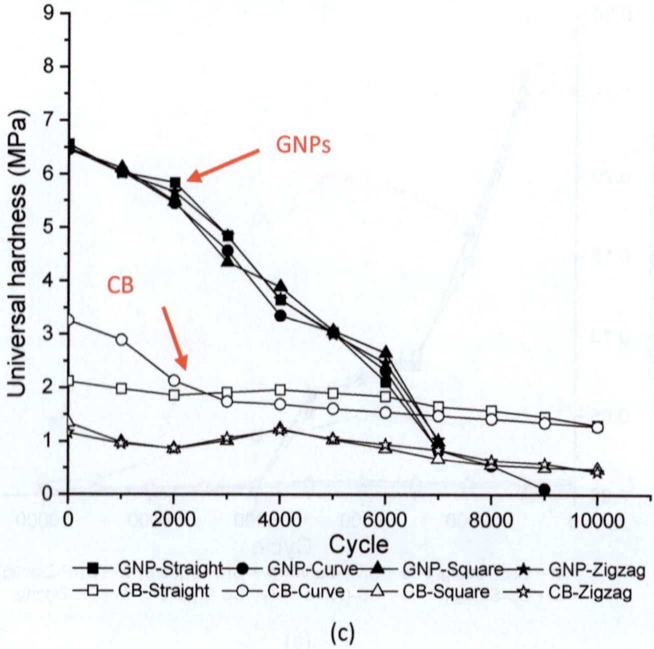

Fig. 7.5 (continued)

also influences the mechanical response. For instance, at a 3 mm width, the straight-line and curve geometries exhibit initially higher hardness values than the square and zigzag patterns. However, over time, their response converges due to uniform material degradation. It is important to note that due to the randomness of the nanoindentation locations, local heterogeneities such as agglomerated filler clusters or voids could cause slight variation in measured hardness, masking any minor differences between geometries.

Moving beyond hardness, Fig. 7.6 presents the evolution of the elastic modulus of conductive ink traces under cyclic bending. This modulus defined as the ratio of stress to strain within the elastic deformation range which is a fundamental property characterising a material's stiffness. In traditional isotropic materials, the elastic modulus remains constant under fatigue until final fracture. However, for nanostructured composite inks, such as GNPs and CB systems, this assumption does not hold true. The elastic modulus here decreases significantly with cyclic loading due to the progressive degradation of the conductive network and matrix integrity [4].

In the case of GNPs ink, after just 1000 bending cycles, a small drop in the elastic modulus (ranging from 4.0 to 9.4%) is observed, attributed to internal stress relaxation and slight plastic strain accumulation. From 2000 to 3000 cycles, a more rapid decline (40% to 55%) occurs due to interfacial debonding, micro-crack formation, and matrix delamination, features typical in fibre-reinforced composites during the onset of fatigue damage [5]. This aligns with fracture mechanics theory, where the

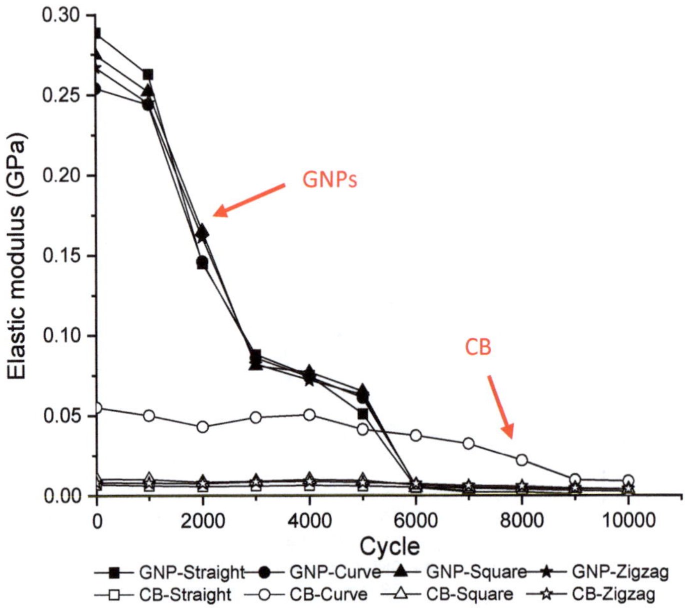

(a)

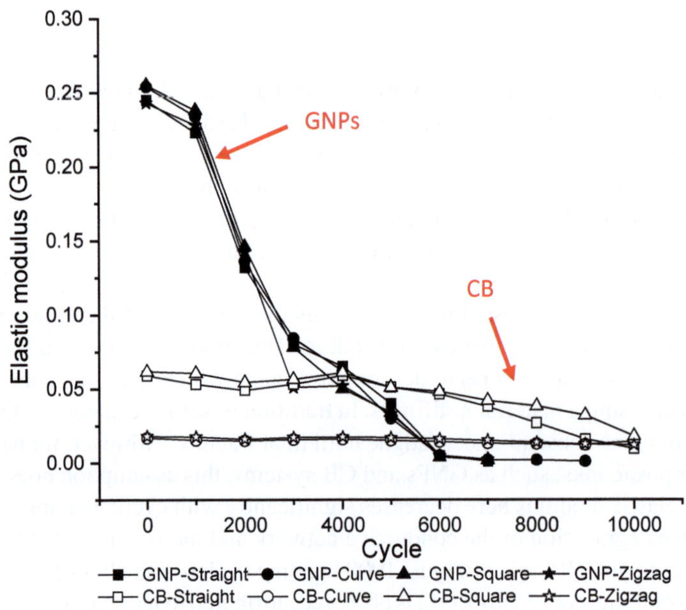

(b)

Fig. 7.6 Graph of elastic modulus of GNPs and CB ink trace with different pattern with **a** 1 mm, **b** 2 mm, and **c** 3 mm widths of cyclic bending

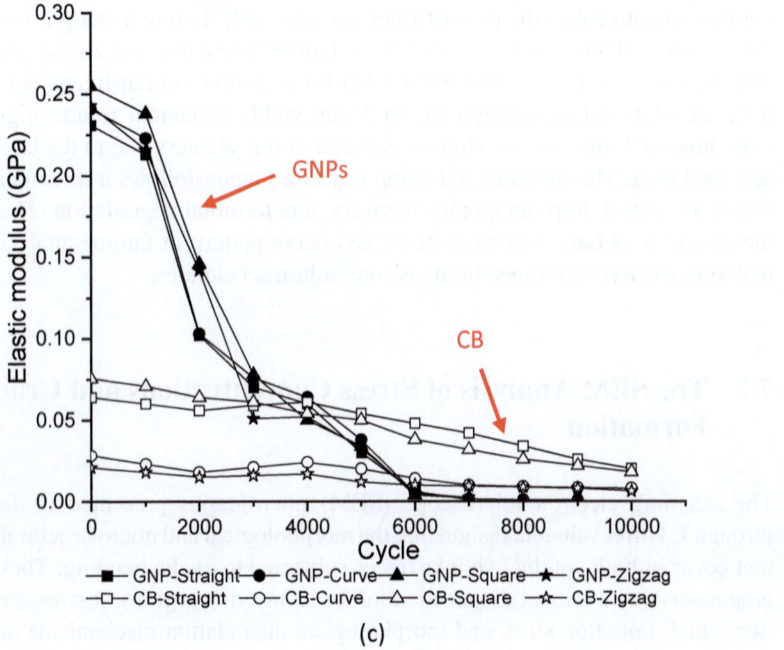

Fig. 7.6 (continued)

formation of micro-voids and debonded zones leads to stiffness degradation and alters the material's load-bearing capability. Between 3000 and 6000 cycles, continued growth of these flaws further reduces the modulus, with final failure characterised by full trace breakage and loss of structural coherence. The reduction in modulus during this phase corresponds with the accumulation of cyclic plastic strain and increased energy dissipation through crack propagation pathways [6].

In CB ink traces, the response differs. For pattern shapes with 1 mm width, initial modulus values lie between 0.007 and 0.0104 GPa, decreasing after 2000 cycles to 0.006–0.0084 GPa, then temporarily rising to 0.01 GPa at 4000 cycles. However, by 10,000 cycles, modulus values fall sharply to 0.0026–0.004 GPa. This non-linear behaviour may be attributed to partial healing of filler contacts or compaction of the network during intermediate stages, before degradation processes dominate again. The square pattern for the inks shows a higher initial modulus (0.0551 GPa), which fluctuates similarly before declining to 0.0088 GPa at 10,000 cycles. Its high initial stiffness can be ascribed to material concentration in geometric corners, enhancing local rigidity. However, these same features promote stress localisation, accelerating fatigue-induced damage.

For traces with 2 mm width, the straight-line and square patterns begin with modulus values of 0.059 and 0.0617 GPa, which decrease to 0.049 and 0.0543 GPa after 2000 cycles. A temporary increase follows at 4000 cycles due to potential rearrangement or densification of filler particles [10], before a sharp drop to 0.01–0.018 GPa at 10,000 cycles due to fatigue failure. The curve and zigzag geometries, initially lower in stiffness (0.0169–0.0181 GPa), follow similar trends but maintain lower absolute values throughout. At 3 mm width, behaviour trends align closely with those at 2 mm, even if slightly elevated initial stiffness due to the larger cross-sectional area. The modulus reduction trajectory again follows a three-phase trend which are initial drop, temporary recovery, and terminal degradation. This reflects the classic S–N (stress-number of cycles) curve pattern in fatigue analysis, where materials exhibit endurance limits before failure accelerates.

7.7 The SEM Analysis of Stress Concentrations and Crack Formation

The scanning electron microscope (SEM) micrographs presented in Tables 7.1 through 7.4 offer valuable insight into the morphological and microstructural changes that occur in both conductive inks traces subjected to cyclic bending. These micrographs serve as a critical diagnostic tool for identifying regions of stress concentration, crack initiation sites, and morphological degradation mechanisms influenced by the geometry and width of the conductive patterns. The SEM images clearly reveal that stress concentrations primarily occur at geometric discontinuities, such as sharp corners, junctions of directional change, and irregularities at the interface between the filler and the polymer binder. These discontinuities induce strain localisation, in line with classical elasticity theory, where abrupt changes in geometry act as singularities that amplify local stress.

In the case of sharp-edged circular patterns, especially those with angular or jagged geometries, the SEM analysis demonstrates elevated stress concentration at internal corners. These points act as primary sites for crack nucleation. During cyclic loading, stress is repeatedly concentrated at these points, initiating micro-cracks that gradually propagate through the ink matrix. This is consistent with fracture mechanics principles, whereby the energy release rate and the stress intensity factor increase at defect sites, thereby driving crack propagation under Mode I (opening) or mixed-mode loading. In contrast, curved or sinusoidal meandering patterns exhibit more gradual transitions in geometry. The smoother curvature promotes a more uniform stress distribution, which delays or suppresses crack formation. This is in line with finite element analysis simulations in structural mechanics, which consistently show reduced von Mises stress values in curved designs as opposed to sharply angled ones. Therefore, from a structural durability perspective, smoothly curved patterns are inherently more resilient to cyclic deformation.

Post-failure SEM imaging of GNPs-based ink traces indicates significant microstructural changes influenced by both trace width and fatigue damage. Cross-sectional images reveal a clear rearrangement of GNPs, particularly in narrower traces where high aspect ratio flakes tend to align preferentially under mechanical

Table 7.1 Cross sectional images of straight-line pattern of GNPs and CB ink trace microstructure with line width of 1, 2, and 3 mm for cyclic bending

GNPs ink trace	CB ink trace
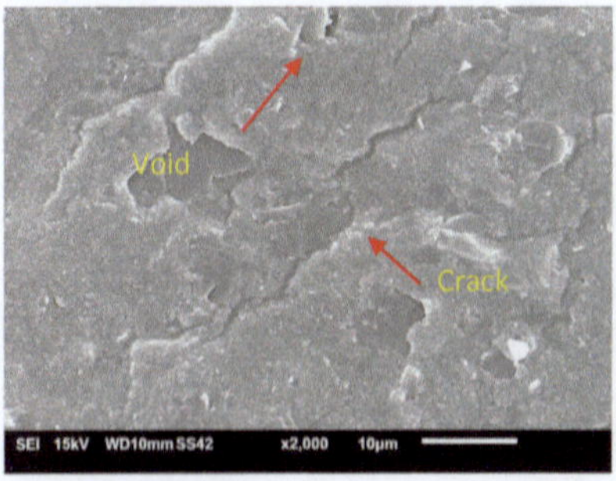	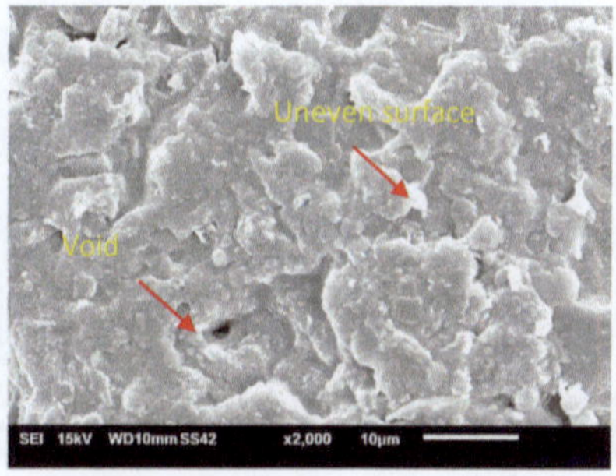
1 mm trace width	
GNPs ink trace	CB ink trace
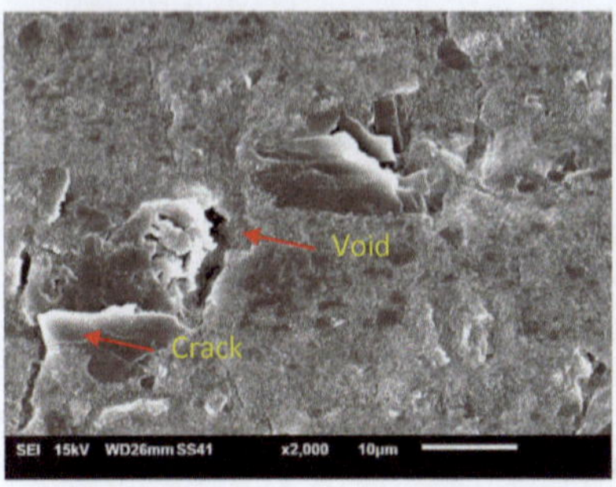	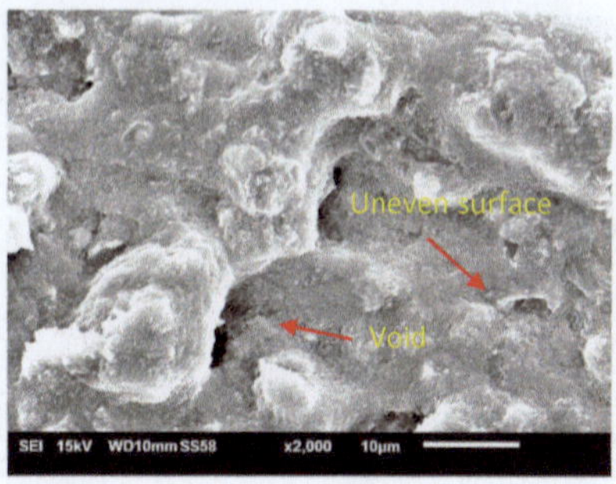
2 mm trace width	
GNPs ink trace	CB ink trace
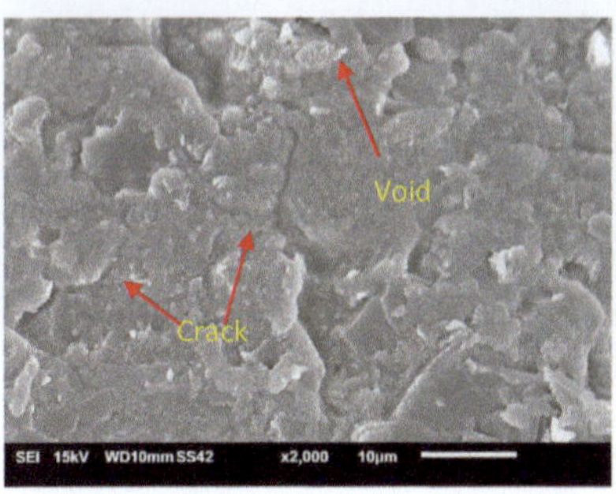	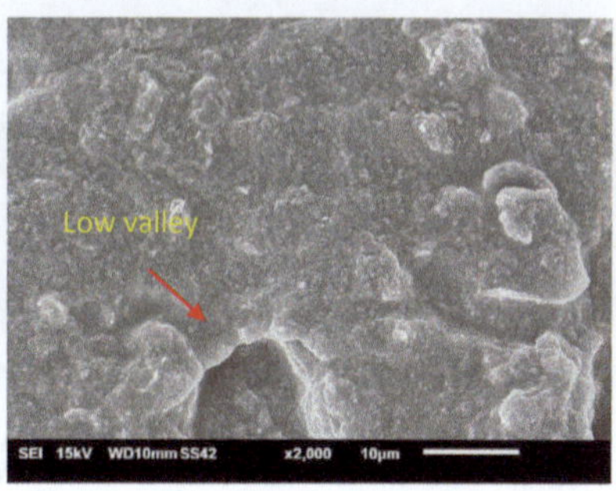
3 mm trace width	

Table 7.2 Cross sectional images of curve pattern of GNPs and CB ink trace microstructure with line width of 1, 2, and 3 mm for cyclic bending

GNPs ink trace

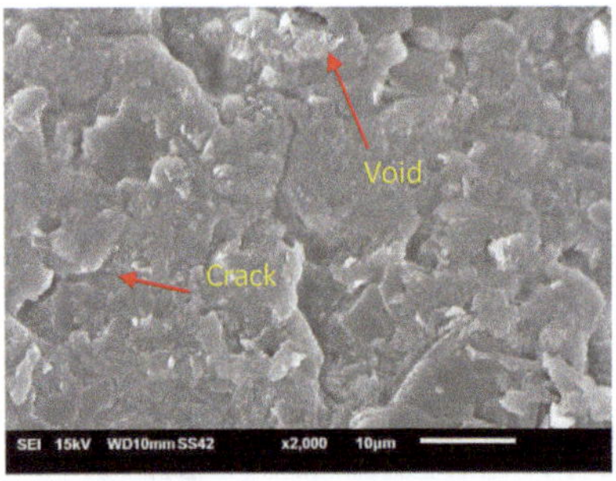

CB ink trace

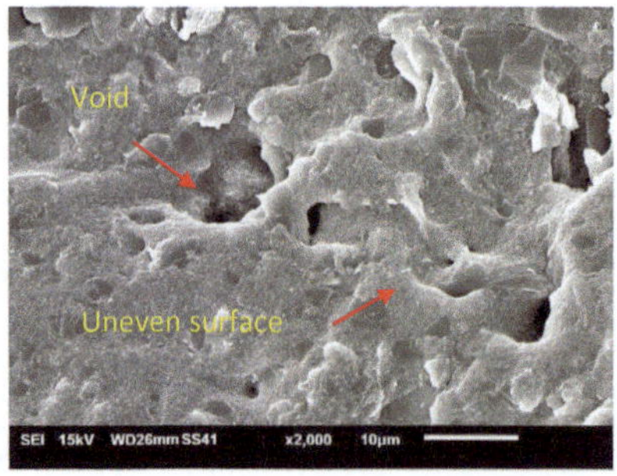

1 mm trace width

GNPs ink trace

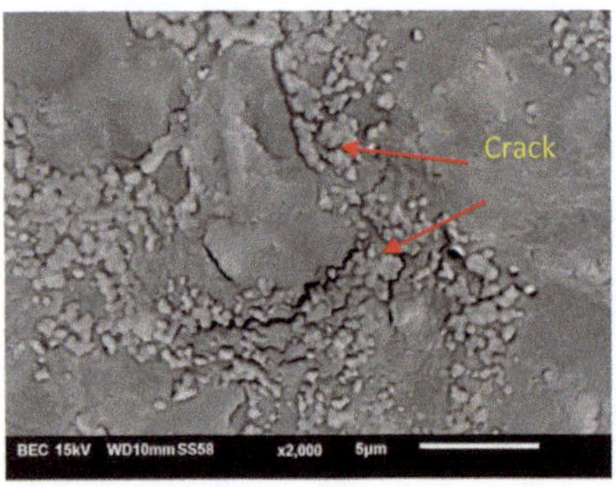

CB ink trace

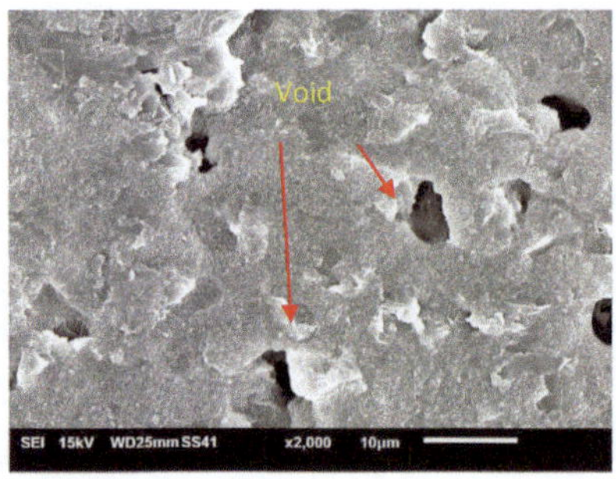

2 mm trace width

GNPs ink trace

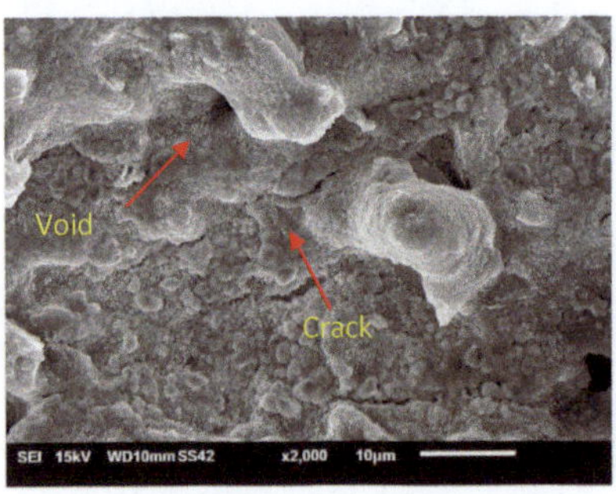

CB ink trace

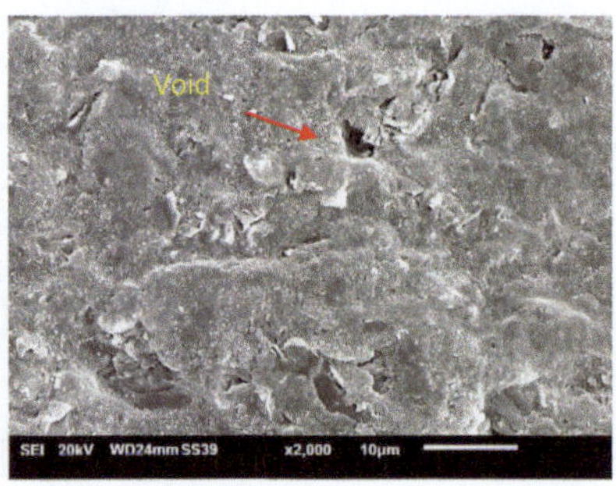

3 mm trace width

Table 7.3 Cross sectional images of square pattern of GNPs and CB ink trace microstructure with line width of 1, 2, and 3 mm for cyclic bending

GNPs ink trace

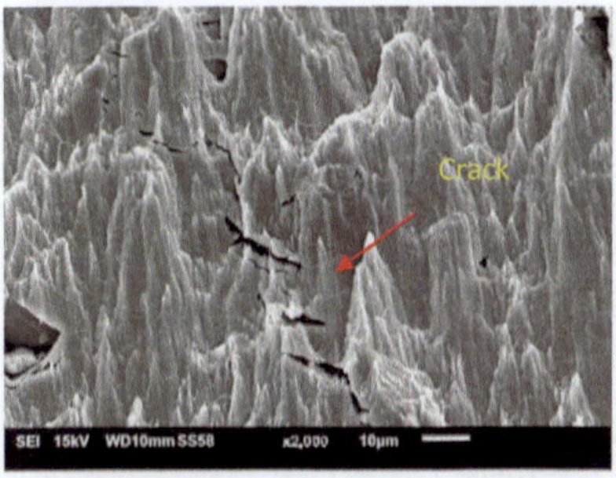

CB ink trace

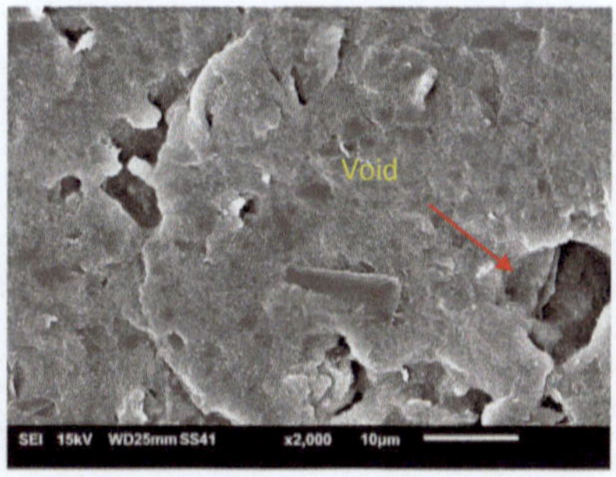

1 mm trace width

GNPs ink trace

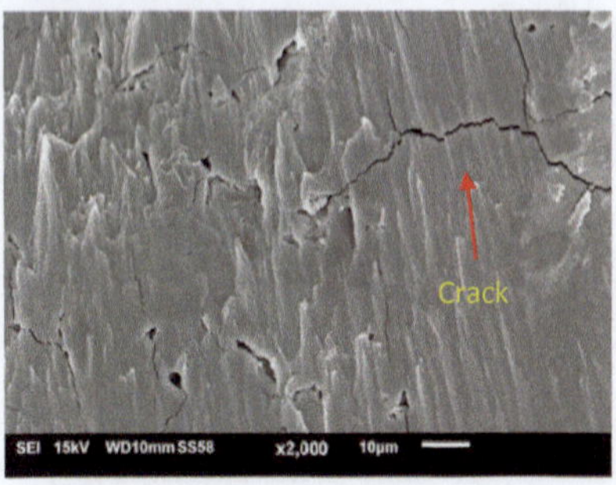

CB ink trace

2 mm trace width

GNPs ink trace

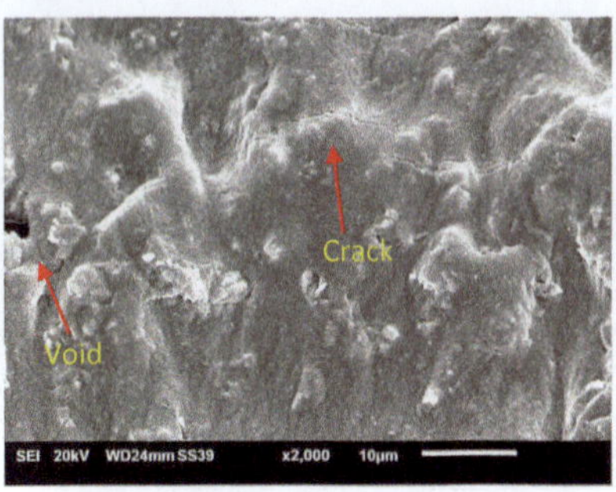

CB ink trace

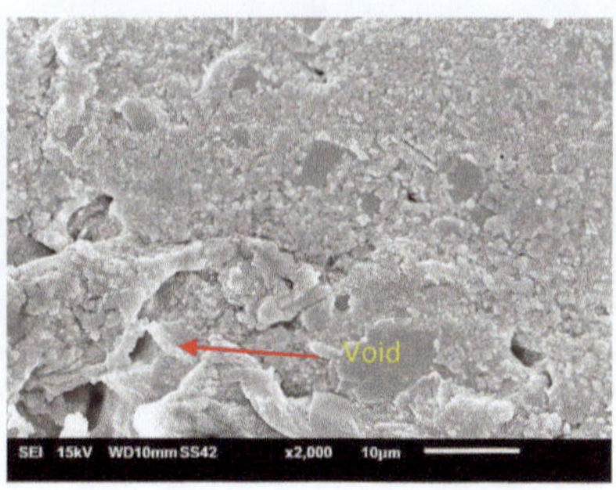

3 mm trace width

Table 7.4 Cross sectional images of zigzag pattern of GNP and CB ink trace microstructure with line width of 1, 2, and 3 mm for cyclic bending

GNPs ink trace

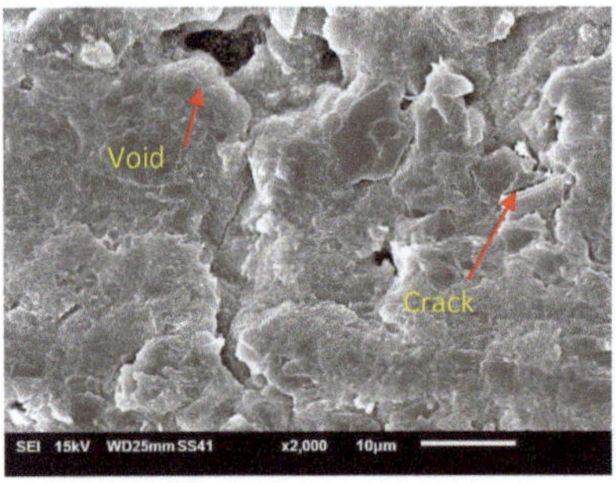

CB ink trace

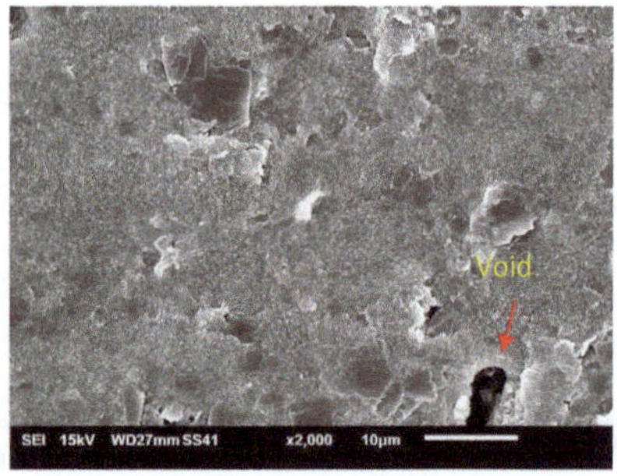

1 mm trace width

GNPs ink trace

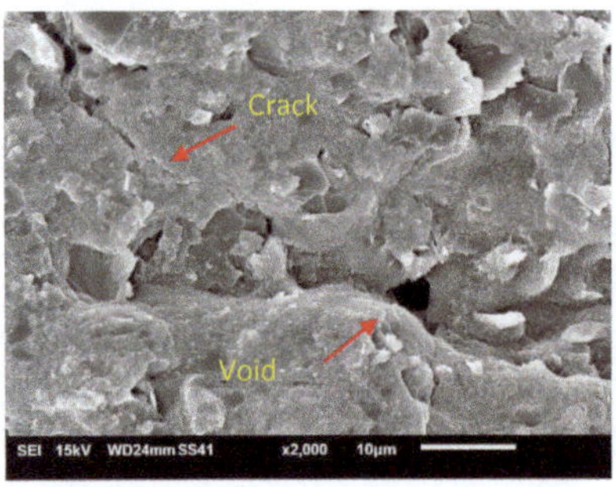

CB ink trace

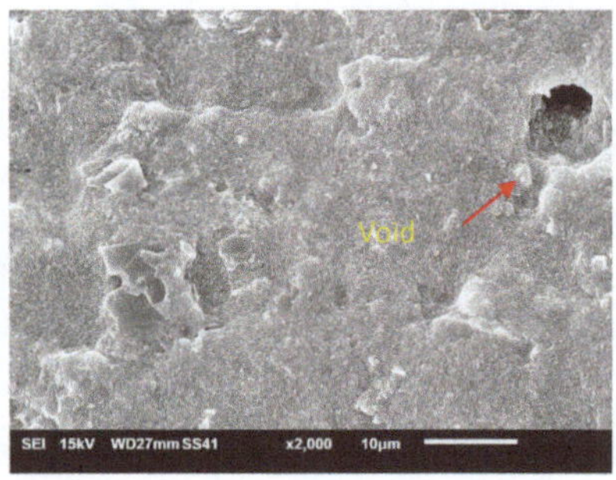

2 mm trace width

GNPs ink trace

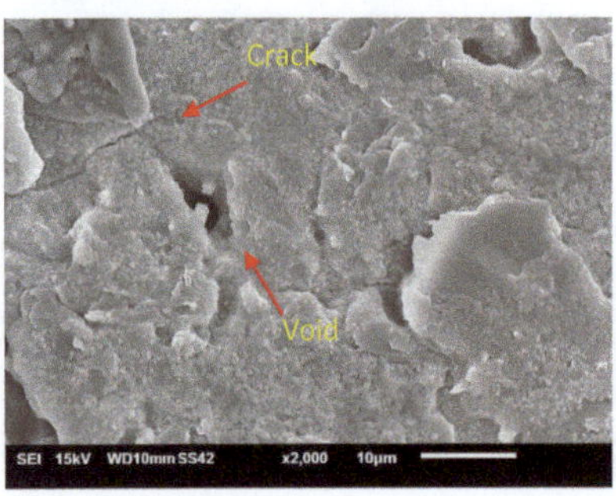

CB ink trace

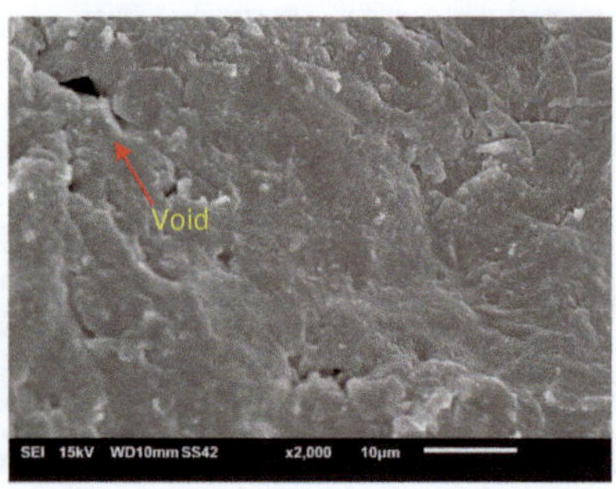

3 mm trace width

strain. After extensive cyclic loading, areas of delamination, void formation, and particle pull-out become more pronounced, especially in traces with widths below 2 mm. This behaviour is attributed to the incompatibility of the filler-matrix interface under prolonged strain cycling. The differences in elastic modulus between the GNPs (typically in the order of hundreds of GPa) and the polymer matrix (several orders of magnitude lower) generate stress mismatches at the interface. As cyclic bending progresses, these mismatches lead to interfacial debonding, disrupting the conductive network and compromising mechanical integrity. The redistribution and fracture of GNPs further aggravate localised stress fields, contributing to nonlinear degradation of the composite system.

In contrast to GNPs ink, CB ink traces display different failure morphologies under SEM. The straight-line pattern exhibits a relatively smooth cross-sectional surface, with minimal visible defects or voids. This suggests better mechanical stability under unidirectional loading. However, circular CB patterns particularly zigzag and square shapes exhibit surface roughness, asymmetry, and distinct waviness, along with scattered voids, hackles, and resin-starved regions. These voids are a direct consequence of the resin's viscous limitations during ink deposition and curing. In composite science, voids are classified as unintended third-phase inclusions that occupy space but do not contribute to mechanical or electrical performance. Their presence disrupts the continuity of the filler network, which is essential for charge carrier transport. According to percolation theory, electrical conductivity in such systems depends critically on the formation of an uninterrupted conductive path. Voids effectively interrupt these pathways, forcing current to reroute or terminate, thereby increasing the overall direct current (DC) resistance of the trace.

Moreover, voids are detrimental to both electrical and mechanical performance. As reported by Appinc (2017), voids act as stress concentrators, reducing the effective load-bearing cross-sectional area and serving as crack initiation sites under cyclic stress. The localised increase in stress around void peripheries can be described by Eshelby's inclusion theory, where internal defects in a matrix generate stress fields that can interact and coalesce, further accelerating damage accumulation. Even though SEM analysis shows fewer observable cracks in CB ink after 10,000 cycles, this does not necessarily imply superior fatigue resistance. Rather, it may reflect the viscoelastic and plastic accommodation of stress by the smaller ink particles. Due to their nanometric size and high surface area, the particles are able to pack more densely and fill micro-voids more effectively than GNPs. This results in fewer large voids and less abrupt interfacial failures. However, the smaller particle size also means the CB matrix is more susceptible to diffuse damage such as matrix softening, polymer chain scission, and micro-void expansion over time.

Another critical observation from the SEM analysis is the role of voids in promoting local oxidation. Voids, by virtue of their open volume, expose larger surface areas to atmospheric oxygen and moisture. The oxidation rate around void walls increases due to the enhanced diffusion pathways and thermal sensitivity of the ink matrix during cyclic deformation. Oxidative degradation leads to embrittlement of the matrix, further weakening mechanical performance and reducing long-term reliability. This aligns with previous findings by [7], who noted that oxidation-induced ageing is a dominant failure mechanism in polymer composites containing micro-voids. The presence of these voids also leads to heterogeneous stress distribution, resulting in non-linear load transfer across the material. As cyclic loading progresses, this inhomogeneous stress field leads to unpredictable failure behaviour, often culminating in catastrophic delamination or sudden crack growth, a concern in mission-critical flexible electronics applications.

7.8 The EDX Analysis

An Energy Dispersive X-ray Spectroscopy (EDX) is an indispensable tool for elemental analysis in materials characterisation. As an adjunct to SEM, EDX enables both qualitative and semi-quantitative identification of elements present in a sample, along with their spatial distribution. This capability is particularly critical for understanding chemical stability, material homogeneity, and compositional changes in conductive ink traces subjected to mechanical fatigue such as cyclic bending. The EDX technique exploits the characteristic X-rays emitted by atoms when bombarded by a focused electron beam, offering insight into the elemental make-up with micron-level resolution. In the context of flexible electronics, where polymeric matrices are infused with conductive fillers, EDX serves to track how repeated deformation alters the internal chemical structure of the ink, especially regarding the distribution of conductive fillers and non-conductive components. This analysis becomes particularly important in assessing material degradation and failure mechanisms at the microscale.

According to the data summarised in Table 7.5, a clear inverse relationship is observed between the carbon and oxygen weight percentages and the trace width in GNPs ink samples. As the width of the conductive trace increases, the carbon content consistently decreases, while oxygen content rises. This trend is uniform across all geometric configurations investigated, including straight-line, curved, square, and zigzag patterns. This shift in elemental composition suggests a dilution of GNPs content relative to the polymeric binder as trace width increases. In practical terms, wider traces likely involve more binder accumulation due to the surface-tension-dominated spreading during the printing process. Since polymer binders are typically oxygen-rich and carbon-deficient relative to GNPs, their increasing proportion naturally reduces the overall carbon content and raises the oxygen fraction.

From a physics view, this shift correlates with reduced electrical percolation. Conductivity in GNPs-based inks largely arises from the formation of a percolated network of overlapping nanoplatelets. A decrease in filler concentration below the percolation threshold disrupts this network, leading to significant increases in electrical resistance. Moreover, the reduction in carbon-rich domains reduces the material's ability to carry mechanical loads, as graphene's intrinsic Young's modulus (1

Table 7.5 Carbon and oxygen weight percentage inside GNPs and CB ink trace for each pattern measured from EDX for cyclic bending

	GNPs ink trace						CB ink trace					
Trace Pattern	Carbon weight percentage, %			Oxygen weight percentage, %			Carbon weight percentage, %			Oxygen weight percentage, %		
Width, mm	1	2	3	1	2	3	1	2	3	1	2	3
Straight line	53.2	45.6	44.8	46.8	54.4	55.2	44.9	44.6	43.2	55.1	55.4	56.8
Curve	65.4	61.0	60.9	34.6	39.0	39.1	69.2	60.9	58.8	30.8	39.1	41.2
Square	53.8	54.2	51.1	46.2	45.8	48.9	55.8	54.2	53.1	44.2	45.8	46.9
Zigzag	58.6	58.1	57.8	41.4	41.9	42.2	62.9	58.1	54.3	37.1	41.9	45.7

TPa) and tensile strength (130 GPa) are far superior to those of the polymer matrix. Therefore, the observed compositional shift indicates both electrical and mechanical performance degradation as trace width increases. This observation supports previous findings [8], where the deterioration in ink functionality was attributed to non-uniform filler dispersion and binder dominance at larger scales. It is also indicative of nanoparticle redistribution during mechanical loading, as GNPs may migrate or cluster due to cyclic strain energy input.

In CB inks, the EDX results show a less uniform trend compared to GNPs. For straight-line and sharp-corner meandering patterns, oxygen content is higher than carbon content, whereas for the curved meandering pattern, the reverse is true. This inversion hints at different filler-to-binder ratios influenced not just by trace width, but also by pattern-induced mechanical response and ink flow dynamics during deposition. The higher oxygen levels in sharp-edged traces may arise from two coupled mechanisms: (i) mechanical degradation during bending, which facilitates exposure of internal surfaces to air, increasing the likelihood of oxidative reactions, and (ii) ink recession and binder pooling at corners due to capillary and centrifugal forces during printing and curing, which leave behind oxygen-rich regions.

Regions subjected to high stress concentration, such as acute corners, are face down to crack nucleation and binder fracture, resulting in the formation of voids. These voids expose fresh surfaces to the environment, accelerating oxidation and thus increasing oxygen signal during EDX. As reported by [9], oxygen-rich degradation is a typical outcome of polymer matrix embrittlement, where atmospheric oxygen diffuses preferentially along crack paths and reacts with functional groups within the resin matrix. The relative abundance of carbon in the curved rendering CB patterns suggests a more homogeneous distribution of the particles and less binder pooling. The smoother geometry promotes uniform mechanical deformation and reduced strain localisation, thereby limiting crack formation and oxidation. This aligns with mechanical stress distribution theories, which show that curvilinear paths reduce peak stress values by allowing gradual load redistribution.

A more granular view is provided by EDX elemental mapping in Tables 7.6 and 7.7, which display the distribution of carbon and oxygen across the cross-sectional surface of GNPs and CB ink traces (specifically, the curved pattern with 3 mm width). In the GNPs trace, carbon and oxygen account for 60.9% and 39.1% of the local composition, respectively. In the CB trace, the values are 58.8% carbon and 41.2% oxygen. Although these compositions are broadly similar, the underlying mechanisms differ significantly. In GNPs-based inks, cracks and voids appear more prevalent due to the planar morphology of nanoplatelets and their tendency to stack with weaker van der Waals bonding. These interplatelet gaps serve as diffusion channels for oxygen ingress, particularly under cyclic strain. Repeated deformation exacerbates the formation of micro-voids, which act as sinks for oxygen accumulation. According to Fick's second law of diffusion, the local concentration of diffusing species increases with both exposure time and defect density, explaining the elevated oxygen levels at damaged sites.

In contrast, carbon black fillers possess a spherical morphology, leading to isotropic packing with better inter-particle contact. While the traces may contain inherent voids due to their smaller size and irregular agglomeration, these voids are less likely to form extended crack paths, thus reducing the exposure of inner domains to ambient oxygen. However, oxygen can still accumulate in loosely packed or resin-rich areas, particularly where the ink failed to fully consolidate.

The presence of oxide layers in both systems introduces additional complications. Oxidation of the polymer matrix or surface carbon atoms forms insulating oxides,

Table 7.6 The EDX mapping images for patterned GNPs ink (curve and 3 mm) for cyclic bending

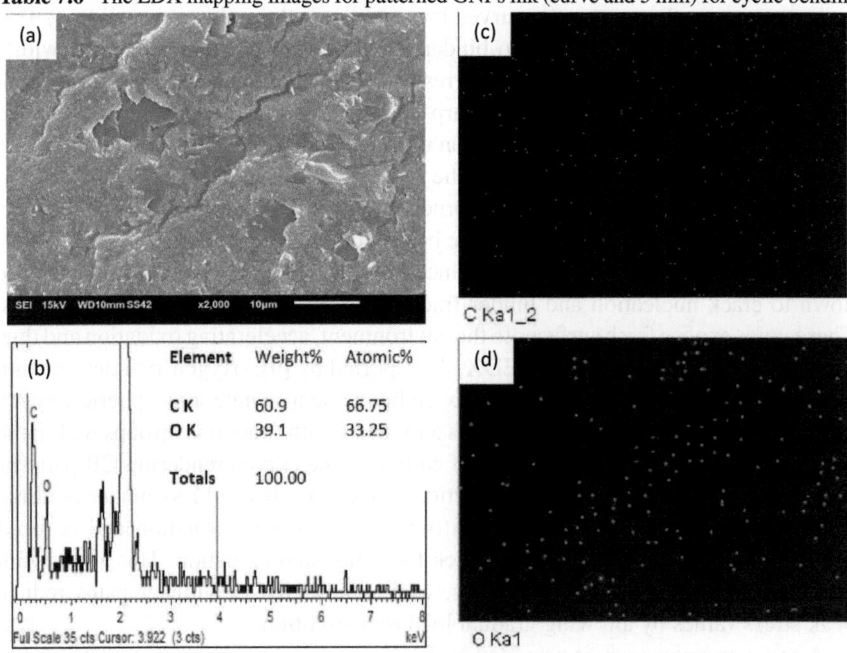

(a) SEM image, (b) EDX element graph, (c) EDX mapping image of Carbon (GNPs), and (d) EDX mapping image of Oxygen

Table 7.7 The EDX mapping images for patterned CB ink (curve and 3 mm) for cyclic bending

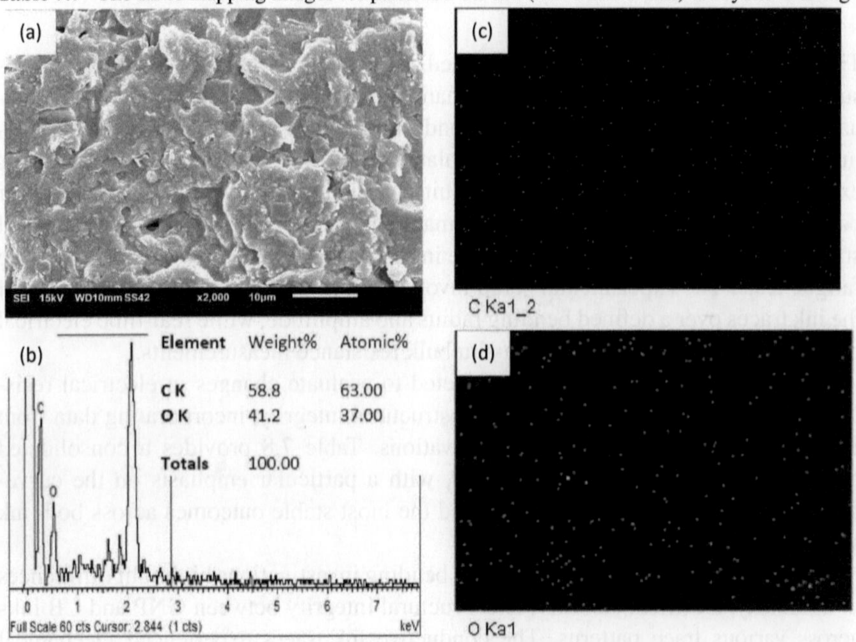

(a) SEM image, (b) EDX element graph, (c) EDX mapping image of Carbon (CB), and (d) EDX mapping image of Oxygen

such as carbonyl or carboxyl functional groups, which degrade conductivity. These oxides may also inhibit carbon diffusion and filler mobility, effectively sealing off conductive pathways and increasing contact resistance. The resultant degradation is not only electrical but also mechanical, as oxide layers can embrittle the interface and promote crack propagation under load.

These findings have profound implications for the design of reliable conductive inks for flexible electronic applications. The pattern geometry, trace width, and filler morphology all influence the distribution of elemental constituents, which in turn governs performance characteristics such as: (i) electrical conductivity (percolation and network integrity), (ii) mechanical resilience (stress distribution and crack resistance), and (iii) environmental stability (susceptibility to oxidation and moisture ingress). Material selection and printing parameters must therefore be carefully optimised to mitigate binder-dominated zones and avoid localised oxygen accumulation. Strategies such as post-curing in inert atmospheres, addition of antioxidant agents, or surface treatments to reduce oxygen permeability may be necessary to enhance long-term stability and maintain functionality under cyclic deformation.

7.9 Summary

This chapter comprehensively addressed the behaviour of conductive ink traces subjected to both static and cyclic mechanical loading, followed by a detailed evaluation of their electrical, mechanical, and morphological performance. The experimental sequence was designed to simulate the real-mechanical fatigue conditions experienced by flexible electronic circuits, particularly under repeated flexing or bending in wearable, foldable, or deformable devices. Following the completion of static loading tests, the same conductive ink traces were subjected to cyclic bending fatigue tests. The experimental setup involved repeated mechanical deformation of the ink traces over a defined bending radius and amplitude, while real-time electrical functionality was monitored via in-situ bulk resistance measurements.

A multi-characterisation was conducted to evaluate changes in electrical resistance, mechanical resilience, and microstructural integrity, incorporating data from SEM, EDX, and morphological observations. Table 7.8 provides a consolidated overview of the resulting performance, with a particular emphasis on the curve-pattern trace, which consistently yielded the most stable outcomes across both ink systems.

The primary findings from the cyclic bending investigations highlight differences in durability, electrical stability, and structural integrity between GNP and CB inks across various trace patterns. The conductive ink traces experienced mechanical failure during the cyclic bending test at varying cycle counts, depending on the pattern geometry. Specifically, failure manifested after approximately 7000 cycles for the straight-line, square, and zigzag patterns. In contrast, the curve-patterned trace endured up to 9000 cycles before structural breakdown was observed. These results suggest that abrupt geometric transitions and angular corners inherent in non-curvilinear patterns generate localised stress concentrations, facilitating early crack nucleation and propagation. In contrast, the curve pattern's continuous and smooth geometry promoted uniform strain distribution, thereby reducing mechanical localisation and delaying failure.

Table 7.8 Summary of electrical and mechanical characterisation of cyclic bending load test for patterned GNPs and carbon black (curve pattern)

		Width, mm		
		1	2	3
GNPs	Volume resistivity, Ω cm	6.230	12.898	13.528
	Hardness, MPa	0.990	0.220	0.110
	Elastic modulus, GPa	0.001	0.001	0.001
Carbon Black	Volume resistivity, Ω cm	0.008	0.009	0.009
	Hardness, MPa	0.670	1.320	1.280
	Elastic modulus, GPa	0.009	0.015	0.008

On the other side, the endurance differences can be attributed to strain energy accumulation and stress intensity factors at geometrical discontinuities. Sharp edges inherently create conditions where the applied stress exceeds the local material strength, promoting micro-crack initiation. In GNPs ink, this issue is exacerbated by the layered structure of the nanoplatelets, which are down to delamination under shear stress, especially when the load cycles are high and the filler alignment becomes compromised.

In contrast, all CB conductive ink traces successfully withstood the full 10,000 cycles test without exhibiting catastrophic failure. The superior performance of the traces can be attributed to several factors: (i) spherical morphology of the particles enables isotropic packing, thereby promoting a more uniform stress response under bending, (ii) smaller particle sizes allow for better dispersion within the polymer matrix, reducing void formation and interfacial debonding, (iii) the flexible and ductile behaviour of the composite enhances its ability to absorb and dissipate mechanical energy, minimising crack propagation pathways. Furthermore, the traces maintained stable electrical performance throughout the cyclic bending process, indicative of a robust percolative conductive network. The reduced inter-particle spacing due to effective packing ensures that electrical connectivity remains largely unaffected, even under substantial mechanical deformation.

Among all geometrical configurations tested, the curve pattern emerged as the most reliable design for both GNPs and CB conductive inks. Its inherent geometry mitigates abrupt directional changes and effectively distributes mechanical stress across a larger surface area during cyclic deformation. This finding reinforces a well-established principle in flexible electronics design: curvilinear paths reduce peak stress and strain gradients, thus enhancing the durability of printed traces under dynamic loading. In addition to mechanical resilience, the curve pattern also supported improved electrical stability, as measured by reduced fluctuations in bulk resistance during testing. This outcome highlights the strong interplay between geometrical optimisation and functional reliability in stretchable or bendable circuit designs.

The comparative analysis between the two conductive ink systems revealed that CB inks consistently outperformed GNPs-based inks in both electrical and mechanical domains under cyclic loading conditions. The mechanical ductility, homogeneous dispersion, and resistance to microstructural damage of carbon black ink contributed to its long-term stability. Conversely, the GNPs ink, while initially offering higher conductivity due to the high aspect ratio and intrinsic conductivity of GNPs, suffered from interfacial debonding, agglomeration, and orientation disruption under repeated mechanical stress. These phenomena led to a progressive increase in electrical resistance and premature mechanical failure.

From a materials science point, this difference highlights the importance of filler-matrix interfacial compatibility and filler morphology. While GNPs possess superior individual electrical and mechanical properties, their planar geometry makes them more susceptible to shear-induced delamination. In contrast, carbon black s granular structure enables better mechanical interlocking and stress transfer, making it a more resilient choice for high-cycle fatigue environments.

This chapter has demonstrated that pattern geometry, filler morphology, and matrix-filler interactions are critical determinants of performance in flexible printed conductive inks. It is evident that mechanical durability under cyclic bending cannot be solely predicted by the electrical properties of fillers, instead, it requires a nuanced understanding of the multi-scale interactions between geometry, microstructure, and material behaviour under dynamic loading. As the field of flexible and wearable electronics continues to advance, such insights will be instrumental in guiding the design of next-generation conductive inks, particularly for applications requiring high mechanical endurance, environmental stability, and long-term functional reliability.

References

1. Znajdek, K., Sibiński, M., Lisik, Z., Apostoluk, A., Zhu, Y., Masenelli, B., Sędzicki, P.: Zinc oxide nanoparticles for improvement of thin film photovoltaic structures' efficiency through down shifting conversion. Opto-Electron. Rev. **25**(2), 99–102 (2017)
2. Graff, K.F.: Ultrasonic metal forming. In: Power Ultrasonics, pp. 377–438 (2015)
3. Bell, T.: Ductility Explained: Tensile Stress and Metals. https://www.thoughtco.com/ductility-metallurgy-4019295 (2018). Retrieved 29 October 2020
4. Lahuta, L.B., Szablińska-Piernik, J., Głowacka, K., Stałanowska, K., Railean-Plugaru, V., Horbowicz, M., Pomastowski, P., Buszewski, B.: The effect of bio-synthesized silver nanoparticles on germination, early seedling development, and metabolome of wheat (Triticum aestivum L.). Molecules **27**(7), 2303 (2022)
5. Claypole, A., Claypole, J., Kilduff, L., Gethin, D., Claypole, T.: Stretchable carbon and silver inks for wearable applications. Nanomaterials **11**(5), 1200 (2021)
6. Padmaraj, N.H., Vijaya, K.M., Dayananda, P.: Experimental investigation on fatigue behaviour of glass/epoxy quasi-isotropic laminate composites under different ageing conditions. Int. J. Fatigue **143**, 105992 (2021)
7. Elbishari, H., Silikas, N., Satterthwaite, J.: The effect of filler size on the presence of voids within resin composite. J. Int. Dent. Med. Res. **11**(2) (2018)
8. Phillips, C., Al-Ahmadi, A., Potts, S.J., Claypole, T., Deganello, D.: The effect of graphite and carbon black ratios on conductive ink performance. J. Mater. Sci. **52**, 9520 (2017)
9. Zhang, H., Li, Y., Zhao, Y., Li, G., Zhang, F.: Carbon black oxidized by air calcination for enhanced H_2O_2 generation and effective organics degradation. ACS Appl. Mater. Interfaces. **11**(31), 27846–27853 (2019)
10. Spahr, M.E., Rothon, R.: Carbon black as a polymer filler. In: Polymers and Polymeric Composites: A Reference Series, pp.1–31 (2016)

Chapter 8
Influence of Pattern Geometry on Fatigue Life of Conductive Ink Under Cyclic Tensile

8.1 Introduction

According to the book Fatigue and Fracture (2019) by ASM International, cyclic tensile loading refers to the repeated application of tensile stress followed by relaxation over time. This loading system induces cyclic deformation within a material, which, if sustained over an extended number of cycles, may result in fatigue failure. This form of failure originates from the nucleation of microscopic cracks within the material matrix, typically at regions of stress concentration, such as material defects, inclusions, grain boundaries, or interfaces between different phases. With continued cyclic loading, these micro-cracks propagate incrementally, ultimately absorption into larger cracks that compromise the structural integrity and lead to catastrophic failure.

A practical example of this phenomenon is demonstrated in the work by Zhang et al. [1–4], which explored the fatigue behaviour of graphene nanoplatelet (GNPs) conductive ink deposited on a flexible substrate under cyclic tensile loading. The investigation revealed a progressive degradation in both electrical conductivity and mechanical integrity of the conductive film. The primary mechanisms attributed to this deterioration include the initiation of micro-cracks within the ink matrix and debonding at the ink-substrate interface. These microstructural changes disrupt the continuity of the conductive network formed by the GNPs, thereby increasing electrical resistance and reducing overall material performance.

From a materials science view, such behaviour can be attributed to the cumulative plastic deformation and strain localisation that occur during repeated loading. Each loading cycle subjects the material to stress reversals, during which dislocations within the crystal structure become mobile and multiply. Over time, these dislocations interact, entangle, and form persistent slip bands, regions of localised strain, which serve as precursors to micro-crack nucleation. In composite or multilayer systems, such as conductive ink on a polymer substrate, differential thermal expansion coefficients and modulus mismatch between the ink and the substrate

M. A. Salim et al., *Nanotechnology in Conductive Inks*, Engineering Materials,
https://doi.org/10.1007/978-981-95-5708-0_8

further exacerbate interfacial stresses, promoting delamination or interfacial crack initiation. By mitigating the risk of fatigue failure in such systems, it is essential to adopt a holistic engineering approach encompassing informed material selection, design optimisation, and various testing protocols. Material selection should prioritise high fatigue resistance, strong interfacial adhesion, and mechanical flexibility, especially in applications involving flexible electronics, wearable devices, or stretchable sensors. In addition, design features that minimise stress concentrations, such as smooth transitions, rounded corners, and appropriate layer thicknesses, and it can help delay the onset of fatigue damage.

Furthermore, accurate characterisation of fatigue performance under realistic loading conditions is critical. The Standard Test Method for Strain-Controlled Fatigue Testing (ASTM E466-20), issued by ASTM International, provides a comprehensive framework for evaluating the fatigue behaviour of materials under cyclic axial loading. This standard outlines procedure for subjecting specimens to cyclic tensile and compressive loads, recording the number of cycles to failure, and interpreting data to determine fatigue strength, endurance limit, and S–N (stress–life) curves. By conducting such tests, engineers can gain insights into the durability and service life of materials under specific operating conditions.

In the context of advanced functional materials like GNPs-based inks, fatigue testing also offers valuable information about the evolution of electrical performance under mechanical stress. This is particularly relevant for applications in flexible and stretchable electronics, where the material is expected to maintain functionality despite undergoing frequent mechanical deformation. Monitoring changes in electrical resistance in situ during cyclic loading provides a direct correlation between mechanical fatigue and electrical degradation, enabling the development of predictive models for failure and informing improvements in formulation or processing. Therefore, understanding fatigue behaviour under cyclic tensile loading is essential for ensuring the long-term reliability of materials used in flexible electronic systems. Through a combination of theoretical understanding, empirical testing, and engineering design, it is possible to develop robust materials and structures that can withstand the demanding conditions of cyclic mechanical stress without premature failure.

8.2 Bulk Resistance Characterisation

The characterisation of bulk electrical resistance in conductive inks, particularly those formulated with GNPs and CB under continuous cyclic tensile loading provides valuable insight into their functional durability and structural integrity. This investigation is essential in evaluating the performance of such inks in flexible and stretchable electronic systems, where materials are routinely subjected to mechanical deformation during operation. By facilitating this analysis, a tensile testing apparatus was employed, enabling the simultaneous application of cyclic tensile loading and real-time monitoring of electrical resistance. Measuring the bulk resistance in situ, rather than post-mortem, is critical for capturing dynamic changes in conductivity as a function of mechanical strain and fatigue-induced damage. This approach offers a more realistic and sensitive assessment of functional degradation, as it mimics the real-operational conditions of flexible electronics.

The results, summarised in Fig. 8.1, reveal that the bulk resistance of both GNPs and CB ink traces generally increases with the number of loading cycles. This trend reflects the progressive deterioration of the conductive network within the ink matrix due to the accumulation of mechanical damage. Repeated tensile stress promotes the initiation and growth of micro-cracks, both within the ink layer and at the ink–substrate interface. These discontinuities disrupt the percolation pathways formed by conductive fillers, thereby increasing electrical resistance. The extent of this degradation is influenced not only by the material formulation but also by geometrical and mechanical parameters, including pattern shape, trace width, and the applied tensile condition.

From a physics theory, the increase in bulk resistance under cyclic loading is consistent with fracture mechanics and percolation theory. As micro-cracks initiate, the conductive network becomes fragmented. The effective conductive cross-section reduces, leading to increased electrical path length and higher resistance. Moreover, with each cycle, stress concentration zones, particularly at sharp geometrical features or material defects, and an intensify localised strain, accelerating damage propagation.

A detailed comparison of different ink trace geometries demonstrated significant variation in bulk resistance behaviour. Among the GNPs patterns, straight-line and curved traces exhibited the lowest initial resistance values. This suggests that these patterns facilitated more efficient current flow due to their relatively undisturbed conductive pathways. In contrast, square and zigzag patterns exhibited higher initial resistance, likely due to increased path tortuosity and sharper directional changes that introduce localised stress concentration points even before mechanical loading. Interestingly, as cyclic loading progressed beyond 3000 to 4000 cycles, the zigzag patterns, particularly those with 1 and 2 mm widths displayed the highest resistance values among all geometries.

This indicates that stress-induced damage accumulates more severely in these patterns, especially at the corners where abrupt directional changes exacerbate stress localisation. The square pattern with 3 mm width similarly exhibited a marked increase in resistance after prolonged cycling.

Trace width was also found to play a significant role. Across all pattern types, samples with narrower widths (1 mm) consistently demonstrated higher resistance, both initially and after cyclic loading, followed by 2 and 3 mm widths respectively. This trend can be attributed to the limited conductive cross-sectional area in narrower traces, which makes them more susceptible to damage accumulation and loss of conductivity. Wider traces are inherently more robust due to greater redundancy in conductive paths, thus better tolerating crack development without substantial loss in electrical performance.

The CB ink traces exhibited a somewhat different resistance profile. Initially, a modest increase in resistance was observed up to approximately 2000 cycles. This

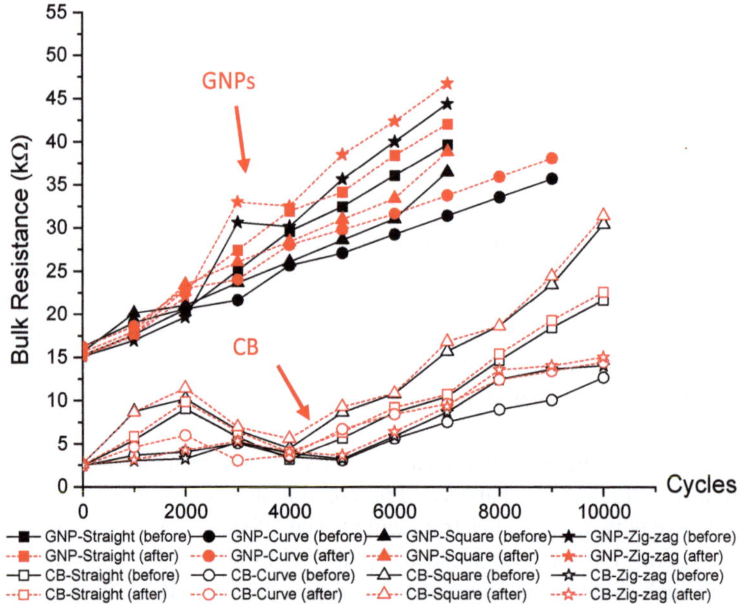

(a)

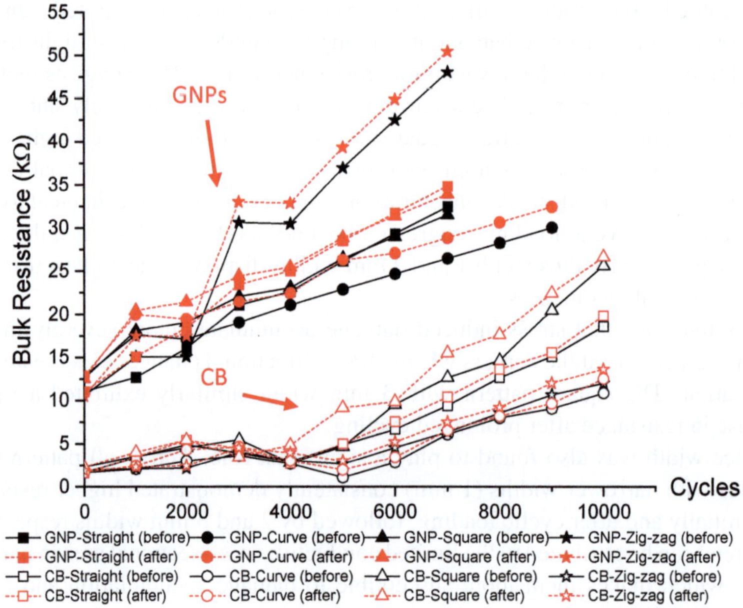

(b)

Fig. 8.1 Graph of bulk resistance value for GNPs and CB ink trace with **a** 1 mm width, **b** 2 mm width, and **c** 3 mm width for 10,000 times of cyclic tensile

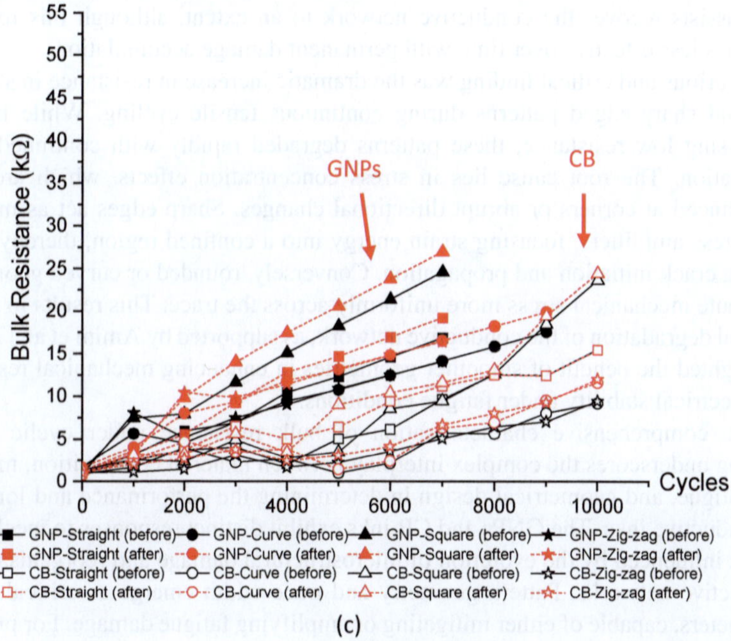

Fig. 8.1 (continued)

early-stage degradation aligns with typical fatigue-induced damage mechanisms. However, an extraordinary deviation occurred thereafter, which is the resistance began to decrease progressively up to 4000 to 5000 cycles. This counterintuitive behaviour is attributable to the mechanical breakdown of CB agglomerates, which, upon disintegration, reorient and reconnect with previously unlinked conductive clusters. As noted by [5], this phenomenon enhances the percolation network by creating new electrical pathways. The dispersion of particles previously hindered by agglomeration facilitates the formation of fresh, more effective conductive bridges. This reorganisation temporarily improves electrical conductivity, manifesting as a reduction in bulk resistance. However, beyond 5000 cycles, the cumulative mechanical fatigue overrides this beneficial effect. Resistance begins to rise again as the formation of new cracks and eventual structural degradation dominate the ink's response.

Another critical observation was the comparison between the resistance values measured in the fully extended (tensile) state and the flat (relaxed) state of the samples. Predictably, the resistance was higher in the tensile state due to the mechanical separation of conductive particles and the opening of fatigue-induced micro-cracks. In the stretched condition, these cracks remain open, resulting in disrupted pathways and higher resistance values. Upon relaxation, many of these micro-cracks close partially or completely, re-establishing some degree of conductivity. This reversible behaviour aligns with the concept of elastic deformation in soft substrates and reversible contact resistance in composite systems. As [6] observed, the reclosure of cracks in the flat state assists recover the conductive network to an extent, although this recovery becomes less effective over time with permanent damage accumulation.

A curious and critical finding was the dramatic increase in resistance in straight-line and sharp-edged patterns during continuous tensile cycling. While initially possessing low resistance, these patterns degraded rapidly with continued stress application. The root cause lies in stress concentration effects, which are more pronounced at corners or abrupt directional changes. Sharp edges act as mechanical stress amplifiers, focusing strain energy into a confined region, thereby accelerating crack initiation and propagation. Conversely, rounded or curved geometries distribute mechanical stress more uniformly across the trace. This results in a more gradual degradation of the conductive network, as supported by Amini et al. [7], who highlighted the benefit of smoother geometries in enhancing mechanical resilience and electrical stability under fatigue conditions.

The comprehensive characterisation of bulk resistance under cyclic tensile loading underscores the complex interplay between material composition, mechanical fatigue, and geometrical design in determining the performance and longevity of conductive inks. The GNPs and CB inks exhibit distinct responses to mechanical stress, influenced by the evolution of microstructural damage and reorganisation of conductive networks. Pattern geometry and trace width emerge as critical design parameters, capable of either mitigating or amplifying fatigue damage. For practical applications in stretchable and flexible electronics, these insights provide a foundation for optimising ink formulations and design strategies to achieve improved electrical reliability and mechanical robustness under cyclic loading environments.

8.3 Sheet Resistance Behaviour

The evaluation presented in Fig. 8.2 focuses on analysing the evolution of sheet resistance in the both conductive inks subjected to continuous cyclic tensile loading. This investigation is critical for assessing the electrical durability and structural reliability of printed conductive ink traces, which are fundamental components in the development of flexible and stretchable electronic systems. Such systems are increasingly applied in wearable electronics, biomedical sensors, and soft robotics, where mechanical deformation is inevitable during usage. By conducting this investigation, a tensile testing apparatus was utilised to apply controlled cyclic mechanical loading while simultaneously measuring the changes in electrical resistance of the ink samples. Specifically, the sheet resistance which a parameter that defines the resistance per unit square of a thin film was used as an indicator of electrical conductivity in the plane of the film. It is especially relevant for thin, planar conductive structures, as it captures the lateral flow of electrical current across the surface, which is the dominant mode of conduction in printed electronics.

Sheet resistance reflects how efficiently charge carriers can move across a conductive network. In composite films made from conductive fillers embedded in a polymer matrix, such as GNPs or CB inks, electrical conductivity is primarily governed by

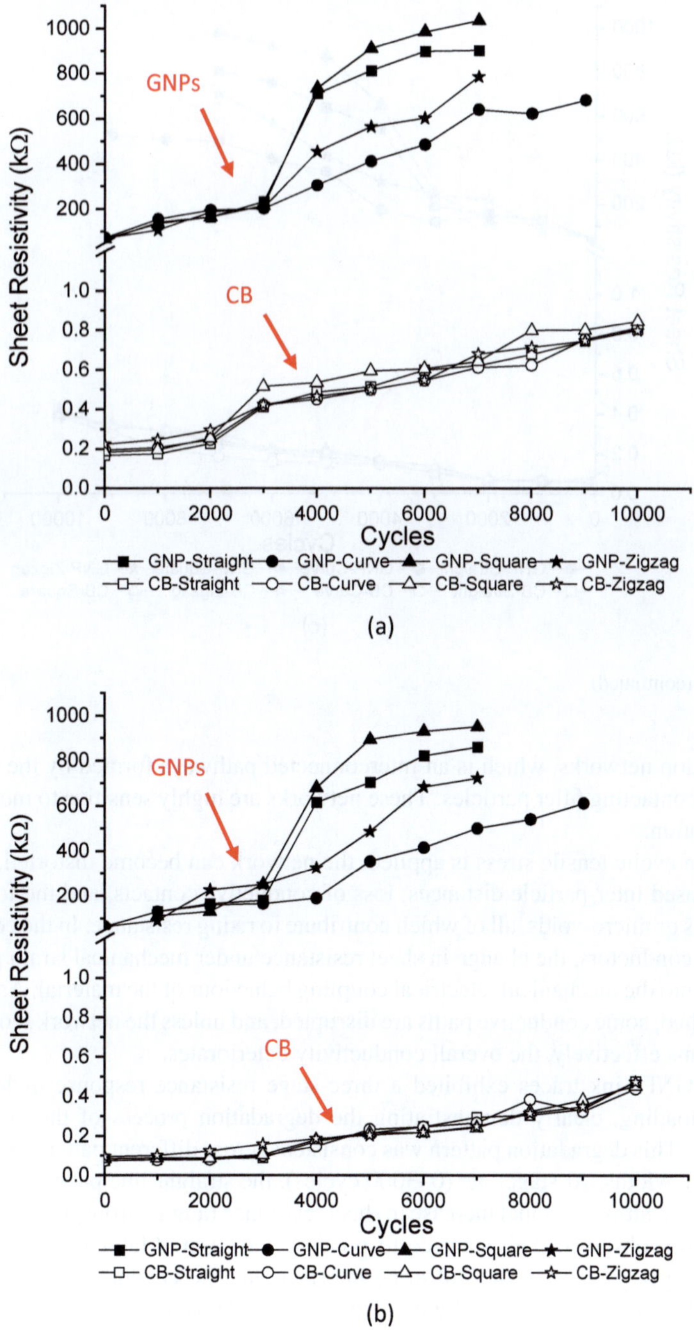

Fig. 8.2 Graph of sheet resistance value for GNPs and CB ink trace with **a** 1 mm width, **b** 2 mm width, and **c** 3 mm width for 10,000 times of cyclic tensile

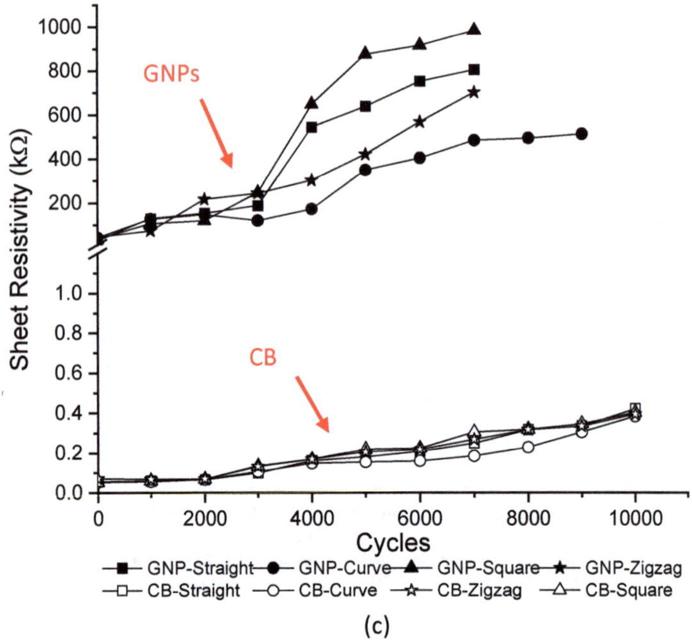

Fig. 8.2 (continued)

percolation networks, which is an interconnected pathways formed by the overlapping or contacting filler particles. These networks are highly sensitive to mechanical deformation.

When cyclic tensile stress is applied, the network can become distorted, leading to increased inter-particle distances, loss of conductive contacts, and the formation of cracks or micro-voids, all of which contribute to rising resistance. In the context of flexible conductors, the change in sheet resistance under mechanical strain provides insight into the mechanical–electrical coupling behaviour of the material. As the film is stretched, some conductive paths are disrupted, and unless the network reorganises or reforms effectively, the overall conductivity deteriorates.

The GNPs ink traces exhibited a three-stage resistance response under cyclic tensile loading, clearly demonstrating the degradation process of the conductive network. This degradation pattern was consistent across different pattern geometries and trace widths. At stage one (0–3000 cycles), the straight-line patterns with 1, 2, and 3 mm widths, a gradual increase in sheet resistance ranging from approximately 8 to 32% was observed. This stage corresponds to the onset of fatigue damage, where debonding begins at the ink–substrate interface and the conductive filler particles within the ink matrix start to slightly displace. The damage is still relatively minor, but it disrupts the percolation network enough to cause measurable electrical degradation. For stage two (3000 to 4000 cycles), a sharp increase in sheet resistance up to approximately 101% was recorded, signalling the propagation of cracks within the ink trace.

This sudden escalation is indicative of a transition from microstructural weakening to macroscopic damage, where the conductive pathways are severely compromised. Crack formation becomes more dominant, resulting in fragmentation of the network and partial loss of function. Finally, for stage three (4000 to 7000 cycles), the ink trace experiences progressive failure. The sheet resistance continues to increase, admitting at a reduced rate (1 to 13%). This behaviour is consistent with the terminal phase of fatigue damage, where cracks continue to expand and connect, progressively severing remaining conductive paths until electrical failure becomes imminent or total.

This progression was mirrored across other geometries which are curve, square, and zigzag patterns with some variation in the number of cycles to failure and the magnitude of resistance change. These variations can be attributed to the distinct stress distribution and current path complexity in each pattern. For instance, zigzag patterns inherently have more junctions and directional changes, which may help dissipate mechanical stress more evenly, delaying crack propagation. In contrast, square patterns introduce high localised stresses at corners, making them more susceptible to earlier crack initiation.

In contrast, a CB ink traces displayed a similar three-stage degradation profile but with distinct resistance trends and lower failure severity. At first stage (0 to 2000 cycles), a modest increase of approximately 20% in sheet resistance was observed in straight-line patterns across all widths. This indicates initial micro-cracking and debonding similar to the GNPs case but suggests that CB networks are initially more resilient or perhaps more deformable due to their different filler morphology. At stage two (2000 to 4000 cycles), a more pronounced increase of about 120% in sheet resistance occurred, corresponding to the expansion and merging of cracks within the film. Unlike GNPs, CB particles tend to form larger agglomerates with less uniform dispersion, making their conductive networks more sensitive to sudden disruption once critical damage accumulates. Finally, at stage three (4000 to 7000 cycles), the resistance continued to rise by another 2 to 20%, reflecting continued degradation. However, no total failure was observed in the ink traces. Instead, electrical performance deteriorated steadily due to distributed micro-cracking and filler disconnection, without a sharp transition to complete loss of conduction. One outstanding aspect of CB ink behaviour is the absence of a catastrophic failure threshold within the testing range. This may be due to the more granular and isotropic nature of the network, where conduction can persist through alternative paths despite damage, albeit at reduced efficiency.

As with bulk resistance characterisation, pattern geometry and trace width again played significant roles in determining sheet resistance behaviour. Wider traces generally exhibited lower resistance and more stable performance, owing to greater redundancy in conductive pathways. Narrower traces (1 mm) were more vulnerable to failure due to their limited conductive cross-section and higher current density, which promotes localised heating and accelerates damage.

Pattern shape influenced both mechanical stress distribution and the continuity of the electrical network. Rounded patterns like curves helped disperse tensile stress more uniformly, delaying crack formation and promoting resistance stability. In contrast, sharp-edged patterns such as squares introduced geometric stress concentrators, accelerating crack nucleation. This effect was particularly pronounced in GNPs traces, where the stiff, planar nature of the nanoplatelets made them more susceptible to delamination and breakage at these stress points.

The degradation of sheet resistance under cyclic loading can be understood through the lens of percolation theory and fracture mechanics. Conductive inks, being composite systems, rely on percolation networks formed by electrically conductive fillers embedded within a deformable polymer matrix. When subjected to mechanical deformation, especially tensile strain, the filler network becomes stretched, disconnected, or partially destroyed. In low-strain, this may be reversible. However, with repeated cycles, irreversible damage accumulates, leading to a permanent increase in resistance.

The mechanical mismatch between the relatively brittle ink and the more flexible substrate exacerbates this problem. While the substrate can elastically deform and recover, the ink layer undergoes plastic deformation and micro-crack development. This difference in mechanical response introduces strain at the interface and within the ink, gradually decoupling the conductive particles and increasing resistance. Once critical stress levels are reached, crack propagation accelerates, and conductive paths are severed irreversibly. These findings are critical for the design of reliable flexible electronic devices. Engineers must consider not only the intrinsic electrical conductivity of the ink but also its mechanical fatigue resistance and interfacial stability with the substrate. Strategies to improve durability could include altering the filler morphology, optimising the filler-to-binder ratio, engineering multilayer architectures, or designing stress-relieving trace geometries.

8.4 Evaluation of Volume Resistivity

Figure 8.3 presents a comparative analysis of the volume resistivity of two types of conductive ink, measured through both bulk resistance and sheet resistance approaches. This investigation provides essential insight into the electrical integrity and degradation behaviour of conductive ink patterns under cyclic tensile loading, which is of particular importance in the development of robust and reliable flexible electronic components.

Volume resistivity, a fundamental material property, quantifies a substance's opposition to current flow throughout its volume and is typically expressed in terms of resistance per unit length and cross-sectional area. It is influenced not only by the inherent properties of the conductive material but also by microstructural integrity, inter-particle connectivity, trace geometry, and the nature of the applied mechanical strain. In this context, both bulk and sheet resistance measurements can be employed to estimate volume resistivity. However, differences in measurement techniques, sample geometry, and probe configuration may yield variations in the calculated values.

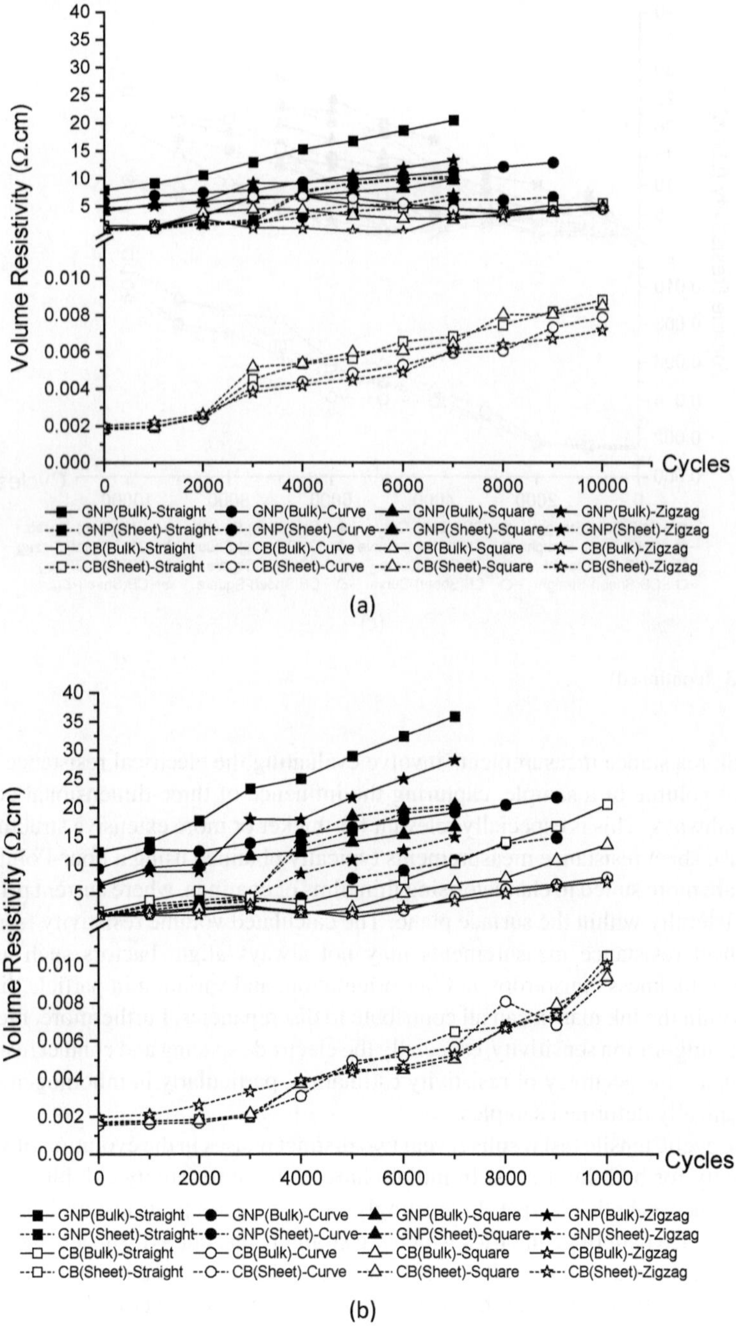

Fig. 8.3 Graph of volume resistivity for GNPs and CB ink trace with **a** 1 mm width, **b** 2 mm width, and **c** 3 mm width for 10,000 times of cyclic tensile

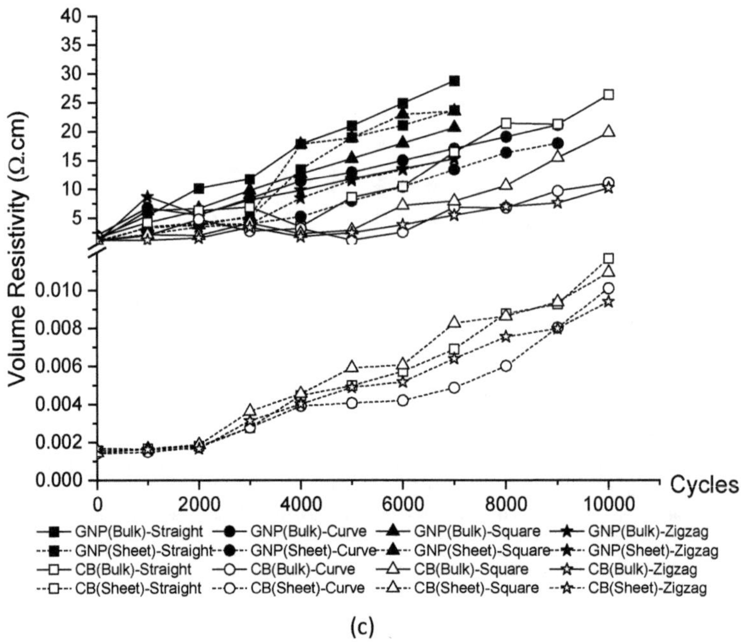

Fig. 8.3 (continued)

Bulk resistance measurements involve evaluating the electrical resistance across the full volume of a sample, capturing the influence of three-dimensional conduction pathways. This is especially relevant for thicker or more extensive structures. In contrast, sheet resistance measurements typically obtained using a Four-Point probe setup are more suited to characterising thin films or coatings, where current primarily flows laterally within the surface plane. The calculated volume resistivity from bulk and sheet resistance measurements may not always align. Factors such as non-uniform thickness, anisotropy in filler orientation, and variation in particle distribution within the ink matrix can all contribute to discrepancies. Furthermore, measurement configuration sensitivity, especially the electrode spacing and contact resistance can impact the accuracy of resistivity estimation, particularly in inhomogeneous or mechanically deformed samples.

The cyclic tensile test results reveal two distinct phases in the evolution of volume resistivity for both ink traces. In initial phase, volume resistivity exhibits a modest increase across both ink types. Interestingly, this increase appears to correlate with the specific trace pattern geometries which are straight-line, curved, zigzag, and square, indicating that trace shape influences how mechanical stress is distributed across the conductive network. During this phase, the conductive ink structure largely remains intact, although slight debonding at the particle or ink–substrate interface begins to occur. The GNPs particles, due to their layered platelet morphology, maintain relatively strong van der Waals bonding and contact area, which delays total network disintegration but does allow for gradual weakening under repeated loading.

In progressive phase or later cycles, which is in the second half of the cyclic test, volume resistivity begins to increase sharply and becomes less dependent on pattern geometry, especially in the case of straight-line patterns. This significant rise indicates the onset of widespread debonding and fracture of conductive pathways. The integrity of the percolation network begins to collapse as mechanical fatigue introduces micro-cracks, particle displacements, and interfacial separations, all of which contribute to the loss of continuous electrical paths. The behaviour observed in this phase reflects the transition from elastic to plastic deformation within the composite ink, leading to irreversible structural changes. For GNPs ink traces, the loss of electrical connectivity in this phase is more pronounced due to the brittle nature of the ink matrix when exposed to cyclic tensile strain. The sharp edges of GNPs flakes are prone to initiating cracks within the surrounding polymer, and once these micro-cracks consolidate, they accelerate network fragmentation. The observed sharp rise in volume resistivity supports the approach of percolation threshold breach, where the number of effective conductive connections falls below the minimum required to sustain electrical conduction.

In CB ink traces, volume resistivity exhibited a somewhat more complex pattern. Initially, the resistivity increased, reflecting the early stages of damage and stress-induced debonding. However, this was followed by a temporary decrease in volume resistivity in subsequent cycles, an effect attributed to the unique behaviour of the filler network. The particles from zero-dimensional (0D) random networks composed of loosely packed agglomerates. Under mechanical stress, these agglomerates can break apart, enabling finer particles to redistribute and re-establish conductive paths. This particle redistribution improves connectivity temporarily, resulting in a decrease in resistivity despite continued mechanical loading. As tensile cycling progressed, however, this compensatory behaviour was eventually overcome by accumulated plastic deformation, and volume resistivity began to rise again. The increase, though significant, did not result in catastrophic failure, even after 10,000 cycles, indicating that the inks, despite their lower conductivity, may offer better mechanical tolerance in certain applications. This finding suggests that networks, due to their less structured and more flexible architecture, may be more resilient to mechanical fatigue than the highly ordered GNPs networks.

An unexpected result was that CB ink traces with a 3 mm width, which exhibited the lowest resistivity in static testing, showed the highest resistivity under cyclic loading at later stages. This reversal indicates that wider traces, while beneficial under static conditions due to greater cross-sectional area and redundancy, may be more liable to internal deformation and filler disintegration when subjected to repeated mechanical strain. This underscores the complex interaction between geometry and fatigue response in flexible electronics.

The experimental investigations highlight the critical role of trace width in governing resistivity behaviour. Wider traces (at 3 mm) provide more conductive pathways under initial conditions but may accumulate more damage under repeated loading due to their larger surface area and greater tendency for localised strain accumulation. Narrower traces, while having higher baseline resistivity, may localise damage more narrowly and thus delay overall failure. Pattern geometry also significantly influences stress distribution. Curved and zigzag patterns, which introduce direction changes and redundant paths, may buffer the effects of tensile strain better than straight-line or square patterns. These geometries help distribute stress more evenly, potentially delaying crack propagation and network fragmentation.

Another key finding is the discrepancy between volume resistivity derived from sheet resistance (using the Four-Point probe method) and that obtained by bulk resistance (typically involving Two-Point probe measurements). The Four-Point probe method yielded lower and more stable resistivity values. This is likely due to its minimisation of contact resistance and its sensitivity to lateral conductivity in thin films. As Li et al. (2012) emphasised, Four-Point measurements offer higher precision and reproducibility, especially for thin conductive coatings, as they effectively isolate the intrinsic resistive properties of the sample from extrinsic artefacts such as electrode contact issues. The Two-Point probe method, in contrast, often includes significant contact resistance and is more sensitive to inhomogeneities and anisotropies within the material. This can result in inflated resistivity measurements, especially in samples experiencing mechanical deformation, where contact points may fluctuate during testing. Hence, for precise and consistent characterisation of conductive ink films, particularly under strain, four-point sheet resistance testing is recommended.

The exploration demonstrates that the degradation of volume resistivity under cyclic tensile loading is multifaceted, influenced by filler type, trace geometry, measurement technique, and mechanical strain history. Understanding these interactions is essential for designing conductive inks and printed trace patterns that are not only electrically efficient but also mechanically durable. For GNPs-based inks, efforts to improve mechanical resilience could focus on hybridising the ink with flexible binders or incorporating bridging agents that help maintain particle connectivity under strain. For CB inks, further optimisation of filler dispersion and matrix compatibility may assist enhance both conductivity and fatigue resistance. Moreover, engineering trace geometries to mitigate stress concentrations and accommodate deformation such as incorporating serpentine or arc patterns, and it can significantly improve performance in stretchable applications.

8.5 Investigation Using Nanoindentation Method

Basically, Fig. 8.4 presents a comparative analysis of the maximum force and maximum indentation depth for both conductive ink traces under both static and cyclic tensile loading conditions, up to 10,000 cycles. This assessment offers critical insights into the evolution of mechanical integrity in conductive inks, which is especially important in stretchable and flexible electronic systems subjected to repeated mechanical deformation during use. In mechanical characterisation, maximum force represents the maximum tensile force a material can withstand before yielding or failure, while maximum indentation depth denotes the depth of indentation under load during nanoindentation testing which an indicator of surface hardness, plasticity, and internal cohesion of the material. These metrics are particularly useful for evaluating how mechanical fatigue and microstructural changes impact the mechanical performance of printed electronic materials.

Under static loading, both traces exhibit relatively high maximum force and low indentation depth values. This indicates good mechanical resilience, with both materials initially capable of withstanding tensile stress without significant deformation. The relatively low indentation depth further suggests a dense, well-bonded internal structure with strong inter-particle connectivity and minimal voids or delamination. However, as the ink traces are subjected to cyclic tensile loading, a progressive and notable degradation in mechanical response is observed. Both maximum force and indentation depth show significant changes, reflecting deterioration in the mechanical and structural properties of the ink materials due to cyclic fatigue. The force decreases steadily with an increasing number of cycles, indicating a loss of load-bearing capacity. Besides, indentation depth increases progressively, suggesting that the ink traces become mechanically softer and more compliant, likely due to microstructural damage and reduction in internal cohesion.

The degradation in GNPs traces under cyclic loading is more pronounced than in CB inks. The GNPs, while possessing excellent electrical conductivity and high modulus due to its graphene-based structure, is intrinsically brittle. Its two-dimensional platelet morphology contributes to stiffness but offers limited plastic deformation capability. As a result, GNPs-filled inks have a lower fracture toughness and reduced capacity to absorb and dissipate mechanical energy under repeated stress. After approximately 1000 tensile cycles, the ink trace demonstrates a deeper indentation under a similar load compared to an unloaded sample. This reflects early-stage weakening of the microstructure likely due to micro-crack initiation, interfacial debonding between the platelets and polymer binder, and reduction in inter-particle van der Waals forces.

As cyclic loading progresses, these micro-cracks propagate and consolidate, leading to reduced structural rigidity and increased plastic deformation, which is evident in the steadily increasing indentation depth and decreasing force. The ink's increasing compliance is further exacerbated by stress concentration effects, especially at filler edges or geometric discontinuities within the trace. Over time, the accumulated strain energy leads to irreversible damage and eventual structural failure, confirming the brittle nature of the inks under mechanical cycling.

In contrast, a CB conductive ink exhibits a different degradation profile under cyclic loading. Initially, similar trends are observed a gradual reduction in maximum force and an increase in indentation for force, indicating weakening of the structure. However, around the midpoint of the cyclic test (approximately 5000 cycles), the ink demonstrates an unexpected behaviour, a slight hardening is observed, with a temporary decrease in indentation depth. This phenomenon may be attributed to structural rearrangements within the ink. The particles form isotropic, 0D random networks. When subjected to repeated loading, these networks can compact and redistribute, improving load transfer through localised densification. This compaction may temporarily increase resistance to indentation. However, as cycling continues

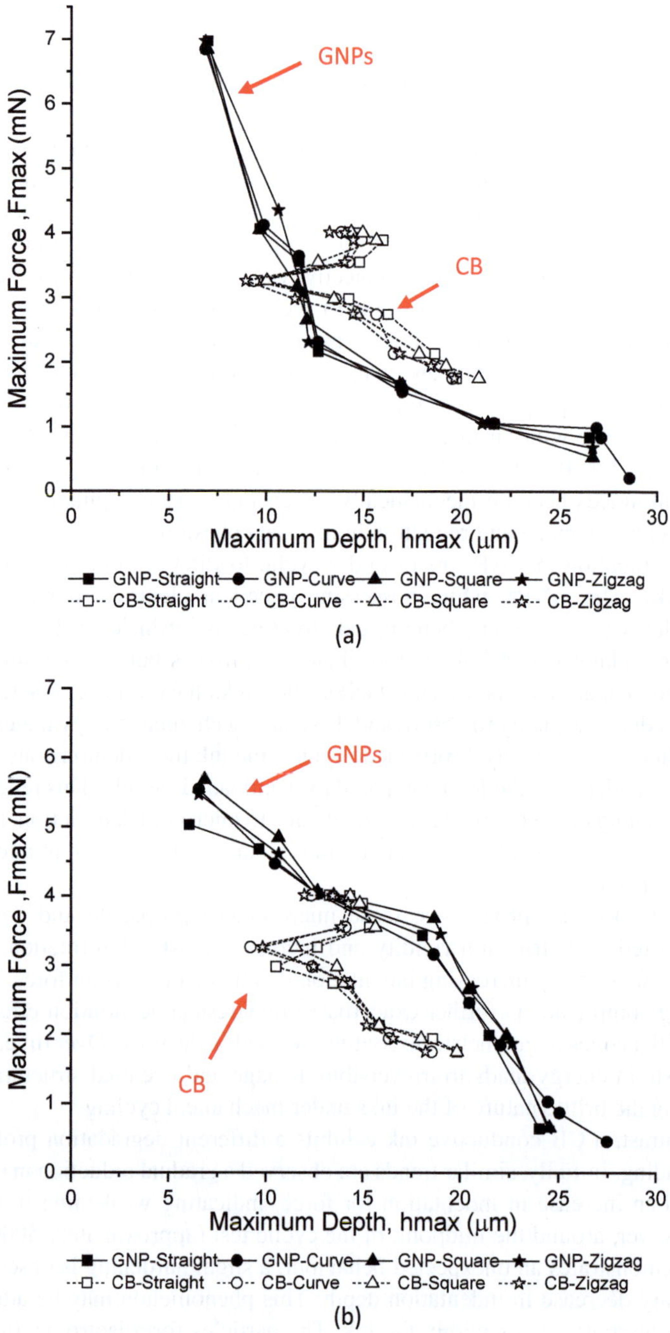

Fig. 8.4 Graph of maximum force versus maximum load of GNPs and CB ink trace with different pattern with **a** 1 mm, **b** 2 mm, and **c** 3 mm widths for cyclic tensile

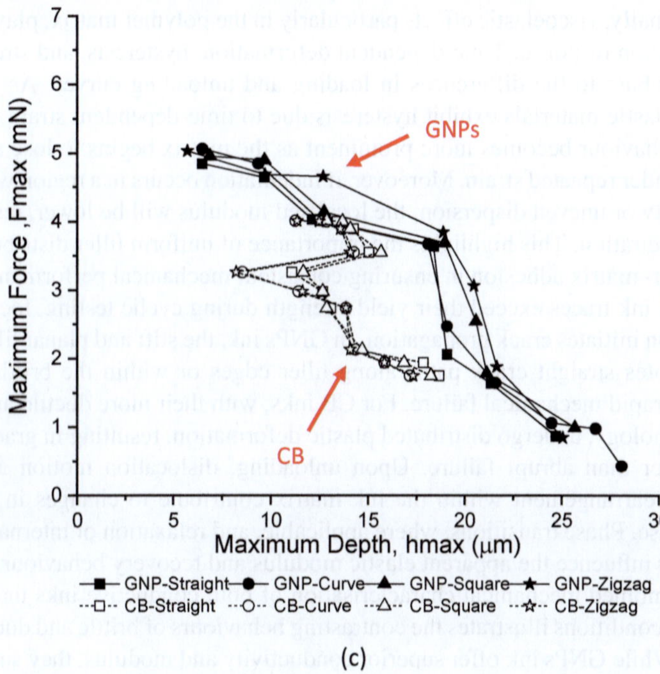

Fig. 8.4 (continued)

beyond 5000 cycles, the fatigue damage accumulates, and the ink once again softens, with indentation depth increasing and force decreasing, reflecting further structural degradation and plastic deformation. Importantly, CB inks do not exhibit complete failure within 10,000 cycles, suggesting a higher tolerance to cyclic stress. This is largely due to the ductile nature of the it network and the greater flexibility of its agglomerates, which allows for more energy dissipation and strain accommodation compared to the stiff and brittle GNPs network.

Nanoindentation testing serves as an effective method to probe the elastic–plastic behaviour of ink traces. During indentation, both elastic (recoverable) and plastic (permanent) deformation occurs. The loading phase captures the total deformation behaviour, while the unloading phase reflects the degree of elastic recovery. In materials with a strong filler–matrix interface, elastic recovery is typically higher, and the indentation depth is lower. As cyclic loading progresses, the indentation curves for both GNPs and CB traces shift, reflecting a shift from predominantly elastic to increasingly plastic behaviour. The increasing indentation depth under constant or reduced loading conditions reveals that the material becomes more compliant, allowing deeper penetration, which is symptomatic of mechanical softening and internal failure.

Additionally, viscoelastic effects particularly in the polymer matrix, play a role in the indentation response. Time-dependent deformation, hysteresis, and stress relaxation contribute to the differences in loading and unloading curves. As noted by [7], viscoelastic materials exhibit hysteresis due to time-dependent strain recovery, and this behaviour becomes more prominent as the matrix begins to lose structural integrity under repeated strain. Moreover, if indentation occurs in a region with lower filler density or uneven dispersion, the localised modulus will be lower, resulting in deeper penetration. This highlights the importance of uniform filler distribution and strong filler–matrix adhesion in ensuring consistent mechanical performance.

As both ink traces exceed their yield strength during cyclic testing, local plastic deformation initiates crack propagation. In GNPs ink, the stiff and planar filler structure promotes straight crack paths along filler edges or within the brittle matrix, leading to rapid mechanical failure. For CB inks, with their more ductile and entangled morphology, undergo distributed plastic deformation, resulting in gradual softening rather than abrupt failure. Upon unloading, dislocation motion and grain boundary rearrangement within the ink matrix contribute to changes in mechanical response. Phase transitions, where applicable, and relaxation of internal stresses can further influence the apparent elastic modulus and recovery behaviour.

The combined mechanical characterisation of both conductive inks under static and cyclic conditions illustrates the contrasting behaviours of brittle and ductile filler systems. While GNPs ink offer superior conductivity and modulus, they suffer from rapid mechanical degradation under cyclic loading. In CB inks, though less conductive, exhibit greater toughness and durability under repeated mechanical stress. From an engineering perspective, the selection of conductive ink for flexible electronics must consider both electrical performance and mechanical endurance. In applications where devices are subjected to repeated bending or stretching, ductile inks with stable mechanical properties under fatigue like CB materials may be more appropriate. Where high conductivity is paramount and mechanical cycling is limited, GNPs-based inks can offer a better solution, with additional design strategies to mitigate fatigue effects, such as encapsulation, geometric reinforcement, or hybridisation with ductile binders.

8.6 Hardness and Elastic Modulus Degradation

Figure 8.5 degradation illustrates the progressive changes in mechanical hardness of both ink traces subjected to cyclic tensile loading. The work highlights how repeated mechanical deformation affects the structural integrity of the conductive ink films, particularly in terms of their hardness response over time. Hardness is a key mechanical property that reflects a material's resistance to permanent surface deformation. It is inherently linked to the material's internal cohesion, microstructural integrity, and its ability to withstand mechanical stress without yielding or fracturing.

The findings from cyclic loading reveal that the mechanical hardness of GNPs ink traces decreases consistently with increasing tensile cycles, regardless of the

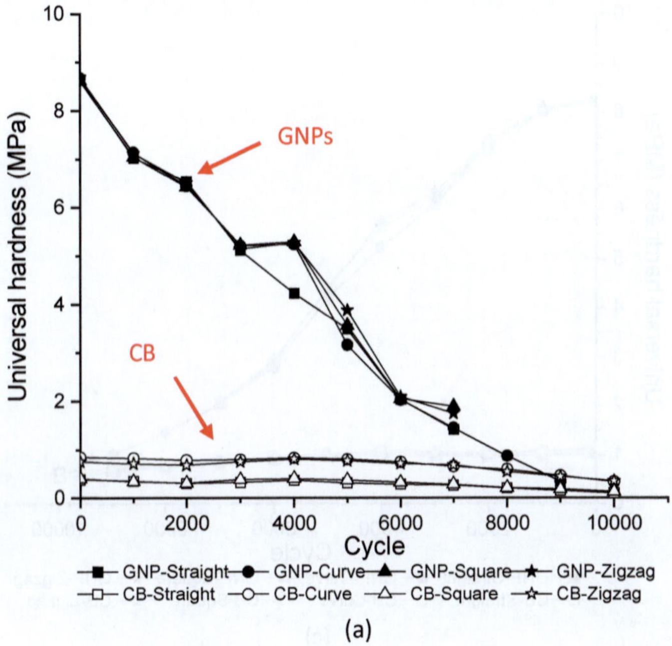

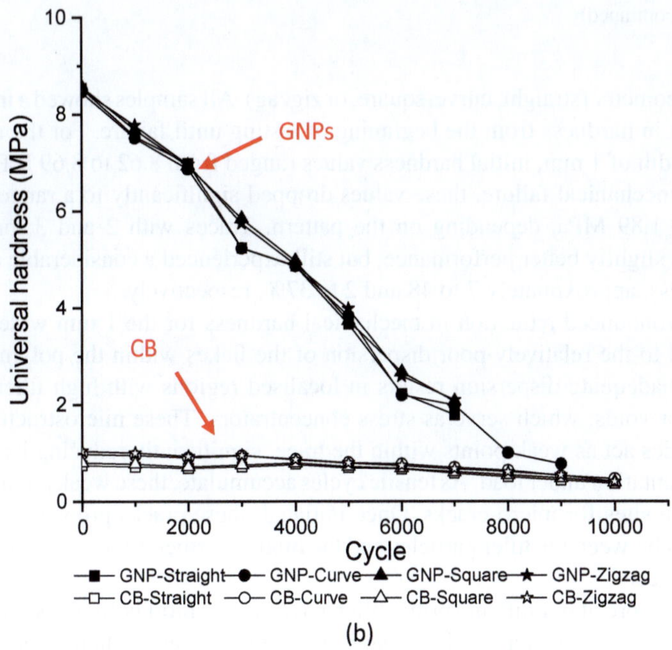

Fig. 8.5 Graph of hardness of GNP and CB ink trace with different pattern with **a** 1 mm, **b** 2 mm, and **c** 3 mm widths for cyclic tensile

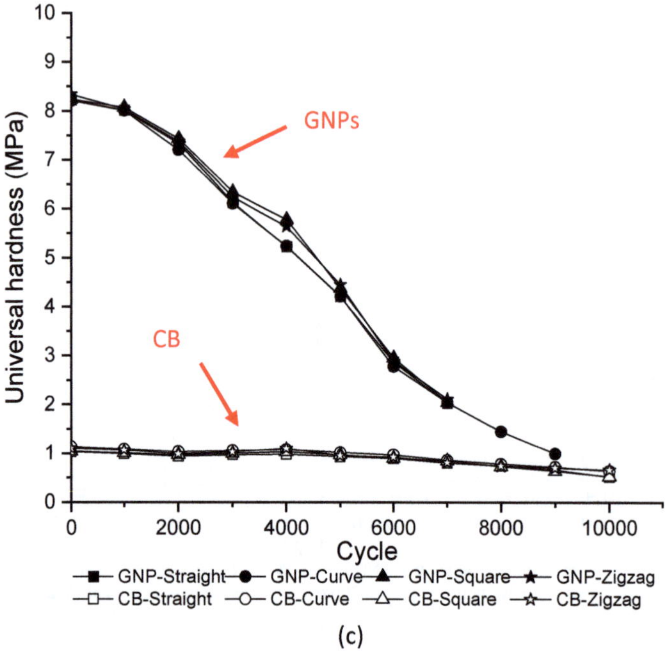

Fig. 8.5 (continued)

pattern geometry (straight, curve, square, or zigzag). All samples showed a monotonic reduction in hardness from the beginning of testing until failure. For the ink traces with a width of 1 mm, initial hardness values ranged from 8.62 to 8.69 MPa. By the point of mechanical failure, these values dropped significantly to a range between 0.34 and 1.89 MPa, depending on the pattern. Traces with 2 and 3 mm widths exhibited slightly better performance, but still experienced a considerable reduction in hardness, approximately 7 to 48 and 2 to 37%, respectively.

The pronounced reduction in mechanical hardness for the 1 mm wide traces is attributed to the relatively poor dispersion of the flakes within the polymer binder matrix. Inadequate dispersion results in localised regions with high filler concentrations or voids, which serve as stress concentrators. These microstructural inhomogeneities act as weak points within the trace, significantly reducing its ability to resist indentation under load. As tensile cycles accumulate, these weak points become nucleation sites for micro-cracks. Once initiated, these cracks propagate along the interfaces between the filler particles and the binder, further diminishing mechanical performance.

Furthermore, it is a stiff and brittle filler, its plate-like morphology does not readily accommodate strain. Repeated mechanical loading causes gradual delamination at the filler–matrix interface, resulting in internal void formation and loss of cohesion. The polymeric matrix, already under mechanical fatigue, is less able to constrain the movement of the flakes, thus contributing to a progressive decline in overall mechanical hardness. This trend also suggests that the width of the trace plays a more influential role in mechanical performance than the pattern shape. Wider traces (2 and 3 mm) allow for more uniform distribution of stress and strain, and better dispersion of filler particles, reducing the severity of stress concentration and micro-crack formation.

For CB ink traces display a more complex and pattern-dependent mechanical response. At the beginning of the tensile cycle, the straight-line and square patterns with 1 mm width exhibited mechanical hardness values of approximately 0.38 and 0.40 MPa, respectively. These values increased with trace width up to 1.05 MPa for 3 mm-wide square patterns. Over the course of 10,000 cycles, the hardness of these patterns gradually decreased, reaching as low as 0.1 to 0.53 MPa depending on the trace width. Conversely, the curved and zigzag patterns initially exhibited higher hardness values, 0.85 and 0.75 MPa respectively for 1 mm traces, and experienced a slight increase in hardness up to around 4000 cycles. This transient increase can be attributed to the reorganisation of the particle network. As mechanical cycling progresses, loosely packed the ink agglomerates may compact and improve inter-particle contact, temporarily enhancing mechanical rigidity. However, with continued cycling, plastic deformation accumulates, leading to softening and eventual decline in hardness values, reaching 0.33 to 0.35 MPa by 10,000 cycles.

This unique behaviour reinforces the influence of pattern geometry on mechanical durability. Curved and zigzag patterns inherently redirect and distribute mechanical strain more uniformly, reducing localised stress concentrations. The direction changes inherent in these patterns promote redistribution of mechanical loads, improving the ink trace's ability to maintain hardness over more cycles. In contrast, straight-line and square geometries impose linear strain paths that favour the development of longitudinal micro-cracks, resulting in more pronounced mechanical degradation.

Then, Fig. 8.6 presents the evolution of elastic modulus in both conductive ink traces under the same cyclic tensile loading conditions. The elastic modulus, a measure of a material's stiffness, reflects the relationship between stress and strain within the elastic region of deformation. It is a critical parameter in understanding how a material responds to mechanical loading and its ability to recover its original shape after stress removal.

In the case of GNPs ink traces, the elastic modulus exhibits a continuous decline as the number of cycles increases. During the early phase up to 3000 cycles, the modulus drops significantly, by 40 to 70%, indicating rapid degradation of stiffness. This is followed by further reduction between 27 and 45% up to 6000 cycles, as the material transitions from the early fatigue stage into more advanced microstructural failure. This reduction in elastic modulus is indicative of accumulated damage at the filler–matrix interface and within the polymer phase. The onset of delamination, combined with matrix cracking and interfacial debonding, leads to a breakdown in the effective load transfer between the particles and the surrounding matrix. The disruption of this network reduces the ability of the ink trace to bear elastic loads and recover post-deformation.

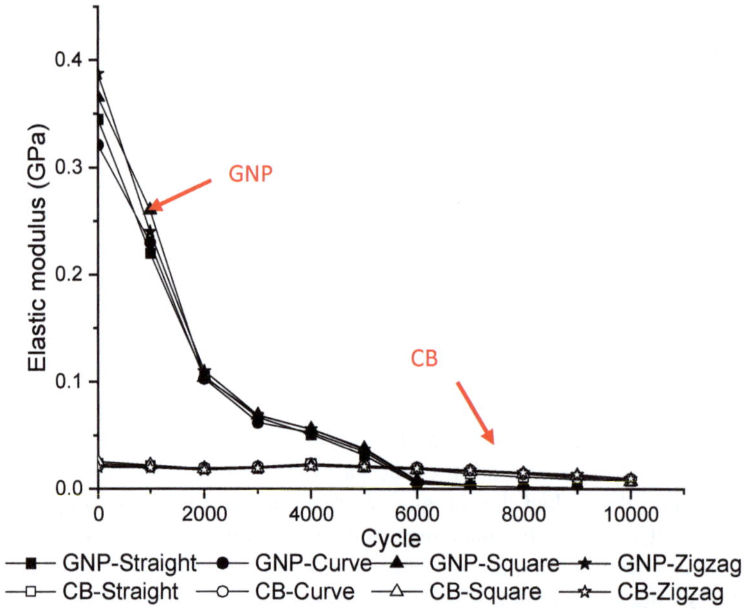

(a)

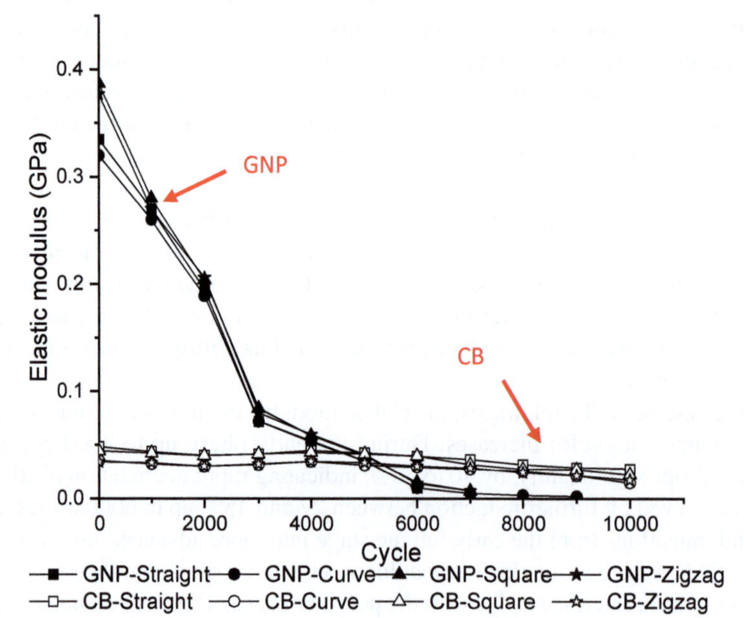

(b)

Fig. 8.6 Graph of elastic modulus of GNPs and CB ink trace with different pattern with **a** 1 mm, **b** 2 mm, and **c** 3 mm widths for cyclic tensile

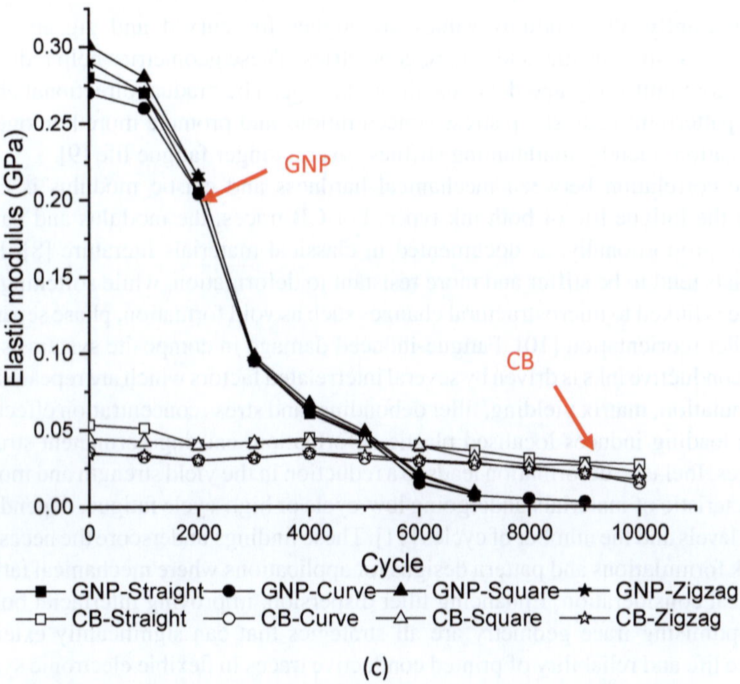

Fig. 8.6 (continued)

The high modulus of the GNPs flakes themselves is insufficient to compensate for the weakening binder, especially when the flakes begin to detach or reorient under stress. As cracks propagate and coalesce, the material becomes increasingly compliant and less structurally cohesive, culminating in full mechanical failure.

In CB ink traces exhibit a different modulus behaviour, reflecting their more ductile nature. The elastic modulus initially decreases during the first 2000 cycles from 0.0253 GPa to around 0.0178 GPa, consistent with the onset of plastic deformation and micro-crack formation. However, between 2000 and 4000 cycles, the modulus increases again, peaking around 0.0235 GPa. This temporary hardening effect is attributed to strain-induced densification of the particle network. As cyclic loading progresses, the random ink agglomerates may become more compacted, improving inter-particle contact and enhancing mechanical rigidity. This is a classic case of work hardening, where the internal microstructure becomes more resistant to further deformation due to reconfiguration of particles and realignment of polymer chains. However, this effect is short-lived. Between 4000 and 10,000 cycles, the elastic modulus decreases sharply, reaching values as low as 0.0074 GPa. At this stage, cumulative plastic deformation and damage at the particle–matrix interface dominate. The particle network begins to fracture, and the matrix yields under accumulated strain, resulting in a softer, less elastic material.

Importantly, the modulus values are higher for curved and zigzag patterns compared to straight-line and square geometries. These geometries help redistribute stress more uniformly and delay localised damage. The gradual directional changes in the pattern mitigate sharp stress concentrations and promote more homogeneous deformation, thereby maintaining stiffness over a longer fatigue life [9].

The correlation between mechanical hardness and elastic modulus is evident across the fatigue life of both ink types. For CB traces, the modulus and hardness emerge proportionally, as documented in classical materials literature [8]. Harder materials tend to be stiffer and more resistant to deformation, while softening under fatigue is linked to microstructural changes such as void formation, phase separation, and filler reorientation [10]. Fatigue-induced damage in composite systems such as these conductive inks is driven by several interrelated factors which are repeated strain accumulation, matrix yielding, filler debonding, and stress concentration effects. The cyclic loading induces localised plastic deformation, causing permanent structural changes. Inelastic deformation leads to a reduction in the yield strength and modulus, characteristic of materials undergoing low-cycle or high-cycle fatigue, depending on stress levels and the number of cycles [11]. These findings underscore the necessity of the ink formulations and pattern designs for applications where mechanical fatigue is a critical consideration. Enhancing filler dispersion, improving interfacial bonding, and optimising trace geometry are all strategies that can significantly extend the service life and reliability of printed conductive traces in flexible electronic systems.

8.7 The SEM Analysis of Stress Concentrations and Crack Formation

Nanocomposite materials, comprising nanoscale reinforcements dispersed in a continuous matrix, have garnered substantial attention in recent years due to their unique and enhanced properties compared to conventional materials. These improvements often arise from a high specific surface area of the fillers and the strong interfacial interactions they can establish with the matrix, ultimately imparting superior mechanical, electrical and thermal properties. However, to fully understand and leverage these advantages, it is essential to characterise their morphology, especially after they have been subjected to cyclic mechanical loads, and to establish clear structure, property relationships.

The Scanning Electron Microscopy (SEM) is a powerful and well-established technique that enables to examine the microstructure of nanocomposites at high magnification. In the context of nanocomposite conductive inks, such as those formulated with GNPs or CB SEM imaging provides invaluable insights into the dispersion of the conductive fillers, the integrity of the percolating network and the presence of microstructural defects. The behaviour of these nanocomposites under repeated cyclic tensile loading can also be evaluated in terms of micro-crack initiation, crack propagation, particle debonding and interfacial failure. Tables 8.1, 8.2, 8.3 and 8.4

show the SEM micrographs of the both conductive ink traces printed in various patterns and at different widths. Careful analysis of these micrographs reveals correlations between the microstructural evolution of the traces and their macroscopic electrical and mechanical performance.

For the GNPs-based inks, the cross-sectional SEM images highlight significant microstructural rearrangement after the traces are subjected to cyclic tensile loading. Initially, its tend to align partially along the direction of printing, creating an anisotropic microstructure with good in-plane conductivity. However, upon repeated stretching and relaxing, especially at high strain amplitudes, the nanosheets reorient, separate and agglomerate, leading to disruption of the percolation pathways that originally enabled efficient electron transport. The SEM images clearly show that crack formation is prevalent throughout the microstructure. These cracks initiate at points of stress concentration, typically located at rough or undulating surfaces and at filler–binder boundaries. The binder, often a polymer or resin, acts as a compliant matrix to hold the GNPs together. However, under cyclic tension, localised strain mismatches between the stiff, planar GNPs and the softer binder cause debonding at the interface. This debonding serves as a nucleation site for cracks, which subsequently propagate under cyclic stress.

Moreover, these cracks tend to coalesce into larger assured along the trace, exacerbated by uneven ink distributions that produce weak spots in the network. The cyclic loading further promotes nanosheet pull-out and layer separation, a phenomenon that increases the effective distance between adjacent conductive fillers and consequently decreases the trace's effective electrical conductivity. Such observations are consistent with earlier findings by [12], who demonstrated that cyclic tensile loading degrades the filler–matrix interface and promotes micro-cracking in ink nanocomposites.

By contrast, a CB inks show a markedly different behaviour under similar cyclic loads. The SEM micrographs of straight-line traces exhibit a smooth surface, with a dense and uniform microstructure and minimal surface roughness. The spherical ink particles tend to form percolating chains and clusters, creating a more isotropic structure that can better accommodate cyclic deformation. Even after multiple cycles of tension, these ink traces show no obvious cracks or debonding. This can be attributed to the smaller average size of the particles and their tendency to fill interstitial spaces more effectively than larger planar fillers like GNPs. The smaller particle size also yields a lower local stress concentration and a more homogeneous stress distribution across the trace during loading.

In the case of meandering and serpentine patterns, the ink traces appear rougher, with more indirect surfaces and some micro-voids. These wavy-like structures are deliberate geometric features that help distribute strain more evenly across the trace length. [13] noted that under high strain, the curved shapes can deform elastically by straightening out the meanders before reaching critical strain thresholds for material failure. However, some voids and hackles may still form due to incomplete resin infiltration or filler agglomerates. These features, if left unchecked, can serve as stress risers and may potentially reduce the trace's durability over extended cyclic loading.

Table 8.1 Cross sectional images of straight-line pattern of GNPs and CB ink trace microstructure with line width of 1, 2 and 3 mm for cyclic tensile

GNPs ink trace

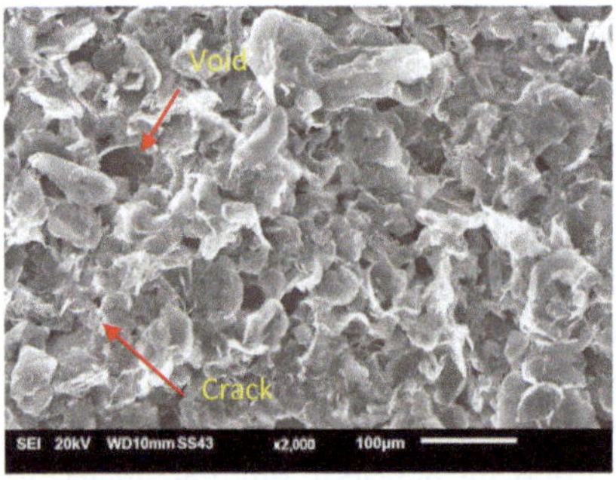

CB ink trace

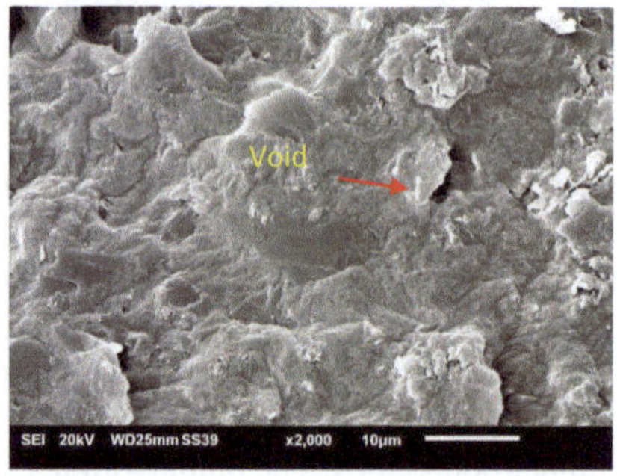

1 mm trace width

GNPs ink trace

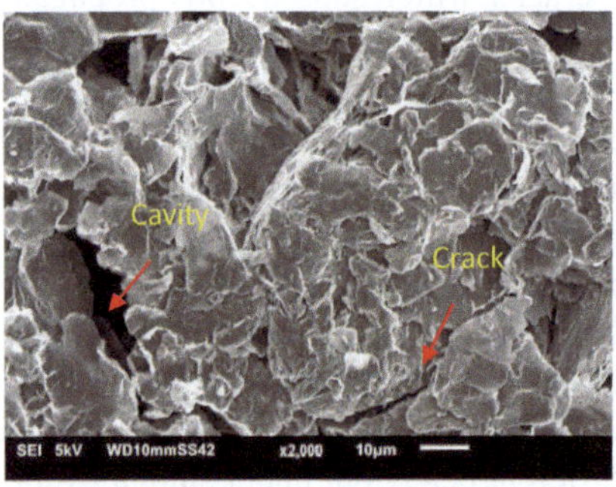

CB ink trace

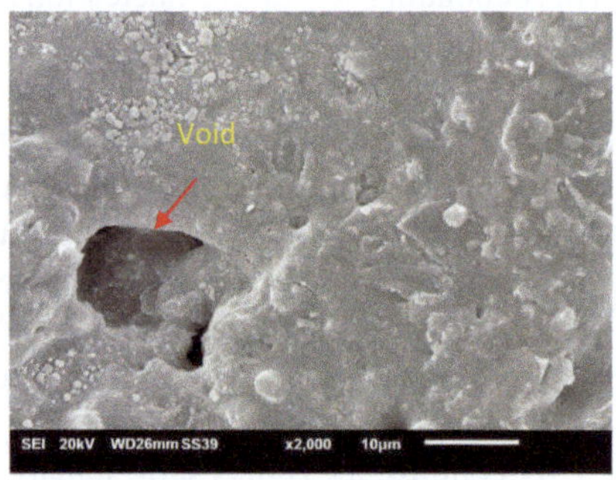

2 mm trace width

GNPs ink trace

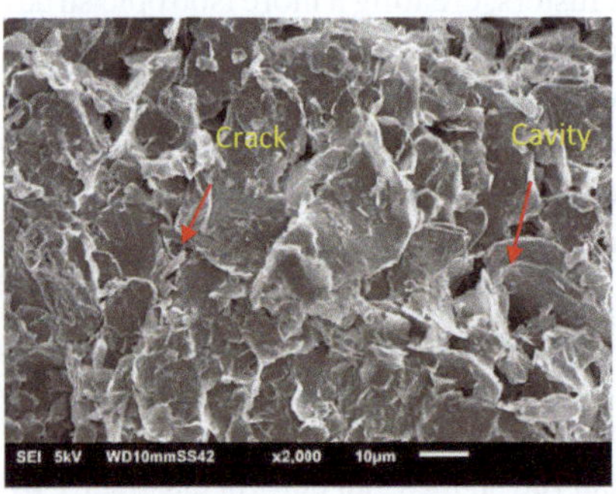

CB ink trace

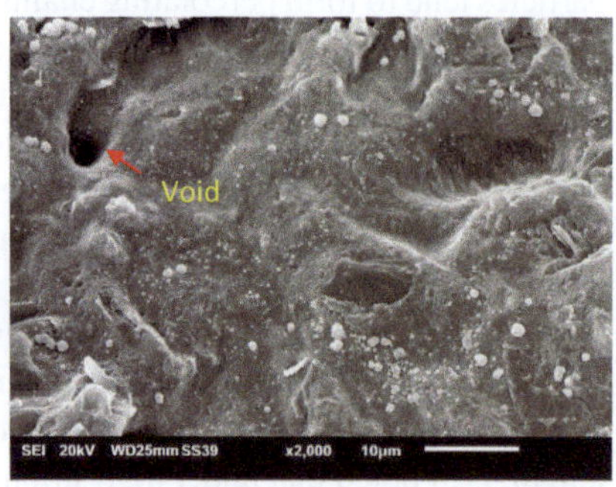

3 mm trace width

Table 8.2 Cross sectional images of curve pattern of GNPs and CBink trace microstructure with line width of 1, 2 and 3 mm for cyclic tensile

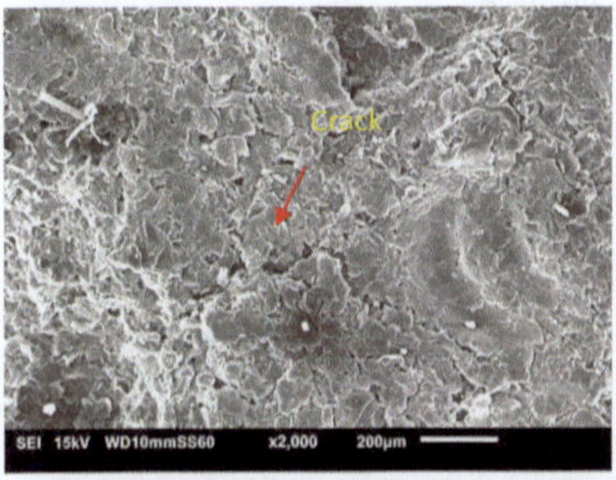

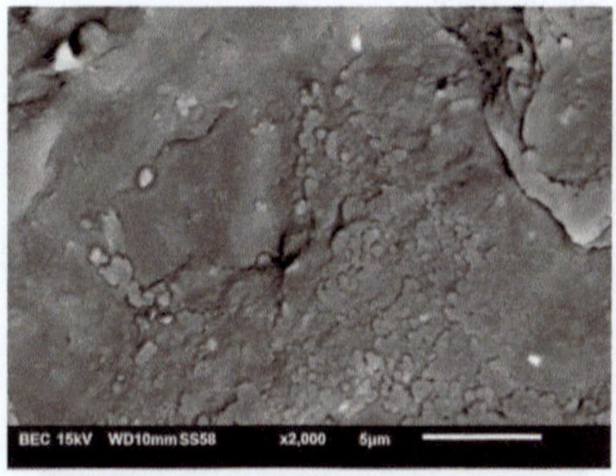

1 mm trace width

GNPs ink trace

CB ink trace

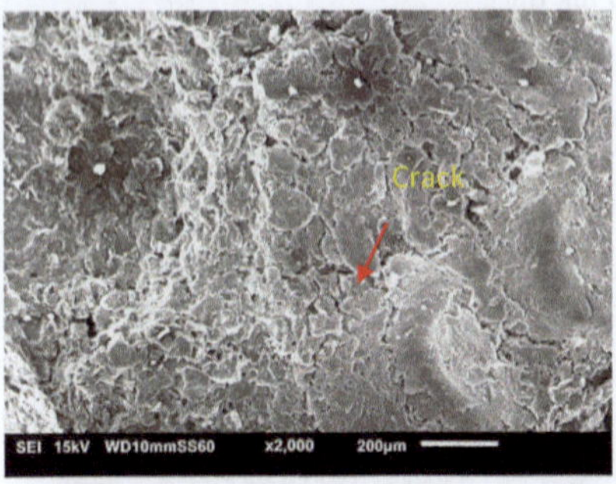

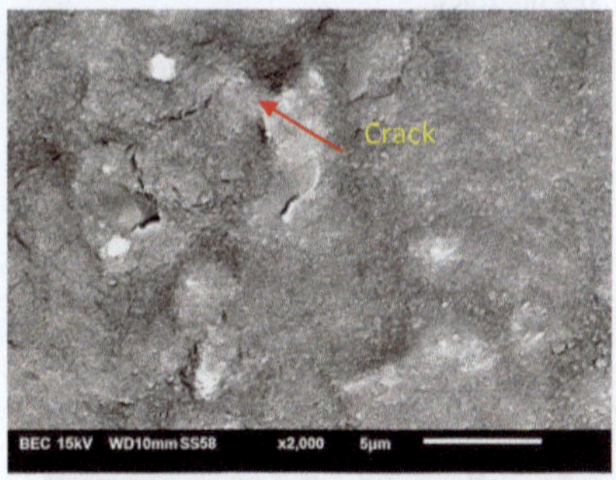

2 mm trace width

GNPs ink trace

CB ink trace

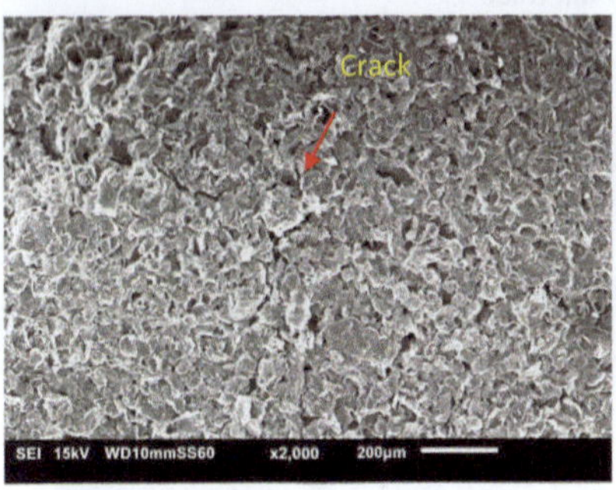

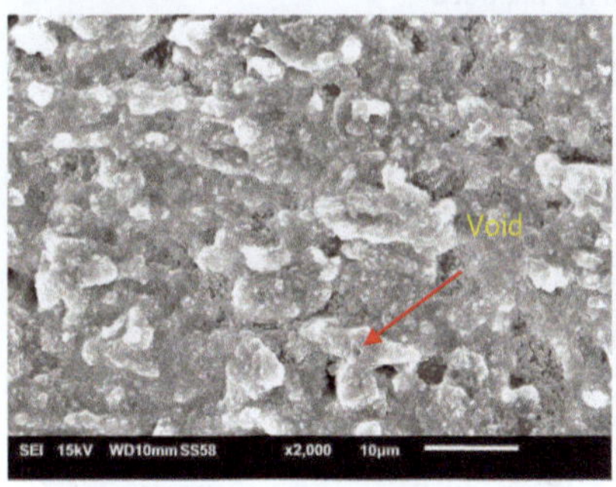

3 mm trace width

Table 8.3 Cross sectional images of square pattern of GNPs and CB ink trace microstructure with line width of 1, 2 and 3 mm for cyclic tensile

GNPs ink trace

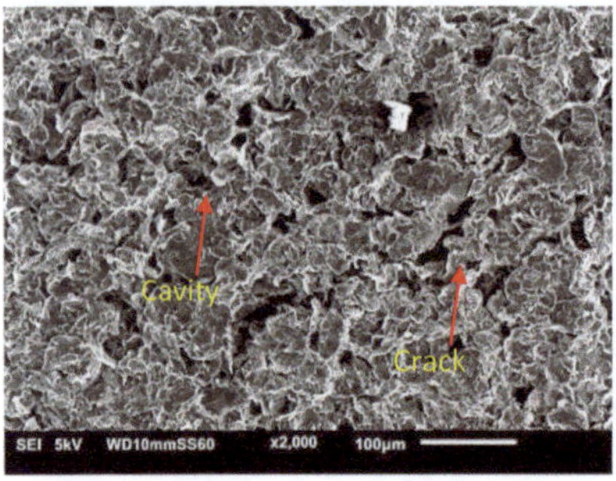

CB ink trace

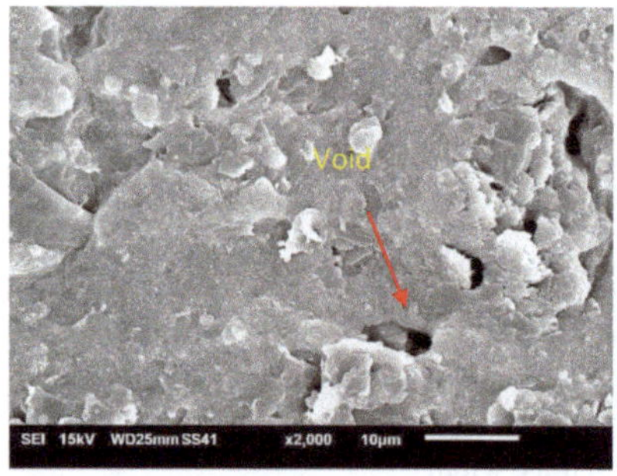

1 mm trace width

GNPs ink trace

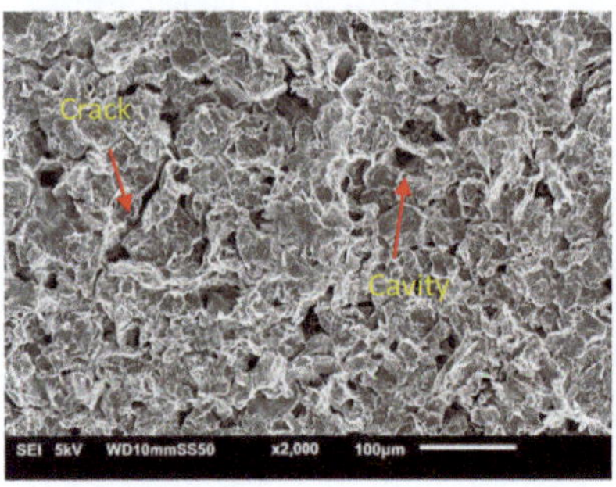

CB ink trace

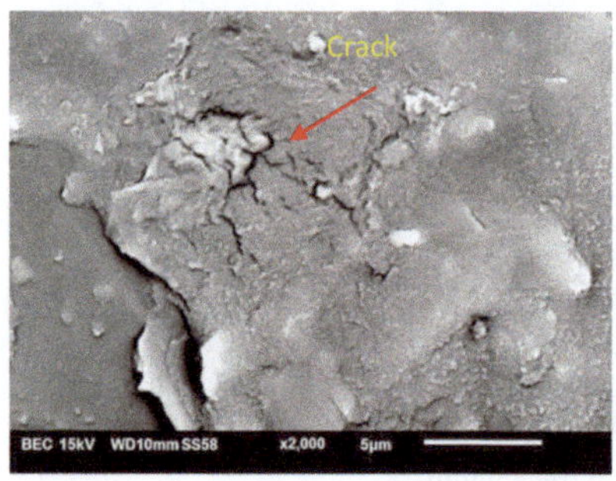

2 mm trace width

GNPs ink trace

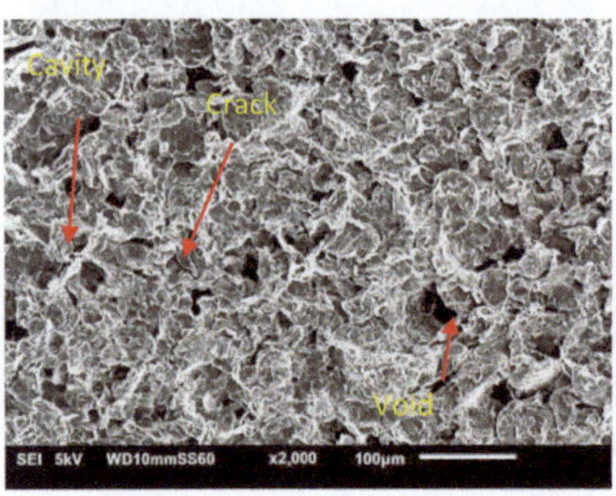

CB ink trace

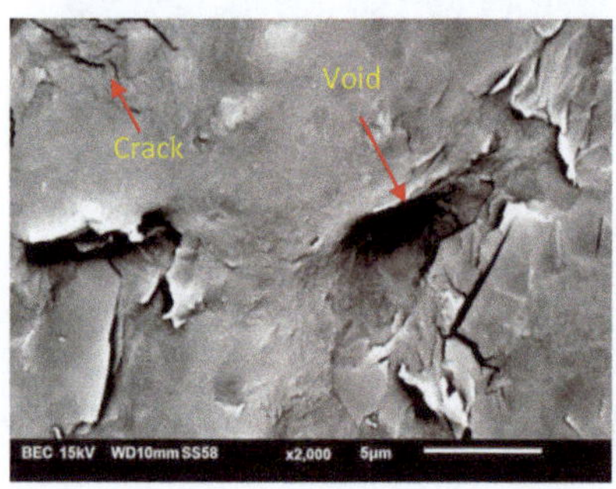

3 mm trace width

Table 8.4 Cross sectional images of zigzag pattern of GNPs and CB ink trace microstructure with line width of 1, 2 and 3 mm for cyclic tensile

GNPs ink trace

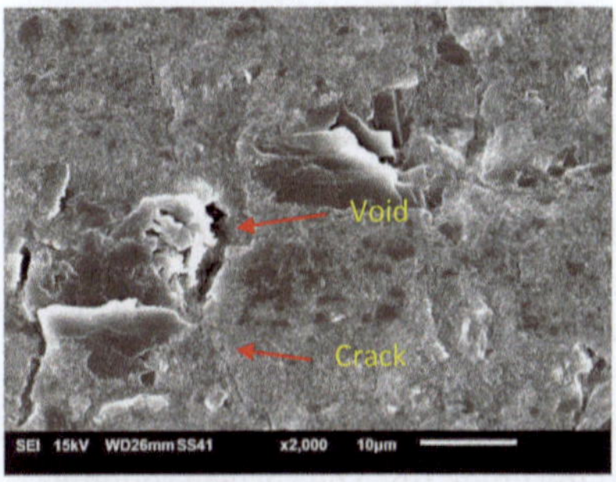

CB ink trace

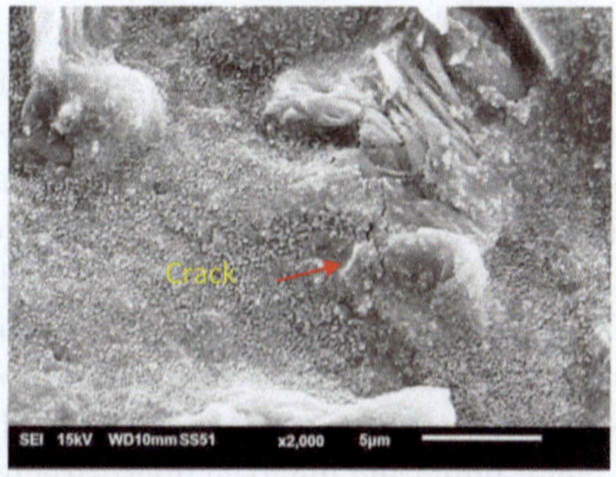

1 mm trace width

GNPs ink trace

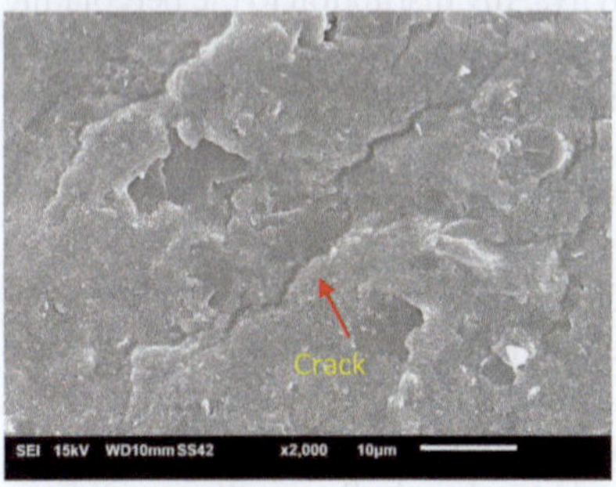

CB ink trace

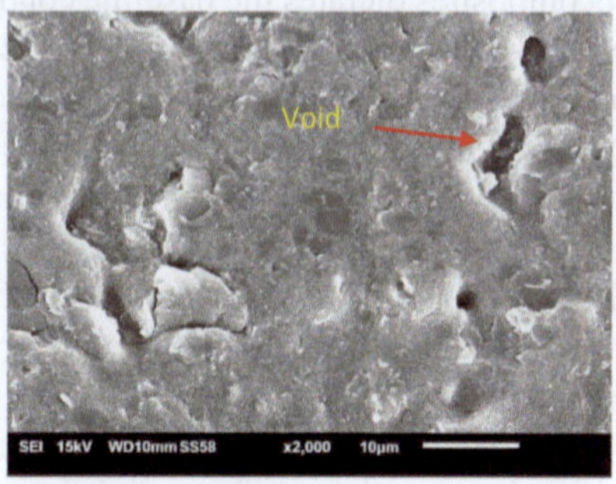

2 mm trace width

GNPs ink trace

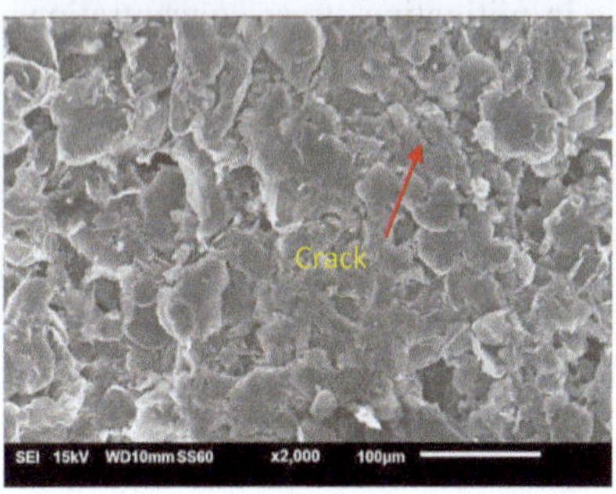

CB ink trace

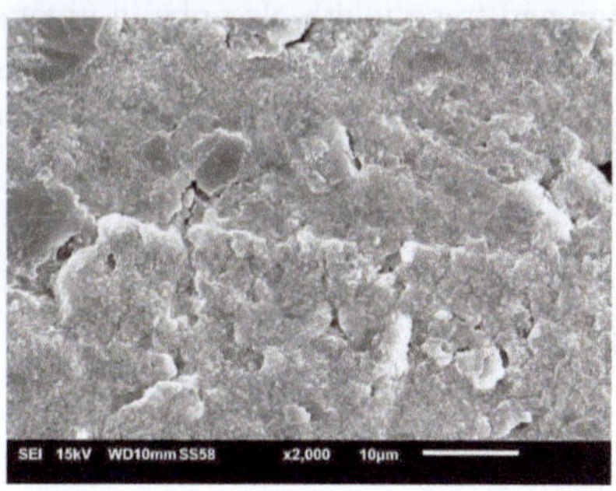

3 mm trace width

Voids is defined as small, empty spaces or unfilled resin pockets, and it can have a substantial impact on both the electrical and mechanical integrity of printed traces. They typically arise from incomplete wetting of the conductive filler by the polymer matrix during the manufacturing process, as noted by Appinc.co (2017). Voids and micro-cracks act as stress concentrators under tensile loading, promoting further crack propagation, filler pull-out and ultimately, failure of the composite. The presence of such defects decreases the effective cross-sectional area capable of carrying current, reducing conductivity, and also acts as nucleation sites for fatigue cracks during cyclic loading [14]. Progressive growth of these cracks and voids eventually compromises the overall performance, reducing the device's lifespan and reliability.

From a mechanical perspective, crack propagation and microstructural rearrangement under cyclic loading can be described by fracture mechanics principles. The rate of crack growth per cycle, governed by Paris' law, depends on the stress intensity factor range and the local microstructure at the crack tip. The surface roughness, filler orientation and particle–matrix adhesion all significantly influence the stress intensity and crack closure behaviour. Physically, the percolation threshold of these nanocomposites dictates the baseline conductivity. Any disruption to the percolating filler network due to cyclic loading will cause a nonlinear decrease in conductivity as fewer conductive pathways remain intact. Engineering this morphology to reduce stress concentrations. For instance, by choosing optimal trace width, employing smoother patterns and enhancing filler–matrix adhesion through coupling agents, it can significantly enhance the fatigue resistance and performance stability.

Therefore, the SEM imaging reveals that GNPs-based inks tend to exhibit more significant microstructural damage under cyclic tension due to their planar, high-aspect-ratio fillers and corresponding crack sensitivity at filler–matrix interfaces. By contrast, CB inks benefit from the spherical morphology and better filler dispersion, resulting in smoother surfaces and fewer cracks under similar cyclic conditions. Understanding these microstructural differences is critical for designing more robust, durable conductive traces for flexible and stretchable electronics, especially as pattern design and trace width also play a pivotal role in mitigating stress concentrations. Reducing void formation and ensuring strong filler–matrix adhesion remain key engineering challenges to improve both electrical and mechanical longevity of these nanocomposite conductive inks.

8.8 The EDX Analysis

The Energy-Dispersive X-ray spectroscopy (EDX) is an elemental analysis technique that can be employed to investigate the compositional stability of nanocomposite conductive inks exposure to high-cycle cyclic tensile loading. Understanding how mechanical fatigue influences the elemental makeup of these inks provides important insight into degradation mechanisms and the durability of the printed traces. Specifically, EDX analysis yields semi-quantitative data on elemental composition (typically expressed as weight percent of constituent elements), allowing the work of subtle variations in elemental ratios that result from mechanical deformation, micro-crack formation, interfacial debonding, and surface oxidation.

Table 8.5 summarises the elemental composition of the ink traces as a function of pattern geometry. The primary elements detected are carbon, originating from the conductive and oxygen, derived from functional groups, binder material, surface oxides and atmospheric penetration. Following cyclic tensile loading, the GNPs ink traces exhibit a carbon content ranging from 51.4 to 59.8%, with the curve trace pattern retaining the highest relative proportion of carbon. Conversely, the zigzag pattern exhibits a significant increase in oxygen content (up to 55.4%), resulting in the lowest measured carbon content. The pronounced increase in oxygen for the zigzag traces implies a greater degree of structural disruption under cyclic tensile loading, consistent with the observation that sharp directional changes in a zigzag pattern promote stress concentrations, surface cracking and eventual ingress of atmospheric species such as water vapour and oxygen.

The CB ink traces show a broader carbon-content variation, from approximately 30.8 to 69.2%, and an oxygen range between 30.8 and 58.5%. The square trace pattern contains the highest carbon percentage, possibly due to its more evenly distributed strain path and reduced accumulation of micro-damage under cycling. By contrast, the straight-line traces show the highest oxygen content, suggesting more surface exposure and degradation under repeated tensile strain. This correlates with the simplified strain path in straight-line traces, which concentrates tensile stresses along one axis, resulting in microstructural damage, void formation and increased oxygen uptake into the damaged network.

The observed decrease in carbon content and corresponding increase in oxygen content after cyclic deformation is symptomatic of progressive material degradation. Several key mechanisms can explain these findings which are: (i) particle fracture and debonding, which is repeated tensile stresses can lead to micro-fractures of the filler particles and partial detachment at the filler–matrix interface. Once the percolated network is disrupted, new surfaces become exposed to the surrounding environment. These freshly generated surfaces are highly reactive, allowing for rapid oxidation. Chemical interactions with moisture and atmospheric oxygen cause the formation of oxide species and functional groups, thereby reducing the relative proportion

Table 8.5 Carbon and oxygen weight percentage inside GNP and CB ink trace for each pattern measured from EDX for cyclic tensile

	GNPs ink trace						CB ink trace					
Trace pattern	Carbon weight percentage, %			Oxygen weight percentage, %			Carbon weight percentage, %			Oxygen weight percentage, %		
Width, mm	1	2	3	1	2	3	1	2	3	1	2	3
Straight line	51.6	44.1	41.5	48.4	55.9	58.5	44.8	44.5	43.1	55.2	55.5	56.9
Curve	59.8	55.4	51.2	40.2	44.6	48.8	69.2	60.9	58.7	30.8	39.1	41.3
Square	51.4	48.1	40.1	48.6	51.9	59.9	55.8	54.1	53.0	44.2	45.9	46.0
Zigzag	55.1	51.1	47.8	44.9	48.9	52.2	62.9	58.0	54.2	37.1	41.0	45.8

of elemental carbon in the trace, (ii) micro-cracks and void formation, which as cyclic tensile strain is applied, local stress concentration points can give rise to micro-cracks and voids. These structural discontinuities not only decrease the effective contact area between conductive fillers, reducing conductivity, but also provide direct pathways for oxygen ingress into the trace interior. Oxygen atoms diffusing into the micro-cracks can bond with available reactive sites on the carbon fillers and binder matrix, forming carbon–oxygen functional groups and increasing the measured oxygen content. The result is an elevated O:C ratio in highly damaged regions, (iii) surface oxidation and chain scission, which is the high-cycle mechanical deformation can also promote chain scission in any polymeric binder surrounding the fillers. This example exposes polymer end-groups that may react readily with oxygen, further increasing the oxygen content. The degradation of the organic binder can also decrease its ability to maintain filler–filler contacts, resulting in more pronounced electrical resistance increases and further microstructural deterioration under continued cycling, and (iv) the EDX elemental mapping and microstructural correlates, it is additional to pointwise EDX spectra, elemental mapping (as shown in Tables 8.6 and 8.7) provides spatial information on the carbon and oxygen distributions across the trace cross-section.

Table 8.6 The EDX mapping images for patterned GNP ink (curve and 3 mm) for cyclic tensile

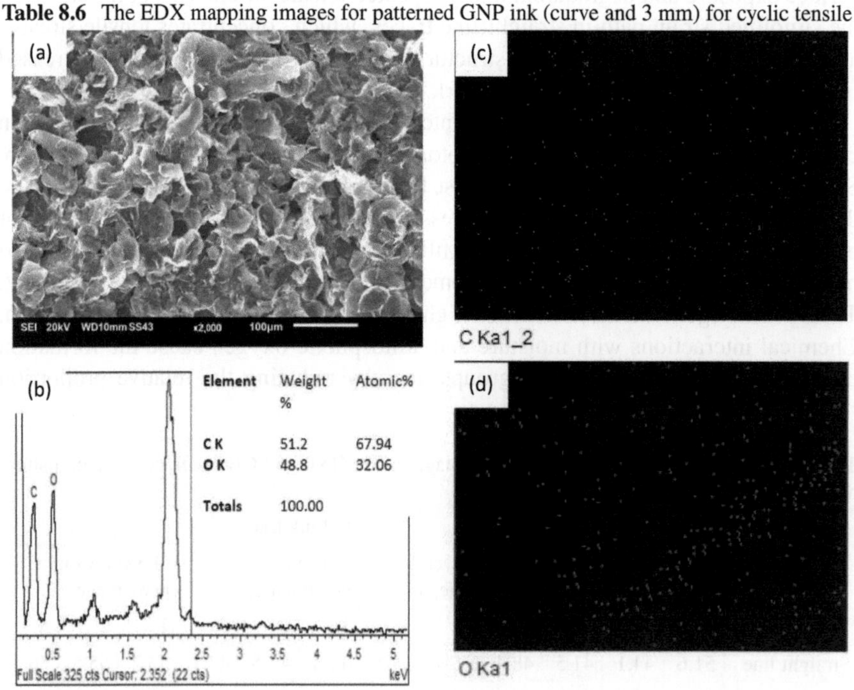

(a) SEM image, (b) EDX element graph, (c) EDX mapping image of carbon (GNPs), and (d) EDX mapping image of oxygen

Table 8.7 The EDX mapping images for patterned CB ink (curve and 3 mm) for cyclic tensile

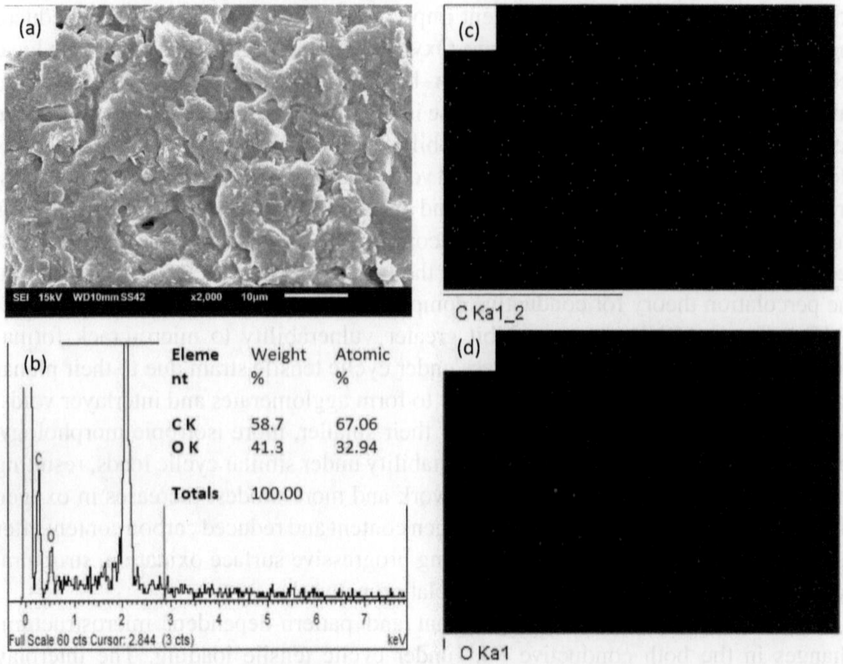

(a) SEM image, (b) EDX element graph, (c) EDX mapping image of carbon (carbon black), and (d) EDX mapping image of oxygen

In the case of GNPs ink, elemental maps reveal an approximately 51.2% carbon and 48.8% oxygen distribution. The higher oxygen fraction is clearly localised around crack surfaces and voids, where disrupted the aggregates expose new surfaces to ambient oxygen. The SEM micrographs corroborate this observation, illustrating prominent micro-cracks and interlayer delamination in ink stacks. The preferential formation of oxide-rich regions near these damaged areas implies that fracture surfaces serve as nucleation sites for oxidation and other degradation reactions, consistent with fracture mechanics and diffusion-limited reaction models.

The CB ink maps show a slightly different picture, with a carbon content of 58.7% and oxygen content of 41.3%. Despite the smoother and more uniform microstructure of the traces compared to GNPs traces, voids and occasional fissures are still observed. The smaller average size and more isotropic shape of the nanoparticles contribute to a denser packing and more uniform percolation network. This more homogeneous filler architecture better distributes mechanical stress, leading to less severe cracking and fewer localised high-oxygen regions. However, small isolated voids persist, acting as preferential sites for oxygen accumulation. Even though major cracks are absent, these voids and minor surface asperities can reduce localised conductivity due to oxidative reactions and disruption of the conducting network.

The engineering consequences of these findings are substantial. Elevated oxygen content and decreased carbon content imply progressive oxidation, which reduces the intrinsic conductivity of the traces. Oxygenated functional groups increase local resistivity at grain boundaries and at filler–binder interfaces. Over sustained cycling, this can manifest as a substantial decrease in the overall percolation efficiency of the network, a reduction in charge carrier mobility and an increase in Joule heating under electrical loading. Furthermore, localised voids and cracks act as stress concentrators, promoting further crack propagation and mechanical instability. From a physics fundamental, as percolation pathways become more tortuous and the effective cross-section for electron transport decreases, the trace resistance increases according to the percolation theory for conductive composites.

Therefore, a GNPs traces exhibit greater vulnerability to micro-crack formation, delamination and oxygen ingress under cyclic tensile strain due to their planar nanoplatelet morphology and propensity to form agglomerates and interlayer voids. Besides, for CB traces, by advantage of their smaller, more isotropic morphology, exhibit relatively better microstructural stability under similar cyclic loads, resulting in less disruption to the conductive network and more modest increases in oxygen content. Both inks display increased oxygen content and reduced carbon content after sustained cyclic tensile loading, indicating progressive surface oxidation, structural damage and disruption of the filler percolation network.

The EDX analysis reveals significant and pattern-dependent microstructural changes in the both conductive inks under cyclic tensile loading. The interplay between mechanical deformation, micro-crack nucleation, filler debonding and surface oxidation profoundly affects the electrical and mechanical integrity of these printed inks. Understanding these degradation mechanisms will inform the design of more resilient ink formulations and trace geometries for flexible and stretchable electronics that must endure cyclic mechanical stresses without compromising their electrical performance.

8.9 Summary

The conductive ink traces that had already been subjected to a static tensile test were further investigated under cyclic loading to evaluate their durability and fatigue life. Following this, a comprehensive characterisation was undertaken, addressing the full spectrum of electrical, mechanical and morphological properties. Concurrently, their operational performance was continuously assessed by monitoring the in-situ bulk resistance under cyclic deformation. This allows for the real-time evaluation of degradation phenomena that might not be evident through in-situ tests alone. A cyclic tensile test was then conducted on the ink traces across different patterns and materials, and the findings pertaining to the electrical and mechanical properties are synthesised in Table 8.8. Importantly, this table underscores the results for the curve-pattern ink trace, which exhibited the most favourable and stable performance across all test parameters.

Table 8.8 Summary of electrical and mechanical characterisation of cyclic tensile load test for patterned GNP and CB (curve pattern)

		Width, mm		
		1	2	3
GNP	Volume resistivity, Ω cm	6.628	14.743	17.976
	Hardness, MPa	0.340	0.780	0.990
	Elastic modulus, GPa	0.001	0.002	0.003
CB	Volume resistivity, Ω cm	0.0079	0.0092	0.0101
	Hardness, MPa	0.330	0.470	0.640
	Elastic modulus, GPa	0.0094	0.015	0.015

Based on the cyclic loading data acquired, the key outcomes of the cyclic tensile tests can be summarised. The GNPs conductive ink traces failed at approximately 7000 cycles for the straight-line, square, and zigzag patterns. However, the curve pattern sustained up to 9000 cycles before electrical and mechanical failure was observed. In contrast, CB conductive ink traces exhibited exceptional resilience, with all pattern geometries (straight-line, square, zigzag, and curve) sustaining the full 10,000 cycles tensile load test without failure. Among all patterns tested, the curve pattern displayed the most favourable electrical and mechanical properties for both inks.

Furthermore, the CB conductive ink traces demonstrated superior performance relative to GNPs ink traces in terms of both electrical conductivity and mechanical durability across all patterns, with the most pronounced advantage observed in cyclic tests. From an engineering perspective, the differences in cyclic durability between the both ink traces can be attributed to variations in filler morphology, percolation pathways, stress distribution and energy dissipation under repeated loading.

The relatively larger lateral dimensions of GNPs and their planar geometry result in a more pronounced sensitivity to tensile stress. Under cyclic loading, micro-cracks tend to nucleate along the edges of the nanoplatelets, especially at sites of filler–matrix debonding and at stress concentrators introduced by sharp pattern geometries (for example at zigzag and straight-line). Progressive delamination of these nanoplatelets restricts further stress transfer across the trace and disrupts the percolated electron transport network. The observation that the curve pattern can endure approximately 2000 more cycles than other GNPs patterns indicates that smoother geometrical profiles distribute cyclic strains more evenly across the trace, mitigating localised stress concentrations and delaying the onset of crack propagation.

By contrast, CB particles have a much smaller mean particle size and are approximately spherical in shape, allowing them to pack more isotropically and establish a more uniform percolation network. The near-isotropic particle arrangement enables the ink trace to better accommodate cyclic strains without significant disruption to the conductive network. Even under high-cycle loading, this particle structure exhibits effective energy dissipation through micro-slippage and slight reorientation without catastrophic debonding. This results in markedly higher endurance, with all CB trace patterns withstanding the full 10,000 cycles.

The impact of trace geometry on fatigue life is also evident. Smooth, gradually varying patterns like the curve trace help disperse tensile stresses more homogeneously along the ink's length. Sharp directional changes, as found in zigzag and square traces, introduce localised areas of higher strain, leading to accelerated damage accumulation. This stress concentration effect is well-described by classical fracture mechanics principles, which is under repeated cyclic loading, cracks tend to initiate at these stress concentrators, propagate incrementally per cycle (as per Paris' law), and eventually coalesce into catastrophic fractures. The curve trace, by avoiding abrupt corners and sharp bends, significantly reduces this concentration, delaying crack initiation and extending the trace's fatigue life.

From a theory of physics, as cracks propagate and filler–matrix debonding intensifies, the continuity of the conductive path is progressively disrupted. This is reflected electrically as an increase in bulk resistance with cycle count. Real-time resistance measurements confirm that the onset of mechanical damage is closely coupled to percolation network breakdown, and hence a rapid rise in electrical resistivity is often a precursor to complete trace failure. This underscores the importance of optimising pattern shapes and filler dispersion to enhance the fatigue resistance of conductive inks in flexible electronics.

The investigations highlight the need for design strategies that maximise the trace lifespan under service-relevant cyclic mechanical conditions. In practical terms, using curve trace patterns combined with well-dispersed CB inks will ensure that the printed traces can reliably sustain large numbers of cycles without a significant loss of conductivity or mechanical integrity. Furthermore, control over trace width and ink deposition can further fine-tune this fatigue performance by ensuring uniform filler density and minimising localised stress concentrations.

References

1. Zhang, H., Wang, W., Shi, L., Zhang, Z.: Mechanical and electrical properties of graphene-based composites: a review. Compos. B Eng. **162**, 63–75 (2019)
2. Zhang, S., et al.: Fatigue behavior of graphene nanoplatelet/carbon black conductive ink on flexible substrate. J. Electron. Mater. **48**(7), 4341–4349 (2019)
3. Zhang, X., Li, L., Niu, L., Hu, H.: Graphene reinforced polymer composites: a review. J. Mater. Sci. Technol. **35**(12), 2831–2844 (2019)
4. Zhang, H., Li, Y., Zhao, Y., Li, G., Zhang, F.: Carbon black oxidized by air calcination for enhanced H_2O_2 generation and effective organics degradation. ACS Appl. Mater. Interfac. **11**(31), 27846–27853 (2019)
5. Ding, T., Wang, L., Wang, P.: Changes in electrical resistance of carbon-black-filled silicone rubber composite during compression. J. Polym. Sci., Part B: Polym. Phys. **45**(19), 2700–2706 (2007)
6. Dziaudin, A.F., Kamarolzaman, A.A., Omar, G., Razali, N., Fadzullah, S.H.S.M.: The effect of graphene nanoplatelets filler size on the electrical and mechanical fatigue properties of conductive epoxy composites. Int. J. Nanoelectr. Mater. **14**(4), 329–344 (2021)

7. Amini, M., Jahanshahi, M., Rahimpour, A.: Synthesis of novel thin film nanocomposite (TFN) forward osmosis membranes using functionalized multi-walled carbon nanotubes. J. Membrane Sci. **435**, 233–241 (2013)
8. Bhushan, B.: Depth-sensing nanoindentation measurement techniques and applications. Microsyst. Technol. **23**, 1595–1649 (2017)
9. Hibbeler, R.C.: Stress and deformation analysis. In: Doe, J. (ed.) Engineering mechanics: statics and dynamics, pp. 87–141. Pearson Education (2013)
10. Callister, W.D., Jr., Rethwisch, D.G.: Materials science and engineering: an introduction, 10th edn. Wiley, Hoboken, NJ (2018)
11. Stanciu, M.D., Draghicescu, H.T., Roşca, I.C.: Mechanical properties of GFRPs exposed to tensile, compression and tensile-tensile cyclic tests. Polymers **13**(6), 898 (2021)
12. Wang, G., Yu, D., Kelkar, A.D., Zhang, L.: Electrospun nanofiber: emerging reinforcing filler in polymer matrix composite materials. Prog. Polym. Sci. **75**, 73–107 (2017)
13. Kim, B.K., Kim, C.J., Cho, D.I.: Tailoring conductive-adhesive interconnects for stretchable electronics. IEEE Trans. Compon. Packag. Manufact. Technol. **7**(4), 625–632 (2017)
14. Elbishari, H., Silikas, N., Satterthwaite, J.: The effect of filler size on the presence of voids within resin composite. J. Int. Dent. Med. Res. **11**(2) (2018)

Chapter 9
Durability and Structural Integrity of Ink Traces Under Cyclic Torsional Loading

9.1 Introduction

Cyclic torsion, often referred to as cyclic twisting, denotes the repeated application of torque, that is, twisting moments to a material or structural component. This kind of loading can accumulate damage over time, ultimately resulting in torsional fatigue failure even when the applied torque is well below the material's monotonic torsional strength. The working principle of cyclic torsion involves imposing alternating shear stresses upon a material's cross-section as the structure experiences repeated twisting and untwisting motions.

Physically, torsional loading induces a state of pure shear in the material. During each cycle of torsion, one side of the specimen is subjected to a clockwise rotation, inducing a positive shear strain, which is subsequently reversed as the rotation is counter-clockwise in the next cycle. Under these fluctuating stress states, microstructural damage accumulates progressively, often beginning at intrinsic flaws or heterogeneities such as voids, inclusions, or grain boundaries, and that act as stress concentrators. These localised sites will experience higher-than-average stress intensities, encouraging dislocation movement and plastic slip on favoured crystallographic planes. The continuous reversals of stress cause cyclic plasticity and dislocation substructure formation, resulting in persistent slip bands and small surface extrusions and intrusions. These, in turn, can nucleate micro-cracks on the material surface.

Over successive cycles, these micro-cracks gradually propagate under mixed-mode loading conditions as the material is twisted back and forth. Crack growth during cyclic torsion is driven by the repeated shearing of the crack tip region, allowing cracks to extend incrementally during each load reversal. The rate of crack growth is influenced by key material parameters, including yield strength, shear modulus, toughness, and ductility as well as by the magnitude of the applied cyclic torque and the geometry of the specimen or component. Furthermore, environmental

M. A. Salim et al., *Nanotechnology in Conductive Inks*, Engineering Materials,
https://doi.org/10.1007/978-981-95-5708-0_9

factors such as temperature, humidity, and corrosive media can exacerbate damage accumulation, for instance through corrosion-fatigue mechanisms.

From a mechanical engineering field, torsional fatigue resistance is fundamentally determined by the material's microstructure and its capacity to accommodate plastic deformation under reversed shear stresses. Ductile metals with high shear ductility tend to endure more torsional cycles before crack initiation and propagation become critical. Conversely, brittle materials including some high-strength composites and ceramics exhibit much lower cyclic torsional endurance due to limited energy dissipation capacity and rapid crack growth once initiated.

The torsional fatigue life is commonly characterised by an S–N curve, relating the applied torsional stress amplitude to the number of cycles to failure. Experimental methods often involve cyclic torsion test machines that apply a controlled torque waveform typically sinusoidal or triangular at a specified frequency. Test parameters can include torsional mean stress, stress range, torque reversal rate, and cycle asymmetry, all of which may significantly influence fatigue life. Microstructural observations using Scanning Electron Microscopy (SEM) or other techniques allow detailed characterisation of crack initiation sites and growth trajectories under cyclic torsion, further enabling the correlation of micro-mechanisms to macroscopic performance. It is also important to note that torsional fatigue behaviour differs from axial or bending fatigue. In torsion, the shear stresses rotate continuously through the material cross-section and produce spiral fracture paths along directions of maximum principal tensile stress. The resulting fracture surfaces often exhibit a characteristic 45° helix aligned with the axis of the specimen, corresponding to the plane of maximum tensile stress under combined torsion and any superimposed axial loads.

The cyclic torsion imposes complex multiaxial stress states that gradually accumulate damage through repeated micro-plastic deformation, dislocation rearrangement and eventual crack nucleation and propagation. The number of cycles to failure is determined by the interplay of material properties which are yield strength, ductility, toughness, structural parameters such as cross-sectional geometry, surface finish, patterning, loading characteristics such as torque amplitude, frequency, mean stress, and environmental effects. A deep understanding of these factors is essential in engineering design, allowing for accurate prediction of service life and the development of more robust materials and geometries for torsional loaded components across a wide range of industries including automotive drive shafts, torsion springs, rotating machinery, and power transmission systems.

9.2 Bulk Resistance Characterisation

Figure 9.1 illustrates a detailed investigation into the evolution of bulk electrical resistance in patterned graphene nanoplateletes (GNPs) and carbon black (CB) conductive inks subjected to continuous cyclic torsional loading. Samples comprising a range of geometrical patterns which are straight-line, curve, zigzag, and square, and different trace widths were tested under a torsional fatigue setup, with real-time resistance monitoring. The cyclic torsion tests were carried out at a rate of 20 cycles per minute to mimic dynamic twisting deformations experienced by practical flexible electronic devices, such as wearable sensors or roll able displays. This section examines the interplay between trace pattern, width, torsional cycle number and the resulting variations in bulk electrical resistance.

The measured data clearly show that for all GNPs ink patterns, the bulk resistance rises progressively with the increasing number of torsion cycles. This is indicative of cumulative microstructural damage typically including the debonding of fillers from the polymeric binder, progressive breakage of inter-particle conductive contacts, and localised micro-cracking under repeated shear and rotational stresses. Initially, straight-line patterns present the lowest resistance due to their shortest and most direct conductive pathways between electrodes. However, as cycling progresses, these straight-line traces accumulate substantial localised strain at the narrowest gauge sections, leading to earlier micro-crack initiation. Beyond approximately 3000 cycles, their resistance exceeds that of the square patterns as cracks and delamination impede electron flow.

By contrast, the curve pattern exhibits lower initial resistance at intermediate trace widths (2 mm) and sustains this relatively low resistance longer than more angular designs. Smooth curves distribute the torsional strain more evenly along the trace length, reducing sharp stress gradients and delaying microstructural failure. Interestingly, zigzag patterns at 1 mm width initially outperform other shapes due to their shape path allowing more uniform strain accommodation at small scales. However, as cycle counts rise, these sharp-angle designs eventually concentrate torsional shear into cornered regions of the trace, expediting debonding and nanoparticle pull-out at the vertices.

The role of trace width is equally evident. Increasing trace width raises the effective cross-sectional area through which current passes, providing multiple parallel pathways that decrease baseline resistance. Wider traces also reduce current density and Joule heating per unit length, which can improve fatigue life. However, under torsion, wider traces can exhibit increased overall stiffness, leading to different stress distributions and potentially higher crack densities at the trace edges. Hence, optimal performance often lies in balancing trace width against pattern design to reduce local torsional stress concentrations.

The response of CB ink to cyclic torsion exhibits a more complex, staged behaviour. Initially, as torsional cycles accumulate up to approximately 3000–4000 cycles, bulk resistance rises across all ink trace patterns. This is attributed to elongation and debonding at the nanoscale, where cyclic shear loads stretch the polymer binder and reduce effective percolation density between CB agglomerates. Mechanical energy imparted by torsion breaks some of the fragile conductive pathways particularly those sustained by weak van der Waals interactions between particles which raises the overall resistance.

However, past this early degradation stage, a counterintuitive decrease in resistance is observed at cycle counts of 3000–4000 and further decreases up to around 5000 cycles. This can be explained by the progressive redistribution and breakage of agglomerates into finer it clusters under continued torsional shearing. Smaller

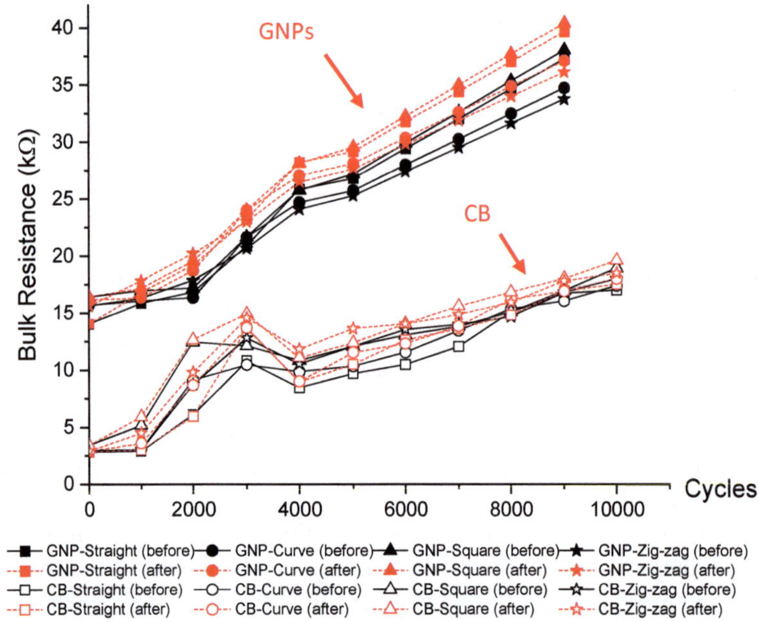

(a)

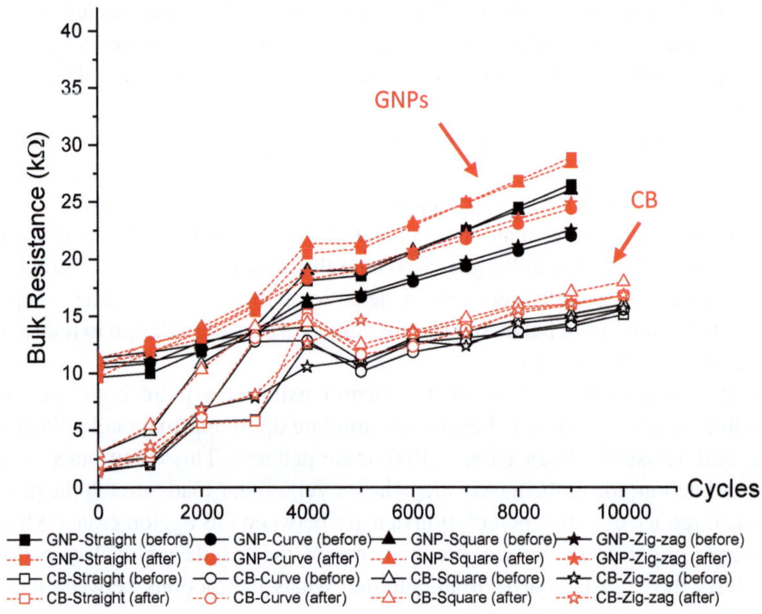

(b)

Fig. 9.1 Graph of bulk resistance value for GNPs and CB ink trace with **a** 1 mm width, **b** 2 mm width, and **c** 3 mm width for 10,000 times of cyclic torsion

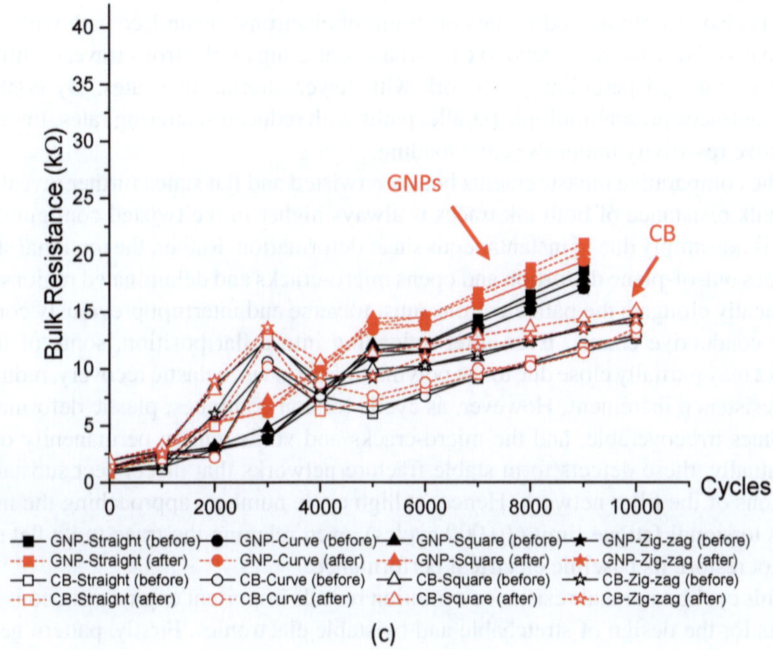

Fig. 9.1 (continued)

particle clusters can reorient, disperse, and fill localised voids created by cracks, creating new percolation routes that reconnect disrupted networks. This "restructuring" or "compaction" phenomenon is sometimes reported in nanocarbon-filled elastomers, where extended cycling allows the agglomerates to "reorganise" into denser, more isotropic configurations that favour electrical conductivity.

Beyond this self-healing behaviour, further torsional cycling eventually dominates by inducing permanent micro-cracks and persistent damage at the filler–binder interface. Once cracks reach a critical density, they cannot close under unloading, leading to irrecoverable increases in resistance up to the maximum of 10,000 cycles tested. This late-stage deterioration is consistent with classic low-cycle torsional fatigue models, where the plasticity of the binder is exhausted, allowing cracks to propagate unchecked.

Trace width plays a crucial role in governing electrical response under torsional deformation. The most significant resistance variations occur in the narrowest 1 mm traces, which present a higher initial resistance due to a limited cross-section and reduced percolating paths for charge transport. Constraining the trace width enhances current density, which in turn intensifies Joule heating and local mechanical strain, potentially promoting earlier micro-crack formation. Wider traces (2 and 3 mm), by contrast, offer lower initial resistance and more stable electrical properties under torsion. They support larger cross-sections and can distribute shear strain over a greater volume, mitigating localised filler debonding and improving strain tolerance.

This is also closely related to the constraint of electrons' mean free path within the ink matrix. Narrow traces tend to exacerbate scattering as electrons traverse through a more confined percolating network with fewer alternative routes. By contrast, broader traces present multiple parallel paths with reduced scattering rates, lowering effective resistivity under dynamic loading.

The comparative measurements between twisted and flat states further reveal that the bulk resistance of both ink traces is always higher in the twisted configuration. This is not simply due to instantaneous shear deformation. Rather, the torsional strain induces out-of-plane distortion and opens micro-cracks and delaminated regions that physically elongate the path electrons must traverse and interrupt previously continuous conductive chains. Even after unloading into a flat position, some of these cracks may partially close due to the polymer matrix's viscoelastic recovery, reducing the resistance increment. However, as cyclic torsion continues, plastic deformation becomes irrecoverable, and the micro-cracks and voids remain permanently open. Eventually, these defects form stable fracture networks that disconnect substantial portions of the filler network. Hence, at high cycle numbers approaching the material's torsional fatigue limit (10,000 cycles), even relaxing the trace to its flat state cannot restore its baseline electrical performance.

This cyclic torsional resistance evolution reveals important engineering considerations for the design of stretchable and twistable electronics. Firstly, pattern geometry plays an important role in mitigating damage localisation, smoother curves or spiral patterns can distribute stress more evenly and delay microstructural damage. Secondly, trace width must be carefully selected to balance electrical requirements and mechanical flexibility, which is overly narrow traces may offer finer resolution for electronics layouts but at the cost of enhanced sensitivity to torsional fatigue. Thirdly, material selection matters considerably, where CB inks show a degree of self-reorganisation under cyclic torsion that GNPs inks do not, demonstrating better long-term stability and suggesting that they may be better suited for highly dynamic or wearable applications. Therefore, this investigation underscores the intertwined relationships between cyclic torsional loading, microstructural evolution and percolation pathways in both inks. Resistance fluctuations over torsional cycling emerge as a complex interplay between stress-induced micro-crack initiation, filler agglomerate disruption and redistribution, strain concentration driven by trace geometry, and the inherent conductive properties of the fillers and binder. Understanding these mechanisms is critical for the design of robust, durable and electrically stable flexible electronic devices that can reliably operate under complex multi-axial strain conditions.

9.3 Sheet Resistance Behaviour

Sheet resistance is a critical parameter in assessing the electrical performance of thin conductive films and printed conductive traces. It reflects the lateral resistive properties of the material distributed over a defined surface area and is conventionally measured in ohms per square. Unlike bulk resistance, which depends on both length and cross-sectional area, sheet resistance normalises the trace thickness so that changes in resistive behaviour can be compared across different geometries. In the present investigation, the sheet resistance of both inks was evaluated under cyclic torsional loading to quantify the degradation in conductive networks across four trace patterns and each produced at three distinct trace widths. This investigation works is shown in Fig. 9.2.

As cyclic torsion was applied, all GNPs traces displayed a pronounced rise in sheet resistance with increasing cycle number. Straight-line traces of 1 mm width increased from approximately 60.54 kΩ/sq up to 1005.98 kΩ/sq by the end of cycling, reflecting severe disruption of the conductive percolation pathways under repetitive shear deformation. The curved and zigzag patterns displayed less extreme degradation rates under equivalent loading, owing to the more evenly distributed torsional strains mitigating extreme localised stress peaks. Meanwhile, square patterns exhibited intermediate resistance progression, where sharper corner features acted as stress concentrators and promoted earlier crack nucleation and propagation across the ink surface.

Physically, this progressive rise in sheet resistance can be attributed to micromechanical damage and progressive fracture at the interface between the flakes and the surrounding polymer binder. It relies on face-to-face stacking and van der Waals forces to establish electrically continuous pathways. Under repeated torsional loading, this delicate particle-particle contact is disturbed by micro-shearing action at the micro- and nanoscale. Hence, with each cycle, some conductive pathways are lost as the filler network is incrementally disrupted which an effect that is compounded by the inherently larger aspect ratios of the ink. Even small relative motions or rotational displacements between these flakes can cause a sharp decrease in tunnelling currents across their surface contacts, particularly where insufficient binder stiffness permits relative sliding and partial separation of these fillers.

Additionally, percolation theory suggests that the conductivity of particulate composites depends on the existence of a percolating network spanning the trace. Once mechanical damage exceeds a critical threshold, often termed the percolation threshold, the sheet resistance skyrockets due to rapid fragmentation of the conductive network into isolated clusters. Experimental observations of sharp rises between 3000 and 4000 cycles in the inks align with this threshold, where the number of intact filler contacts drops unexpectedly.

The tests clearly showed that trace width played an essential role in mitigating damage under torsional loads. Wider traces had comparatively lower absolute sheet resistance and also exhibited more stable resistance profiles over cycling. Physically, this is explained by the larger cross-sectional footprint of the trace, which enables

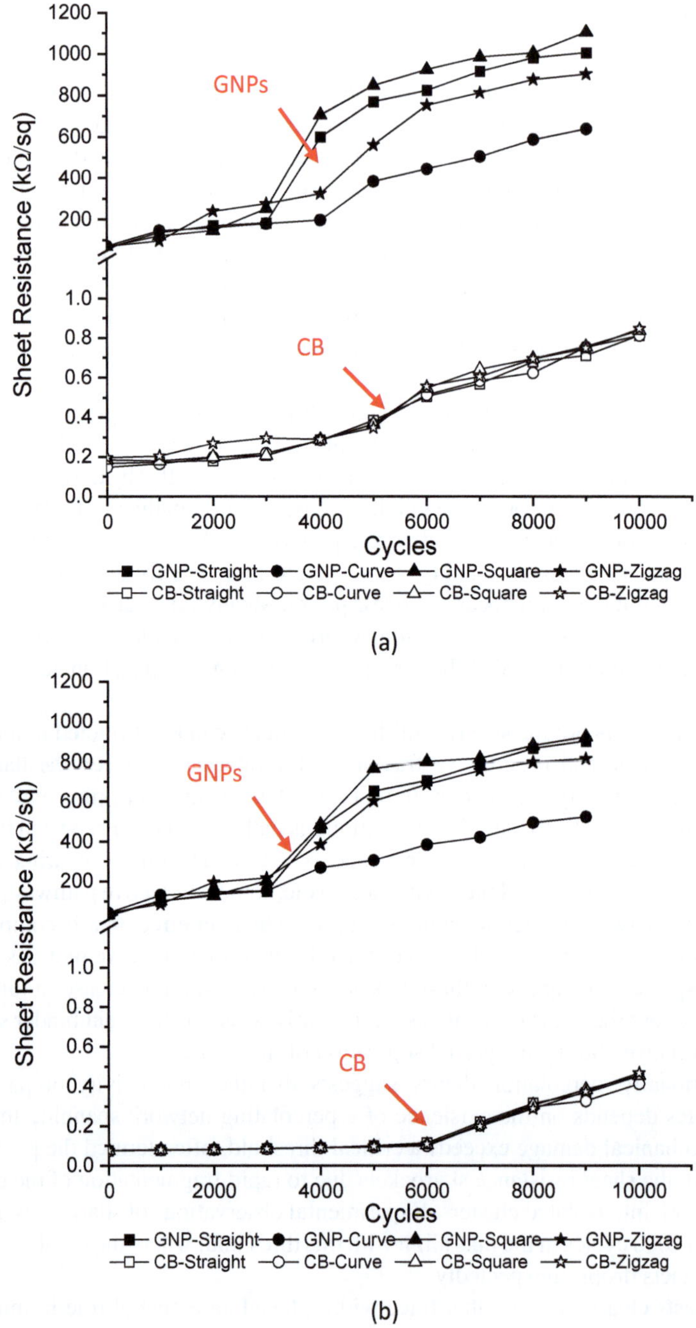

Fig. 9.2 Graph of sheet resistance value for GNPs and CB ink trace with **a** 1 mm width, **b** 2 mm width, and **c** 3 mm width for 10,000 times of cyclic torsion

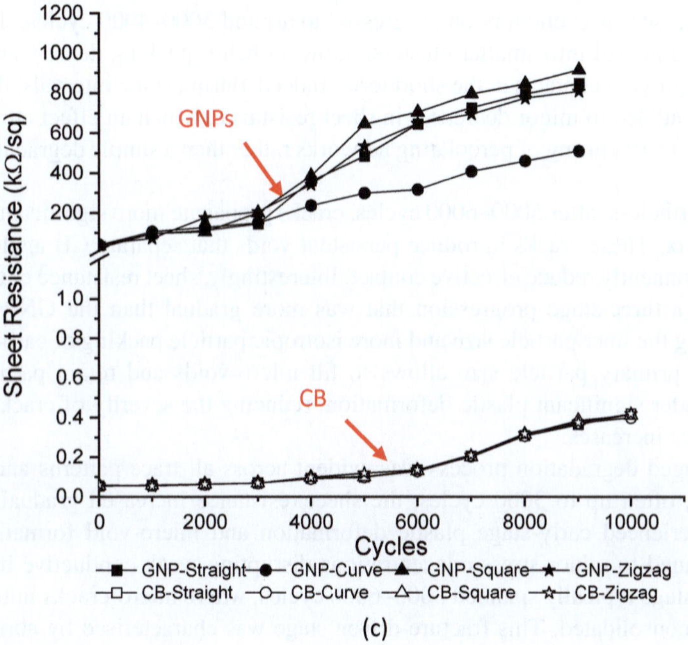

Fig. 9.2 (continued)

multiple parallel conductive pathways to accommodate torsional strains. The torsion-induced curvature in a wider trace is also distributed across a larger area, decreasing localised stresses and reducing the likelihood of early crack formation. Conversely, narrower traces concentrate torsional energy into a smaller conductive footprint, thereby increasing both stress intensity and current density. This drives faster contact degradation and greater electron scattering at filler–matrix and filler–filler interfaces, which all of which contribute to increased sheet resistance.

This width dependency can also be analysed in terms of Ohm's law at the microstructural level. Reductions in effective width cause the current-carrying pathways to funnel into fewer percolation routes, producing higher Joule heating per unit volume. Even under modest current densities, this heating can reduce polymer viscosity and exacerbate micro-crack propagation or particle agglomerate mobility, further destabilising the conductive network under torsional cycling.

In contrast, the CB ink traces exhibited a more stable, progressive rise in sheet resistance under identical torsional fatigue. The sheet resistance of the straight-line traces ranged from approximately 0.1655–0.813 kΩ/sq at 1 mm width and showed a similarly graduated progression across curved, square and zigzag shapes. The inks consist of agglomerated a nanoparticle that form conductive pathways via physical contact and conductive tunnelling. These agglomerates tend to redistribute under shear stresses, as cyclic torsion progressed to around 3000–4000 cycles, the aggregates fragmented into smaller clusters, allowing better packing density and potentially improving contacts in the short term. Indeed, during some intervals, this "rearrangement" led to minor decreases in sheet resistance, which an effect attributed to dynamic restructuring of percolating networks rather than a simple degradation-only model.

Nevertheless, after 5000–6000 cycles, cracks propagate more significantly across the matrix. These cracks introduce persistent voids that separate CB agglomerates and permanently reduce effective contact. Interestingly, sheet resistance of the traces showed a three-stage progression that was more gradual than the GNPs ink, by reflecting the finer particle size and more isotropic particle packing of carbon black. Smaller primary particle size allows to fill micro-voids and retain partial continuity under significant plastic deformation, reducing the severity of crack-induced resistance increases.

A staged degradation process was evident across all trace patterns and widths. Initially, often up to 3000 cycles, the sheet resistance increased gradually as the ink experienced early-stage plastic deformation and micro-void formation. This is attributed to minor stress concentrations disrupting weak conductive links. The second stage typically spanned 3000–6000 cycles, where micro-cracks initiated and rapidly consolidated. This fracture-driven stage was characterised by abrupt resistance jumps of up to 84% for square-pattern GNPs traces, and up to 68% for CB traces, where cracks had traversed substantial portions of the conductive paths. Finally, a third stage took place after approximately 6000 cycles for CB and 4000–5000 for GNPs, representing the terminal failure region. Resistance increases here were slower but continuous, owing to progressive fracture bridging that isolated successive segments of the ink traces.

From a materials engineering field, the observed behaviour underscores the sensitivity of printed inks to dynamic shear, especially at stress concentrators introduced by trace geometry which are the sharp corners in square and zigzag patterns. The CB inks displayed greater resilience attribute to their nanoscale particle size and more isotropic agglomerate shapes that distribute stresses more evenly through the matrix. In contrast, the planar GNPs flake have a directional mechanical response that promotes delamination and slip under repeated torsion. Moreover, the quality of the ink–substrate adhesion plays a critical role which as torsional deformation is imparted through the substrate interface, poor adhesion accelerates debonding and contact loss at the periphery of the traces.

These findings highlight that for applications where flexibility and cyclic torsional loading are expected such as foldable electronics, e-textiles, and biomedical wearable sensors for CB inks, particularly in smoother pattern geometries like curves, provide more stable long-term electrical performance. Meanwhile, pattern design and trace width must be carefully optimised to control localised torsional strains. Combining these strategies with advanced formulations, self-healing polymer binders, or functionalised fillers could greatly enhance torsional endurance and delay micro-crack evolution. Then, the sheet resistance behaviour under cyclic torsional fatigue is governed by multiscale interactions between pattern geometry, trace width, particle–matrix bonding, and microstructural percolation dynamics. Greater understanding of these phenomena supported by careful torsion testing, fracture surface analysis, and finite-element simulations can inform robust engineering design practices for next-generation flexible and stretchable electronic circuits.

9.4 Evaluation of Volume Resistivity

Figure 9.3 presents a comparative analysis of the volume resistivity of both conductive ink traces under cyclic torsional loading, as inferred from both their measured bulk and sheet resistances. Broadly, the graph reveals that GNPs ink traces experienced a progressive increase in volume resistivity as the number of torsional cycles increased, irrespective of the pattern design. Importantly, these increases exhibited pattern- and width-dependencies, underlining the strong interplay between mechanical deformation and the microstructural evolution of the percolating network under cyclic shear.

Physically, this can be explained by the fact that the GNPs ink is comprised of platelet-like nanosheets with high in-plane conductivity. Under torsional cycling, the shearing action promotes misalignment and debonding at particle–particle junctions, increasing electron scattering and reducing the number of continuous percolation pathways. Hence, volume resistivity grows more steeply once the torsional amplitude surpasses a critical value, typically at mid to high cycle numbers, where cracks propagate and the degree of contact between adjacent nanoplatelets is diminished. Moreover, the two fraction of the cycling process revealed distinct behaviours up to approximately 5000 cycles, the networks retain some continuity due to particle rearrangement that can accommodate minor twisting. Beyond this point, increasing plastic deformation induces delamination and rupture of the fragile filler–matrix interface, irreversibly severing conducting paths and allowing the resistivity to escalate.

Conversely, a CB ink traces displayed a different pattern of resistivity change. Initially, during low-cycle torsional testing, their volume resistivity increased modestly as agglomerate networks were gradually disturbed and isolated, similar to GNPs trace. However, after reaching approximately 3000–5000 cycles, the inks displayed a counterintuitive decrease in volume resistivity. This behaviour stems from the fundamentally different microstructural architecture of the particles. The aggregates form random 0D percolating networks within the polymeric binder, which can reorganise under mechanical strain. Cyclic torsional stress can promote localised realignment and agglomerate fragmentation, allowing new contact points to emerge and restoring a functional conductive path which a dynamic self-healing of the conductive network. This process persists until micro-cracks become dominant at very high cycle numbers (10,000 cycles), at which point the volume resistivity of a traces rises steadily as plastic deformation and micro-fractures impair the material's continuous percolating structure.

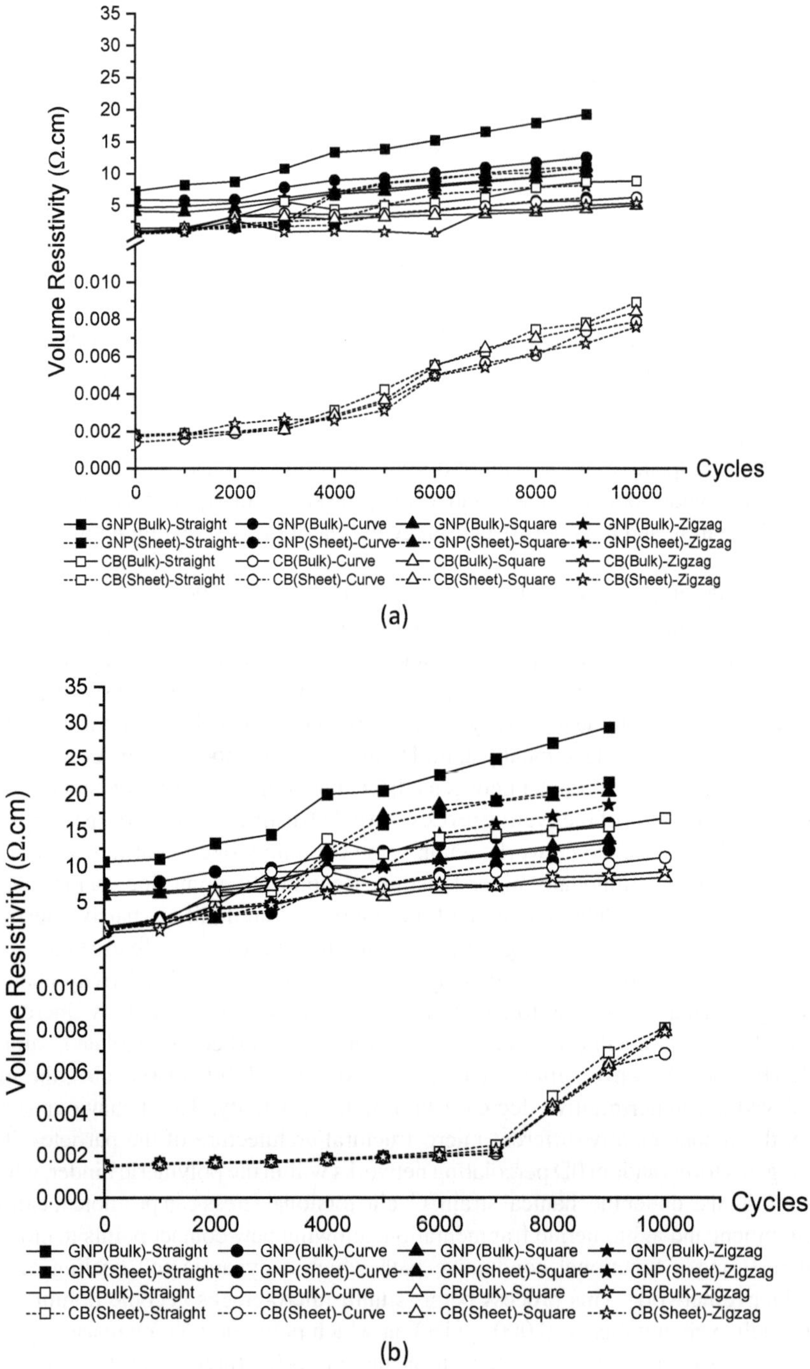

Fig. 9.3 Graph of volume resistivity for GNPs and CB ink trace with **a** 1 mm width, **b** 2 mm width, and **c** 3 mm width for 10,000 times of cyclic torsion

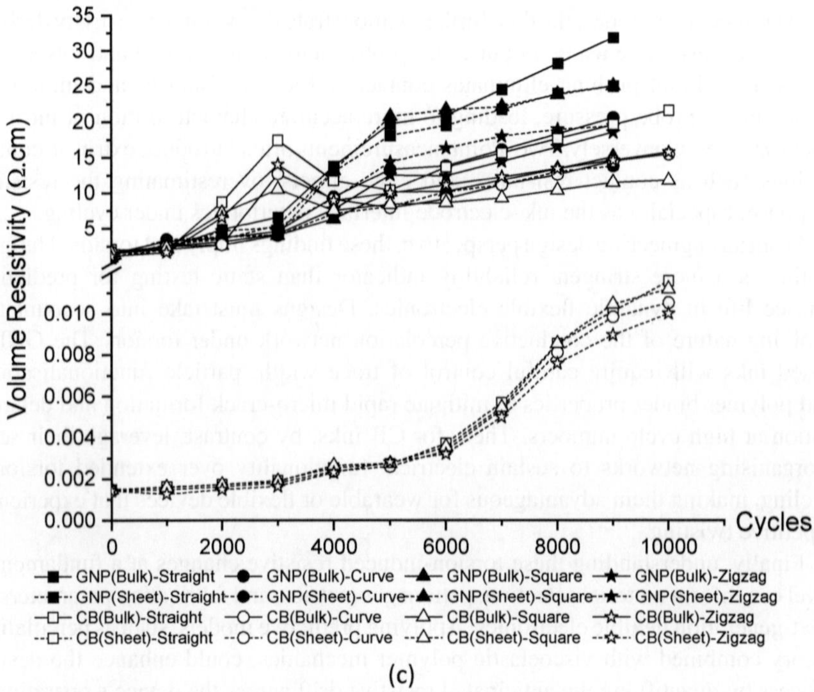

Fig. 9.3 (continued)

The width of the trace played a critical role across all test patterns. Thinner ink sustained fewer continuous pathways and exhibited sharper rises in volume resistivity as torsion-induced cracks readily percolated across the trace width. Thicker traces initially displayed lower absolute resistivity due to the greater cross-sectional area supporting more parallel electron pathways under minimal strain. However, after sustained cycling, these wider traces accumulated more distributed micro-cracks across their width, which led to substantial disruption of the percolating network. Thus, 3 mm traces which had initially the most robust conduction and showed greater relative increases in volume resistivity at high cycle numbers. This finding underscores the engineering challenge of scaling trace width under dynamic torsional stress, which a larger geometries provide more conductive pathways initially but also accumulate greater absolute mechanical damage.

The contrast between GNPs and CB inks also reflects the underlying mechanical properties of the fillers. GNPs, being planar and rigid, tend to fracture under repetitive shear, whereas CB particles are small, spheroidal, and dispersed in agglomerates that can dynamically rearrange under torsion to maintain a degree of conductivity. The polymeric binder also plays a role under torsion it accommodates strain along viscoelastic flow and polymer chain slippage, allowing fillers to relocate to some extent. The higher filler mobility in carbon black inks accounts for the transient recovery of conductivity after a mid-cycle decrease in resistivity. When comparing measurement techniques, the data further demonstrate that volume resistivity derived from sheet resistance using a Four-Point probe yielded more stable and lower estimates. Four-Point probing eliminates contact and lead resistances and minimises errors due to probe pressure, leading to more accurate characterisation of intrinsic ink properties. Conversely, Two-Point measurements often introduce extrinsic contributions such as contact resistance and edge effects, overestimating the resistive response, especially as the ink–electrode interface deteriorates under cycling.

From an engineering design perspective, these findings imply that torsional fatigue testing is a more stringent reliability indicator than static testing for predicting service life in dynamic flexible electronics. Designs must take into account the evolving nature of the conductive percolation network under torsion. The GNPs-based inks will require careful control of trace width, particle functionalisation, and polymer binder properties to mitigate rapid micro-crack formation and delamination at high cycle numbers. Then, for CB inks, by contrast, leverage their self-reorganising networks to sustain electrical functionality over extended torsional cycling, making them advantageous for wearable or flexible devices that experience repetitive twisting.

Finally, understanding these torsion-induced resistive changes at a fundamental level can inform material selection, pattern geometries, and fabrication parameters in next-generation flexible electronics. Applying predictive models, such as percolation theory combined with viscoelastic polymer mechanics, could enhance the design process by quantifying the anticipated resistive drift across the device's operational lifespan under multiaxial stresses.

9.5 Investigation Using Nanoindentation Method

Figure 9.4 illustrates a set of nanoindentation curves describing the variation of maximum indentation force versus maximum penetration depth for both conductive ink traces subjected to cyclic torsional loading up to 10,000 cycles. One of the primary observations is that cyclic torsional loading induces a less pronounced degradation in the mechanical response compared to other cyclic deformation modes, such as tension or bending. This is evident because all traces, regardless of filler composition or pattern width, withstood up to 10,000 torsional cycles without exhibiting abrupt or catastrophic failure. From a materials engineering perspective, torsional loading applies a distributed shear stress across the trace surface rather than a uniaxial stretch or compression. Unlike tensile or bending modes, torsional shear tends to favour energy dissipation through slight rotational movement of filler particles and polymer chains around one another, delaying the onset of micro-cracks that propagate under normal or tensile stresses.

Despite this resilience, indirectly but measurable changes in indentation force and penetration depth are evident. Across all ink traces, penetration depth increased progressively as the number of torsional cycles increased, and the corresponding indentation force required to reach the target indentation depth declined. These

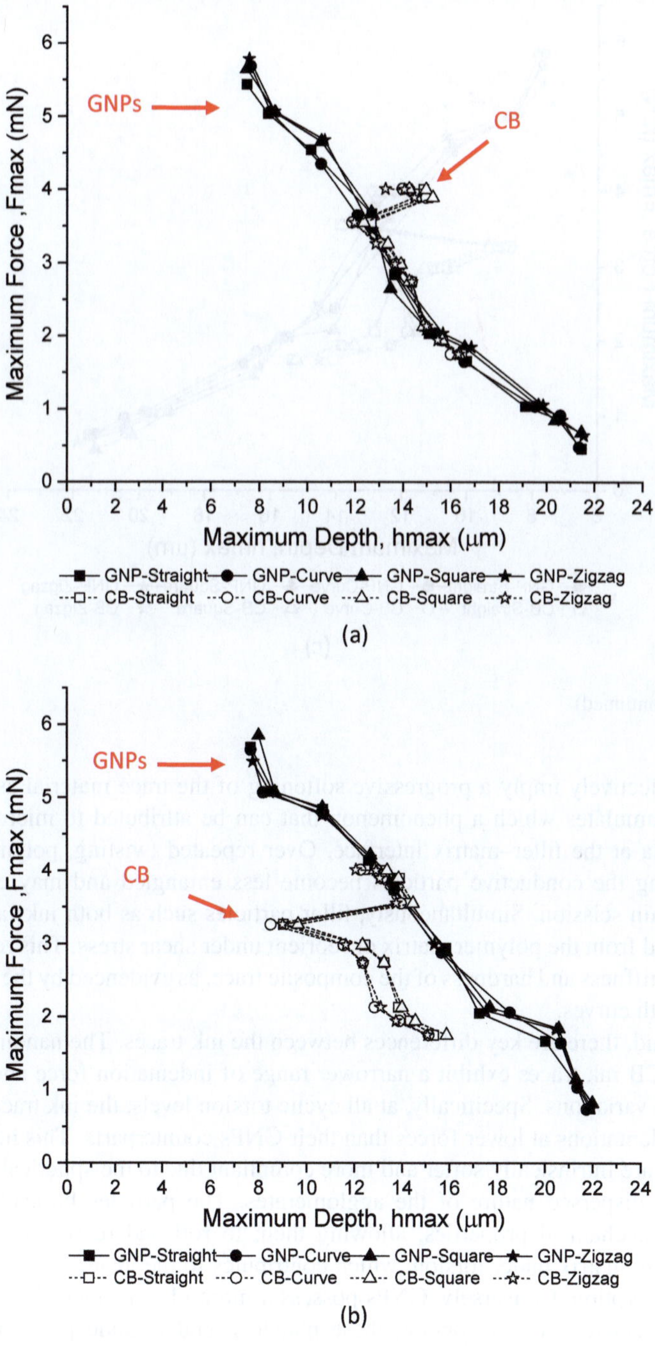

Fig. 9.4 Graph of maximum force versus maximum load of GNPs and CB ink trace with different pattern with **a** 1 mm, **b** 2 mm, and **c** 3 mm widths for cyclic torsion

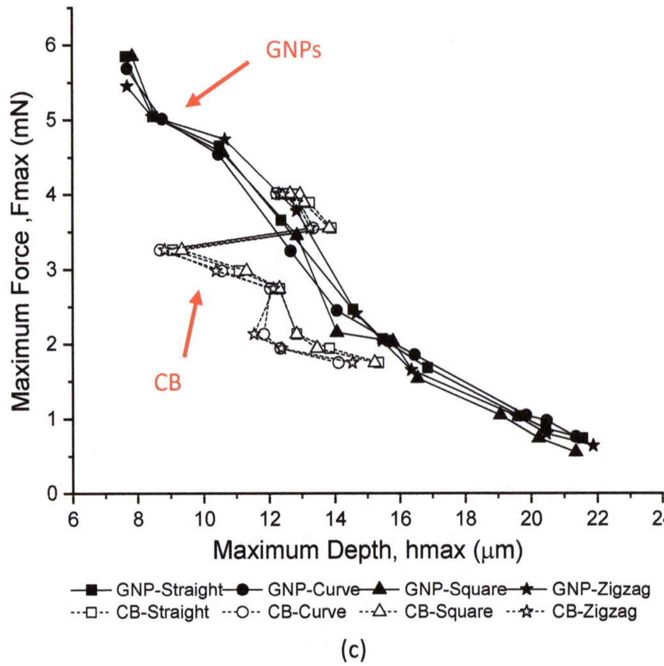

Fig. 9.4 (continued)

trends collectively imply a progressive softening of the trace material as torsional strain accumulates which a phenomenon that can be attributed to microstructural phenomena at the filler–matrix interface. Over repeated twisting, polymer chains surrounding the conductive particles become less entangled and may experience partial chain scission. Simultaneously, filler particles such as both inks aggregates can debond from the polymer matrix or reorient under shear stress. This reduces the effective stiffness and hardness of the composite trace, as evidenced by the shallower force–depth curves.

That said, there are key differences between the ink traces. The nanoindentation plots for CB ink traces exhibit a narrower range of indentation force and penetration depth variations. Specifically, at all cyclic torsion levels, the ink traces yielded deeper indentations at lower forces than their GNPs counterparts. This implies that ink traces are intrinsically softer and more compliant due to the spherical and more uniformly dispersed nature of the agglomerates. The particles basically possess isotropic mechanical properties, allowing them to roll and rearrange within the surrounding matrix under torsion, which contributes to the material's ductility and energy absorption. Conversely, GNPs possess a more planar, stacked morphology. When subjected to cyclic torsion, these platelets tend to slide past one another and can form stress concentrations at platelet edges. This anisotropic behaviour imparts greater initial stiffness to GNPs ink but also leads to a more abrupt stiffness degradation when inter-platelet bonding begins to fail.

The measured indentation force and penetration depth values further quantify these observations. For GNPs trace with widths of 1, 2, and 3 mm, the indentation force ranged between approximately 0.46–5.78, 0.74–5.68, and 0.55–5.85 mN, while penetration depth ranged between 7.47–21.55, 7.66–21.99, and 7.62–21.87 μm, respectively. These data show that trace width exerts only a modest influence on indentation response under torsional fatigue. This is partly because nanoindentation probes only the local microscale deformation response near the surface, which is governed mostly by the filler–matrix interactions rather than the global geometry.

For CB traces, the indentation profile evolved greatly with increasing cycles. Initially, at zero cycles, indentations were limited to approximately 12–15 μm under moderate load. However, at around 3000 cycles, coinciding with the emergence of matrix softening due to partial polymer yielding which is the indentation depth increased slightly (13–15 μm). This corresponds to a decrease in hardness as the polymer network slackens and the agglomerates become more mobile under torsional strain. Interestingly, beyond 3000 cycles, the trace hardened again up to approximately 5000 cycles, as evidenced by reduced indentation depth (8–11 μm). This counterintuitive re-hardening can be attributed to strain-induced densification, under sustained torsion, the aggregates and polymer chains may become more tightly packed due to mechanical settling, which temporarily restricts local deformation.

As torsional cycling continued past 5000 cycles up to 10,000 cycles, the trend reversed, with indentation depths increasing once more to approximately 14–16 μm. This softening phase reflects cumulative damage at the microscale which are micro-cracks coalesce, filler–matrix adhesion deteriorates, and plasticity dominates the response under further indenting. The increased plasticity and reduced elastic recovery manifest as increased for the same indentation force. Thus, nanoindentation reveals a complex interplay of cyclic shear-induced densification and progressive micro-damage in CB traces over the torsion cycle range.

From a mechanics-of-materials standpoint, these findings highlight the enhanced manner in which cyclic torsional deformation influences micromechanical properties of conductive inks. Engineering-wise, this behaviour underscores the need for carefully optimised filler size, shape, and dispersion to control the trade-off between flexibility, resilience, and durability of printed conductive paths. Future designs could leverage surface-modified GNPs or hybrid GNPs–CB formulations to enhance polymer–filler bonding, mitigating torsion-induced debonding and preserving nanoindentation hardness and elastic modulus over extended service life. Overall, nanoindentation after cyclic torsional loading provides an indispensable lens for probing mechanical degradation processes at the microscale, correlating trace hardness and modulus variations with the evolving microstructure under multi-axial loading. This insight enables material engineers to predict long-term reliability and devise strategies such as alternative ink formulations or print pattern geometries, to improve torsional fatigue life of flexible electronic devices.

9.6 Hardness and Elastic Modulus Degradation

Specifically, the repetitive torsional strain induces progressive disruption of particle–particle and particle–matrix interactions, resulting in micro-crack nucleation, debonding at filler–binder interfaces, and localised plasticity, all of which contribute to the observed mechanical softening. Figure 9.5 illustrates the evolution of hardness in both conductive ink traces subjected to sustained cyclic torsional loading up to 10,000 cycles, across different pattern shapes and trace widths. The data reveal a progressive decrease in hardness with increasing torsional cycles for all ink traces, which can be attributed to cumulative microstructural degradation and a reduction in the effective stiffness of the conductive pathways.

For GNPs ink traces across all tested shapes and widths, hardness decreased substantially from the first cycle to 10,000 cycles. The 1 mm width traces began with hardness values of 8.86–8.97 MPa and degraded to 0.43 until 0.57 MPa after 10,000 cycles. The corresponding decreases for 2 and 3 mm width traces followed similar trends, starting from 8.56 to 8.89 and 8.05 to 8.21 MPa, eventually declining to 0.64 until 0.68 MPa and 0.72–0.76 MPa, respectively. The rate of decrease was most pronounced for the narrowest traces owing to their limited cross-sectional stiffness and greater susceptibility to localised plastic deformation. Broadening the trace width enhanced the trace's overall moment of inertia, distributing torsional stresses more effectively across the conductive network and resulting in slightly higher residual hardness after cyclic torsion.

In contrast, the CB traces exhibited more advanced trends. Initially, the straight-line and square patterns at 1 mm width exhibited hardness values of approximately 0.46 and 0.43 MPa. Under low-cycle torsion (up to 2000 cycles), the hardness reduced modestly (0.38 and 0.39 MPa), followed by a transient increase between 2000 and 4000 cycles, reaching approximately 0.40–0.41 MPa. This divergent re-hardening can be explained by a strain-induced densification mechanism under repeated shearing, ink aggregates realign and compact within the polymer matrix, temporarily improving local filler packing and mechanical stiffness. Beyond 4000 cycles, however, sustained damage accumulation outweighs this densification effect, leading to a gradual softening as micro-cracks percolate across the trace.

Curve and zigzag traces followed a similar profile but consistently retained higher hardness than straight-line or square patterns at all cycle numbers. This is attributed to the distributed torsional strain present in smoother geometries. Curvilinear shapes avoid abrupt directional changes and thus exhibit lower stress concentrations, which delays bond rupture at filler–matrix interfaces and reduces micro-crack propagation rates. Even after 10,000 cycles, curve and zigzag ink traces sustained hardness values of approximately 0.46 until 0.50 MPa which a remarkable retention of mechanical integrity under extreme cyclic torsional loading.

This mechanical response underscores the profound influence of pattern geometry. Straight and square patterns introduce sharp directional transitions, which concentrate torsional shear into discrete regions and exacerbate localised debonding and plasticity. This manifests as an accelerated decrease in hardness. By contrast, curve

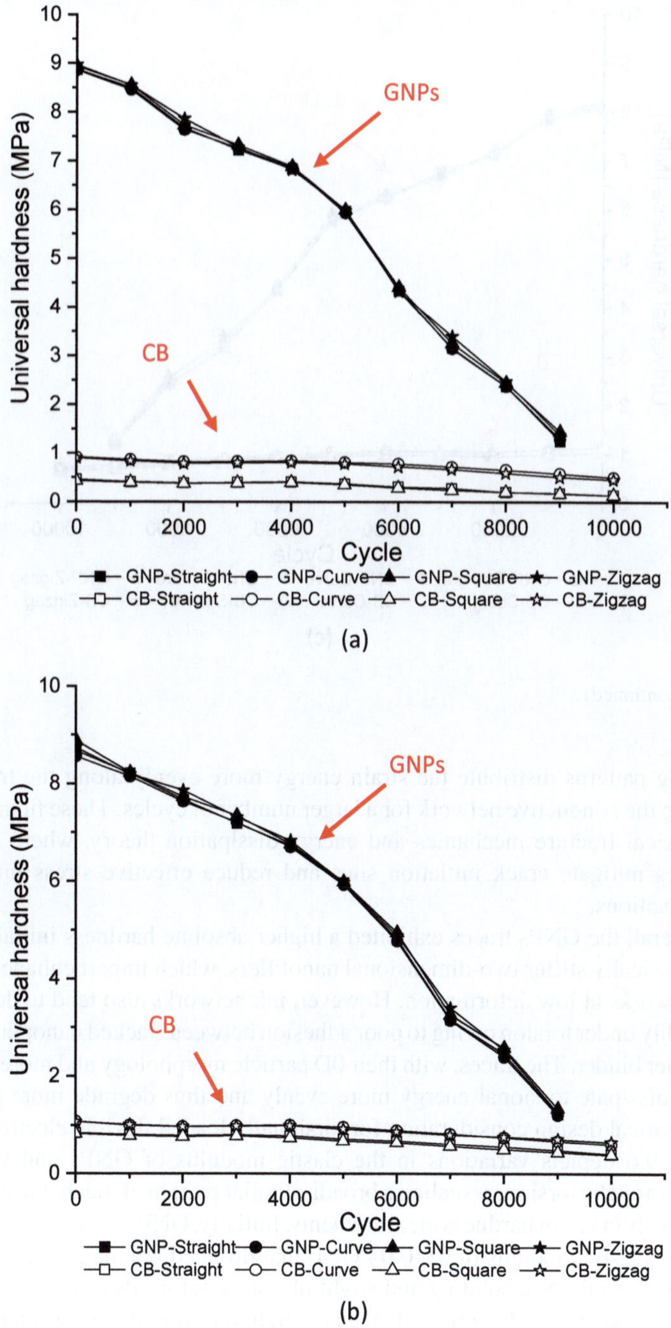

Fig. 9.5 Graph of hardness of GNPs and CB ink trace with different pattern with **a** 1 mm, **b** 2 mm, and **c** 3 mm widths for cyclic torsion

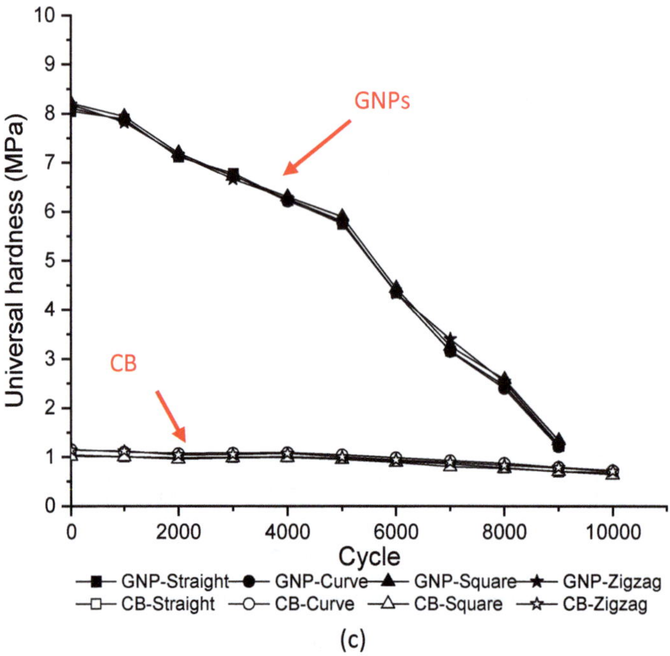

Fig. 9.5 (continued)

and zigzag patterns distribute the strain energy more evenly along the trace path, preserving the conductive network for a larger number of cycles. These findings align with classical fracture mechanics and energy dissipation theory, where smoother geometries mitigate crack initiation sites and reduce effective stress intensity at critical junctions.

In general, the GNPs traces exhibited a higher absolute hardness initially due to their intrinsically stiffer two-dimensional nanofillers, which impart enhanced percolating networks at low deformation. However, ink networks also tend to delaminate more readily under torsion owing to poor adhesion between stacked nanoplatelets and the polymer binder. The traces, with their 0D particle morphology and more isotropic network, dissipate torsional energy more evenly and thus degrade more gradually which a critical design consideration for torsionally loaded flexible electronics.

Figure 9.6 depicts variations in the elastic modulus of GNPs and CB traces subjected to cyclic torsion, revealing a broadly similar pattern of mechanical degradation to that observed in hardness measurements. Initially, GNPs traces possessed relatively high elastic modulus (0.03–0.04 GPa) that dropped gradually during the first 2000 cycles as micro-debonding and slight plasticity set in. Beyond this threshold, the modulus experienced a transient stage or slight recovery due to strain hardening phenomena as filler agglomerates reoriented. Past 4000 cycles, however, modulus degradation became severe as pervasive delamination and interface slippage overcame strain hardening effects, resulting in losses of up to 80% of the initial modulus after 10,000 cycles.

For CB traces, elastic modulus evolution was less pronounced due to the inherently compliant 0D particle matrix. Even at 10,000 cycles, modulus values for curved and zigzag ink traces remained substantially higher (0.45 until 0.50 GPa) than those for straight and square traces (0.01–0.02 GPa), reiterating the advantage of smoother geometries in distributing torsional loads. The traces also displayed a more uniform response across widths, indicating that the higher filler mobility of the particles can compensate for differences in trace cross-section by dynamically adapting the conductive network to torsional strain.

Overall, these findings demonstrate that both the material composition and the geometry of printed patterns strongly influence the torsional fatigue resistance of conductive ink traces. The GNPs ink offer superior stiffness and hardness at low cycle numbers but degrade faster under high-cycle torsion due to particle stacking and interface debonding. Then, for CB inks, despite lower initial stiffness, show greater resilience and capacity for structural recovery owing to dynamic particle rearrangement. Furthermore, pattern shapes that minimise torsional stress concentrations can significantly enhance mechanical endurance which a key consideration for the design of next-generation stretchable and twistable electronics.

9.7 The SEM Analysis of Stress Concentrations and Crack Formation

The ongoing investigation into nanocomposite-based conductive inks is driven by the promise of enhanced electrical, mechanical, and morphological properties when compared to traditional conductive materials. These enhancements originate from the intrinsic nanoscale phenomena governing the interactions between fillers and their polymeric binder matrix. By leveraging these properties successfully, a severe characterisation of the inks' microstructure is indispensable. The SEM serves as a particularly powerful diagnostic technique for this purpose, as it reveals micro- and nano-scale features such as filler dispersion, particle agglomeration, interfacial bonding, porosity, and crack propagation.

As shown in Tables 9.1, 9.2, 9.3, and 9.4, the SEM micrographs of straight-line, curved, square, and zigzag ink traces with different widths highlight great differences in surface topography and fracture behaviour following cyclic torsion. The analysis of these micrographs provides critical insights into the structure–property relationships that control the electromechanical performance of printed ink traces. Specifically, micrographs of GNPs ink traces reveal significant morphological degradation with repeated torsion loading. Crack formation occurs predominantly at the weakest points namely, along boundaries between ink agglomerates and the polymer binder and at

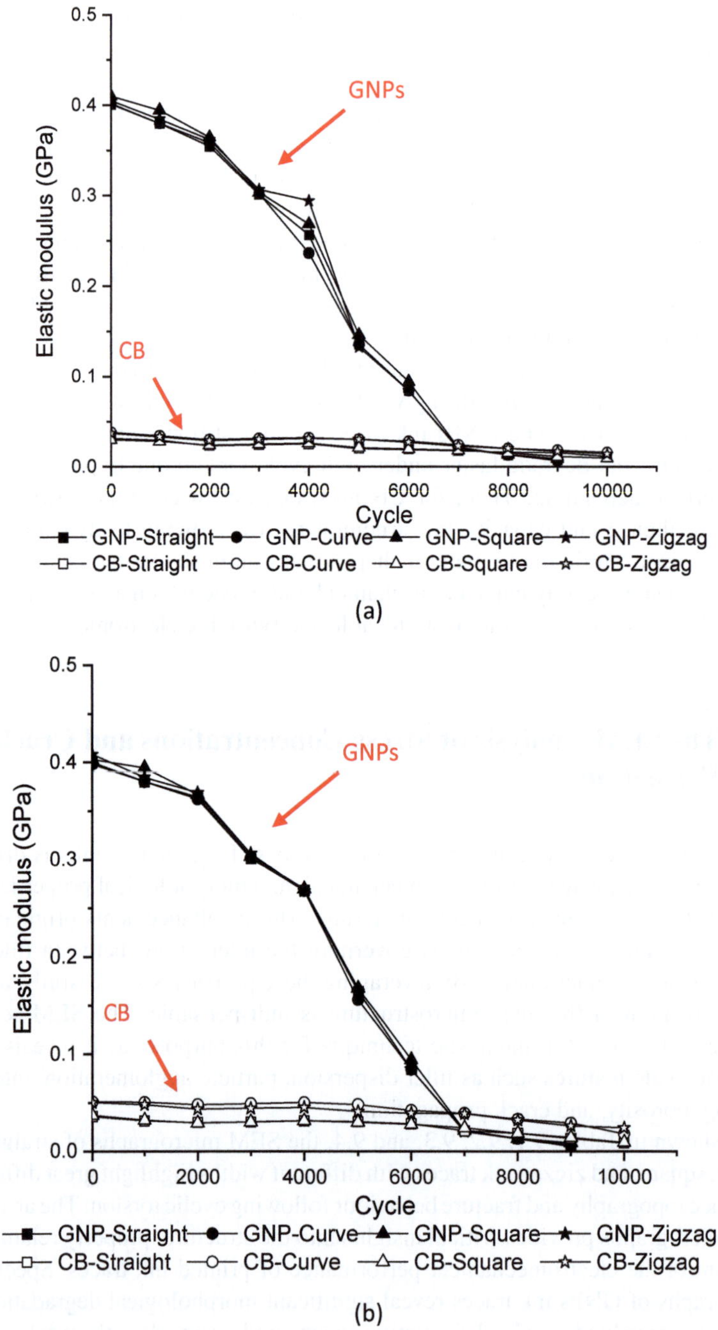

Fig. 9.6 Graph of elastic modulus of GNPs and CB ink trace with different pattern with **a** 1 mm, **b** 2 mm, and **c** 3 mm widths for cyclic torsion

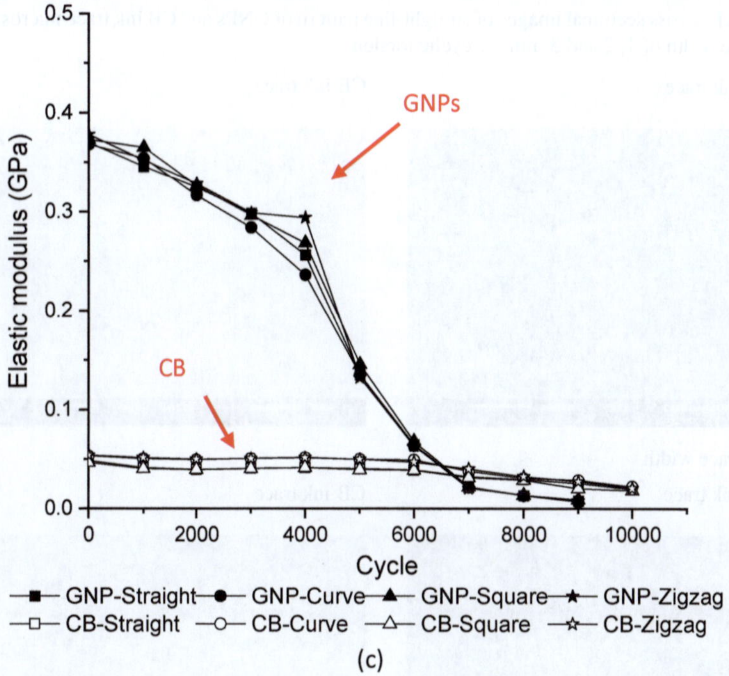

Fig. 9.6 (continued)

existing surface irregularities that act as nucleation sites for delamination and crack growth.

At the microstructural level, GNPs filler tend to organise into plate-like agglomerates due to van der Waals attraction and $\pi - \pi$ stacking interactions between sheets. The SEM images show that these fillers often exist as clustered domains surrounded by the softer polymer binder, leading to a heterogeneous microstructure. These agglomerated clusters introduce local stress concentrators during cyclic torsional loading. Under repeated twisting, the difference in stiffness between the relatively rigid an agglomerates and the surrounding polymer matrix generates interfacial shear stresses. This promotes progressive debonding at the filler–matrix interface, culminating in micro-cracks that propagate along these boundaries. The cracks eventually coalesce, reducing percolation pathways and contributing to an increased sheet and volume resistivity as discussed in earlier sections.

By contrast, the micrographs of CB ink traces exhibit a different morphological response. Ink nanoparticles, which are approximately spherical (0D) and distributed more isotropically within the matrix, lead to a denser and more uniform microstructure under torsional deformation. The SEM images of ink traces show fewer voids, smoother surfaces, and a notable absence of large cracks even after extensive torsional cycling. The inherently smaller particle size and more uniform dispersion of ink

Table 9.1 Cross sectional images of straight-line pattern of GNPs and CB ink trace microstructure with line width of 1, 2 and 3 mm for cyclic torsion

GNPs ink trace	CB ink trace
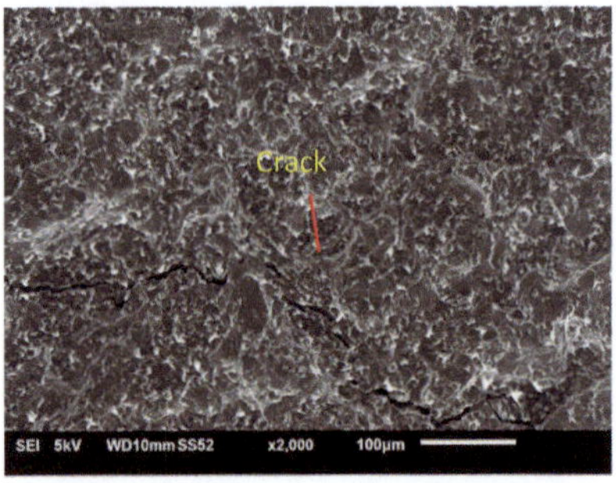	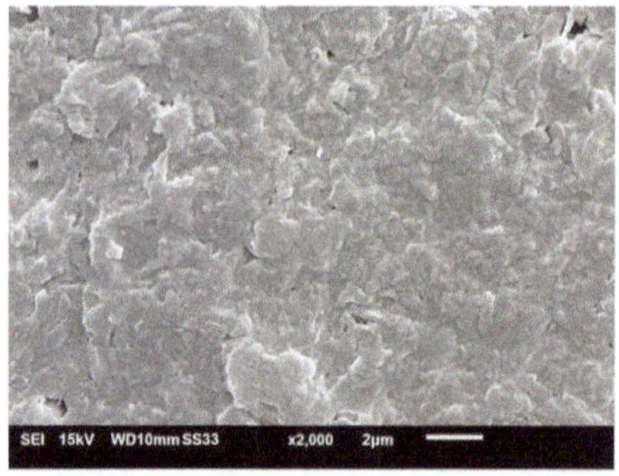
1 mm trace width	
GNPs ink trace	CB ink trace
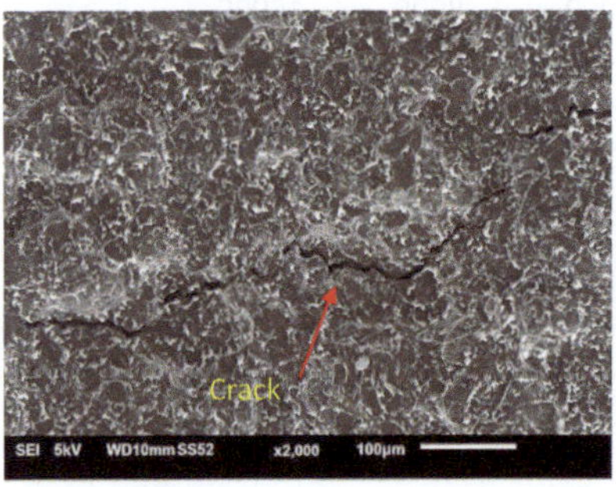	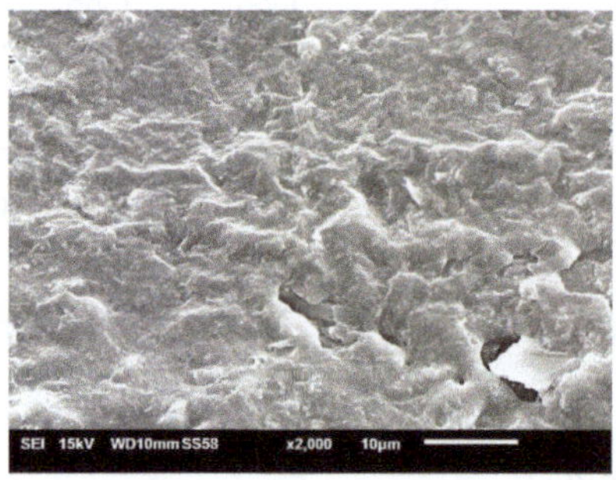
2 mm trace width	
GNPs ink trace	CB ink trace
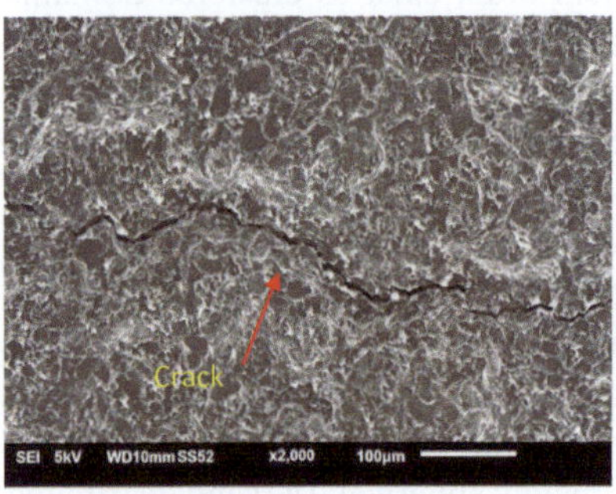	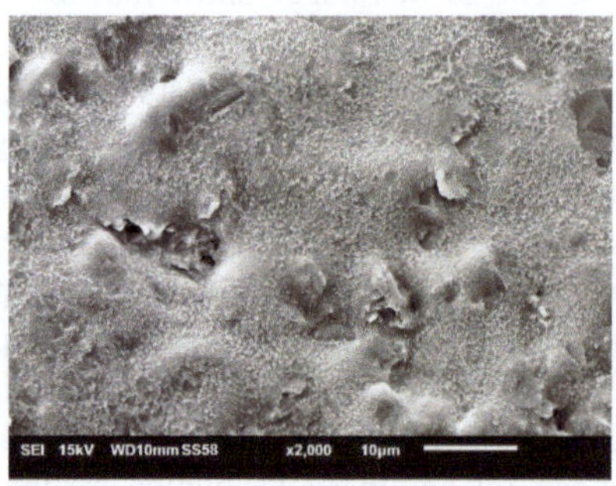
3 mm trace width	

Table 9.2 Cross sectional images of curve pattern of GNPs and CB ink trace microstructure with line width of 1, 2 and 3 mm for cyclic torsion

GNPs ink trace

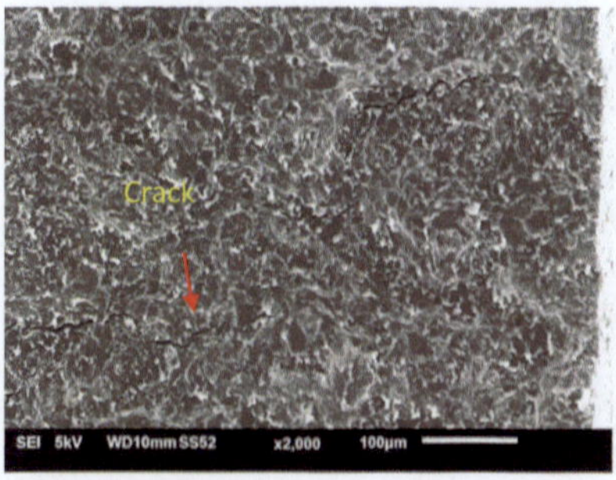

CB ink trace

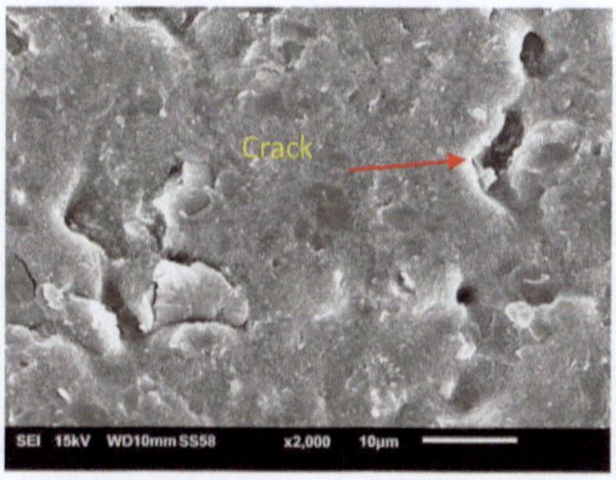

1 mm trace width

GNPs ink trace

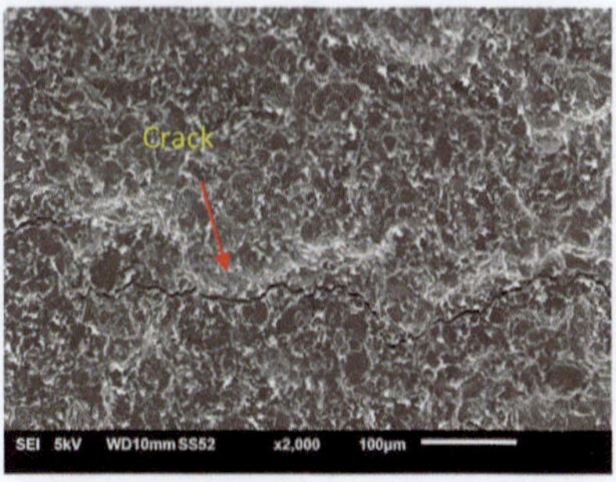

CB ink trace

2 mm trace width

GNPs ink trace

CB ink trace

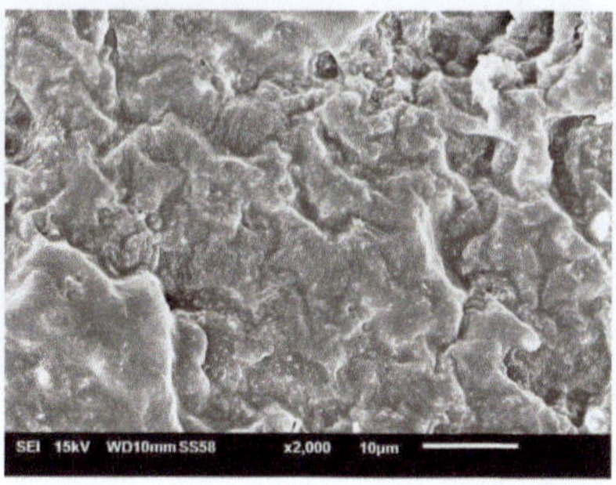

3 mm trace width

Table 9.3 Cross sectional images of square pattern of GNPs and CB ink trace microstructure with line width of 1, 2 and 3 mm for cyclic torsion

GNPs ink trace

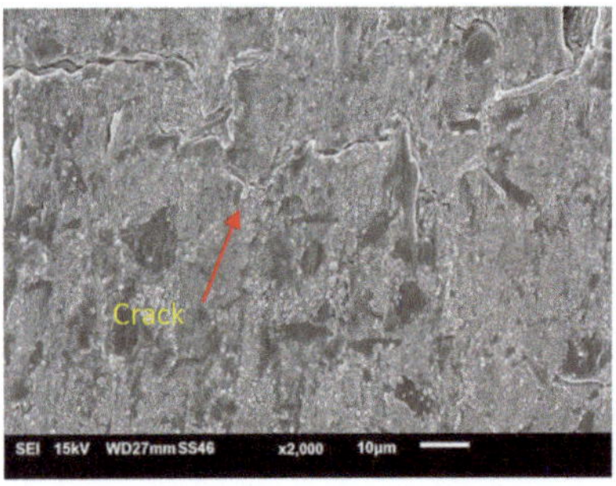

CB ink trace

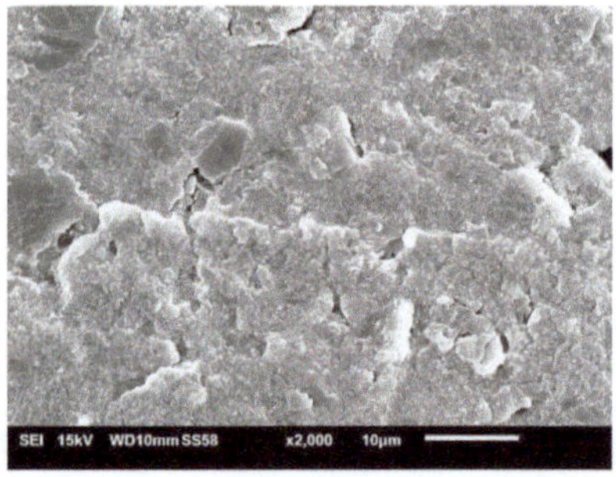

1 mm trace width

GNPs ink trace

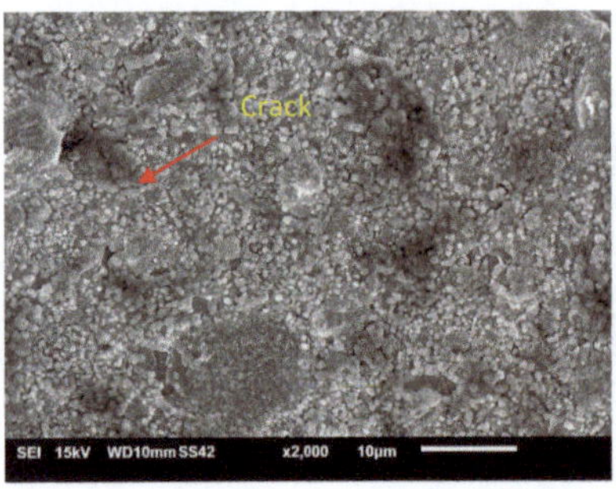

CB ink trace

2 mm trace width

GNPs ink trace

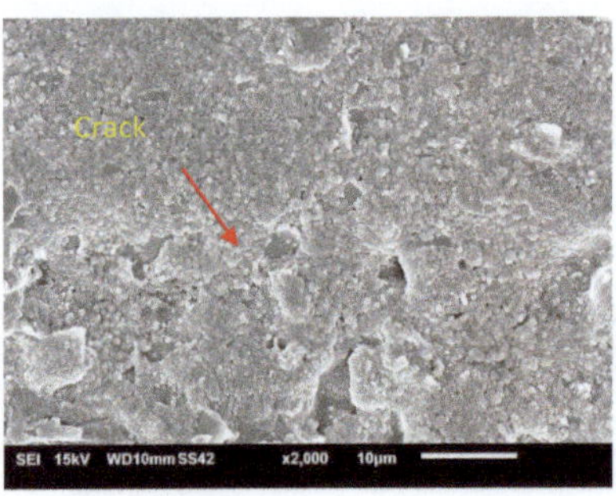

CB ink trace

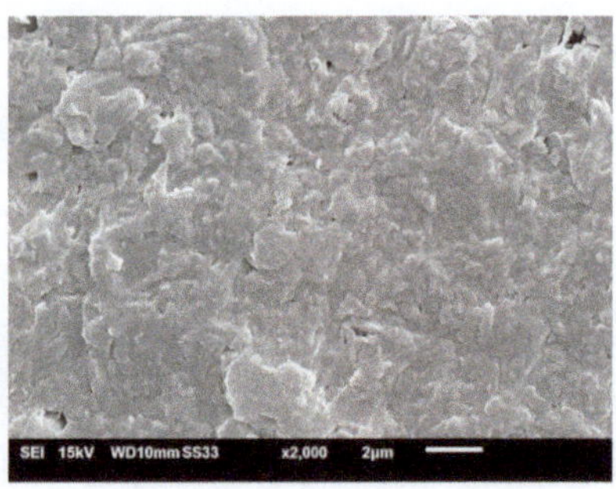

3 mm trace width

Table 9.4 Cross sectional images of zigzag pattern of GNPs and CB ink trace microstructure with line width of 1, 2 and 3 mm for cyclic torsion

GNPs ink trace

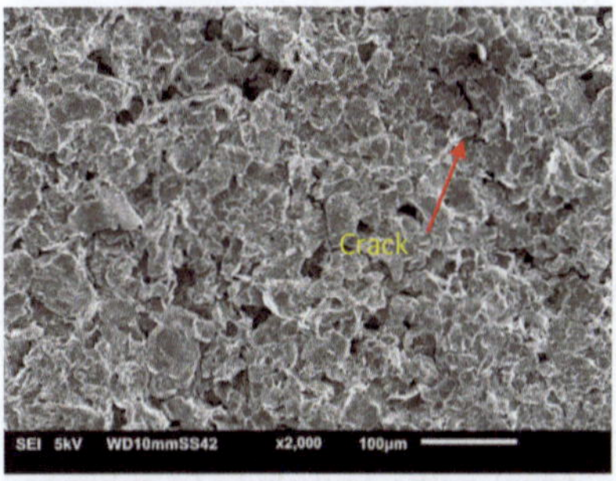

CB ink trace

1 mm trace width

GNPs ink trace

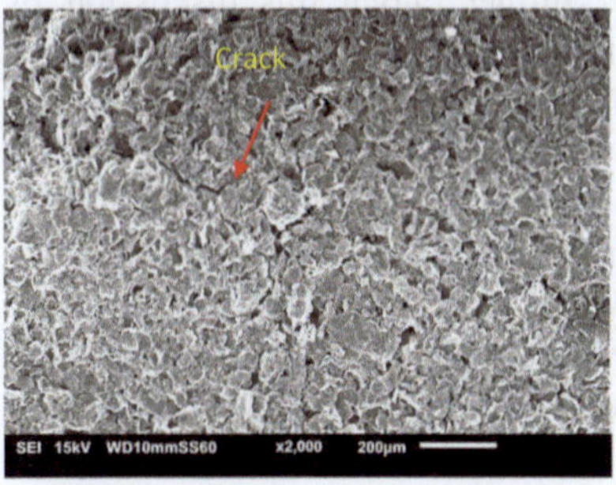

CB ink trace

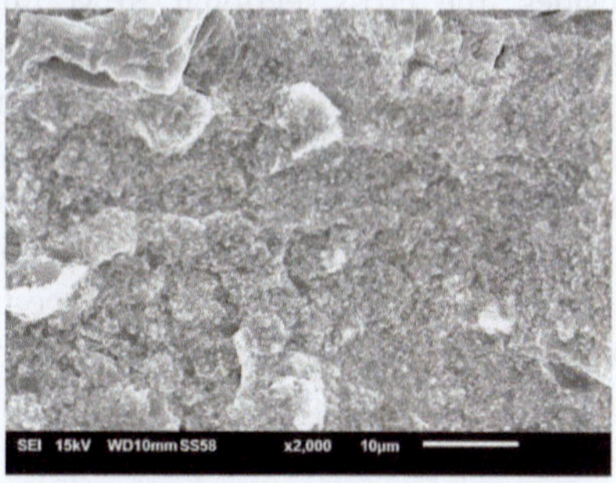

2 mm trace width

GNPs ink trace

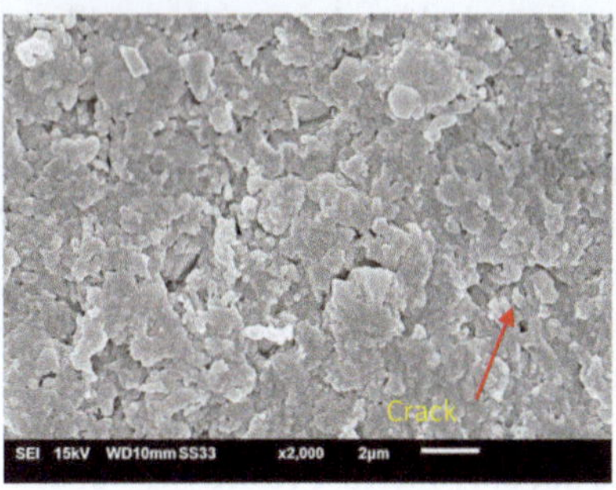

CB ink trace

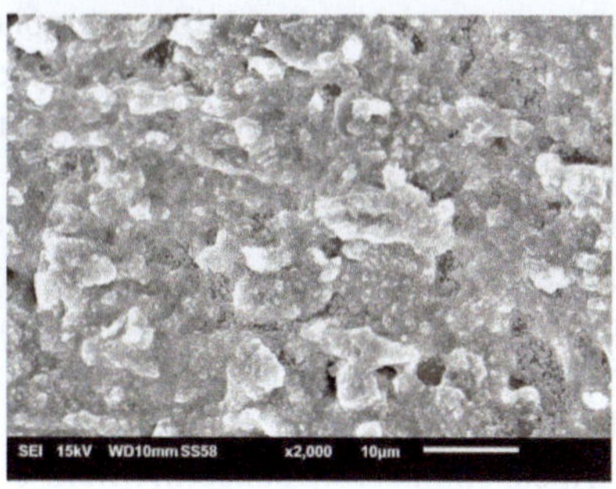

3 mm trace width

produce a network that is less prone to catastrophic delamination, allowing the material to accommodate larger strains without macroscopic failure. This accounts for the superior retention of conductivity and more gradual increases in resistivity in the inks under repeated torsion. The pattern shape plays a central role in the local strain distribution.

A SEM images clearly show that sharp-edged or abrupt directional features, such as those present in square and straight-line patterns, produce micro-cracks aligned with high-stress axes. These cracks initiate at the pattern corners or along straight segments where torsional shear concentrates. Meandering shapes such as curved and zigzag patterns, however, display a smoother surface after testing, with cracks dispersed over larger areas and less evidence of catastrophic failure. The smoother transitions of curved patterns allow for better strain redistribution across the entire trace, mitigating the formation of severe stress concentrations. This observation reinforces earlier findings that smoother geometrical paths contribute to enhanced torsional durability, as they prevent localised material rupture and delay the onset of electrical degradation.

Besides, the cross-sectional SEM images further enrich this discussion by revealing the vertical lamination of the nanocomposites. In GNPs ink, delamination is evident at multiple depth levels, including at filler–matrix interfaces and between ink clusters. These internal delaminations may not propagate directly to the top surface initially but gradually reduce the effective load-carrying area and decrease the number of continuous conductive pathways. In contrast, CB inks present a more homogeneously mixed structure across the trace thickness with fewer distinct boundaries. Even after severe torsional cycling, these traces show comparatively stable interface morphology, supporting the mechanical integrity of the overall film and preserving its electrical properties.

Besides, these observations emphasise the need to adapt both ink formulation and pattern architecture for flexible electronics that must endure torsional stress. Selection of CB as the conductive filler is beneficial where superior torsional resilience is paramount, owing to its uniform percolating network that adapts more readily to mechanical deformation. Conversely, if the application requires the higher intrinsic conductivity of GNPs ink, engineering strategies such as incorporating adhesion-promoting agents, coupling agents, or secondary fillers could improve filler–matrix interfacial strength and decrease crack initiation rates. Additionally, the use of smoother, twisting pattern shapes and appropriate trace widths can distribute cyclic torsional stress more evenly across the surface, preventing premature crack formation and thus preserving device function over extended use.

The progression of micro-cracks and delamination evident in the SEM analyses is of critical engineering concern. Micro-cracks not only increase resistivity but also lead to spatially non-uniform electrical responses under dynamic torsional loads. In a real-application such as wearable sensors, soft robotics, or foldable displays, these variations could produce erratic signal behaviour, reduced sensitivity, and compromised device lifetime. Furthermore, the concentration of cracks at pattern discontinuities reveals potential failure points that require careful attention during design and fabrication. Techniques like optimising cure profiles, using surfactants to improve filler dispersion, and employing multilayer patterning could help enhance both mechanical endurance and long-term signal stability.

Therefore, SEM characterisation of GNPs and CB conductive inks reveals key correlations between nanostructure, pattern geometry, trace width, and mechanical degradation under cyclic torsion. The GNPs trace are more prone to crack initiation and interfacial debonding owing to agglomeration and stiffness mismatch with the binder. In CB traces, by contrast, maintain a more continuous percolating network with fewer intrinsic flaws. Smooth and indirect patterns significantly improve crack resistance and mechanical durability. These findings highlight the importance of holistic engineering of the material–geometry system for durable, torsion-resistant, and stable printed conductive inks which a critical requirement for next-generation flexible and stretchable electronics that must endure dynamic mechanical environments without compromising electrical functionality.

9.8 The EDX Analysis

The Energy-Dispersive X-ray (EDX) spectroscopy provides a quantitative elemental profile of materials and is an important diagnostic technique for understanding how cyclic mechanical stress influences the elemental and compositional structure of nanocomposite conductive inks. Following repeated torsional loading up to high cycle counts, EDX can reveal significant microstructural and chemical alterations within the material. These changes may originate from micro-cracks, surface oxidation, filler pull-out, or binder degradation, all of which profoundly influence the electrical and mechanical performance of printed ink traces in flexible electronics. In this evaluation, EDX was employed to quantify the relative weight percentages of carbon and oxygen in both ink traces with different trace widths and pattern geometries after cyclic torsion. The findings, summarised in Table 9.5, revealed clear differences between the two ink types, providing further insight into their durability under complex multiaxial deformation.

Table 9.5 Carbon and oxygen weight percentage inside GNPs and CB ink trace for each pattern measured from EDX for cyclic torsion

	GNPs ink trace						CB ink trace					
Trace pattern	Carbon weight percentage, %			Oxygen weight percentage, %			Carbon weight percentage, %			Oxygen weight percentage, %		
Width, mm	1	2	3	1	2	3	1	2	3	1	2	3
Straight line	51.0	45.1	43.1	48.0	54.9	56.9	44.8	44.5	43.1	55.2	55.5	56.9
Curve	60.4	57.8	54.2	39.6	42.2	45.8	69.2	60.9	58.7	30.8	39.1	41.3
Square	52.3	50.7	44.1	47.7	49.3	55.9	55.8	54.1	53.0	44.2	45.9	46.0
Zigzag	57.2	52.1	49.8	42.8	47.9	50.2	62.9	58.0	54.2	37.1	41.0	45.8

As expected, GNPs-based ink traces exhibited a higher weight percentage of carbon across all patterns and trace widths than their CB counterparts. This can be attributed to the intrinsic properties of GNPs, which they consist of large, planar graphene sheets with high aspect ratios and a strong in-plane sp2-carbon bonding network. These nanosheets present an extensive surface area and a percolating conductive network that promotes a uniform distribution of carbon. Even under torsional cycling, the ink tends to remain well-dispersed within the polymer matrix due to their lateral size and plate-like structure, mitigating agglomeration and preserving effective load transfer across the trace.

By contrast, CB inks which consist of much smaller, roughly spherical primary particles that form loose agglomerates, exhibited a lower carbon weight fraction in the post-torsion state. The spherical morphology and high surface energy of ink often lead to particle agglomeration, which is further exacerbated by cyclic deformation. During repeated torsional loading, these agglomerated clusters can rupture, creating micro-voids and localised domains of low-carbon density. Furthermore, the smaller size of ink particles renders them more prone to filler migration under stress, leading to less uniform carbon distribution and a relative decrease in overall carbon signal intensity in the EDX spectrum.

Collectively, the oxygen weight fraction was generally higher in CB ink traces than in GNPs-based trace after torsion cycling. This heightened oxygen content is a strong indicator of the susceptibility of CB composites to surface oxidation and polymer degradation at the filler–matrix interface under cyclic strain. The CB is known to possess a greater density of oxygen-containing functional groups which are the hydroxyl, carboxyl, and epoxy groups, as a result of its manufacturing process. These oxygen varieties can serve as stress concentrators and sites for crack nucleation, especially under torsional shear stresses. The cyclic twisting promotes microfibrillation and crazing in the polymer binder, allowing ambient oxygen or moisture to diffuse into newly created micro-voids. The trapped oxygen may participate in localised oxidative reactions, creating further defects in the sp2-carbon network and leading to an increase in electrical resistance.

In contrast, the GNPs trace exhibited lower overall oxygen content due to their lower concentration of intrinsic functional groups. The two-dimensional graphene sheets are also less chemically reactive, so fewer oxygen species are introduced into the conductive network under cyclic torsional strain. Nevertheless, small cracks at the filler–matrix interface and along the GNPs cluster still allow limited atmospheric ingress. This was evident as small patches of increased oxygen signal adjacent to the fracture surfaces in elemental maps.

Both ink traces displayed variations in elemental composition as a function of trace pattern and width. In the case of GNPs trace, more indirect patterns exhibited slightly higher carbon weight fractions than straight-line or square patterns. The shapes redistribute torsional stress across a larger path length, delaying catastrophic failure and maintaining more intact percolation pathways of GNPs. These geometrical features also limit severe delamination and micro-void formation, preserving the local carbon content despite accumulated torsional fatigue. Similarly, trace width played a role in elemental retention. The 3 mm wide traces exhibited more stable ratios after 10,000 cycles relative to narrower traces, which suffered greater localised thinning and bond disruption. Wider traces present a larger cross-sectional area and better capacity to accommodate shear stresses, reducing filler displacement and micro-crack density across the trace.

Elemental mapping (Tables 9.6 and 9.7) further highlighted the spatial variation of carbon and oxygen across the ink microstructure. The GNPs trace displayed more continuous carbon-rich domains with only sparse oxygen-enriched spots distributed along micro-cracks and boundaries, implying that most cracks remained superficial and did not propagate into the deeper layers of the ink. However, in CB traces, elemental maps showed more dispersed and coalescent oxygen-rich pockets surrounding cracks and voids. This points to greater binder rupture and filler debonding under cyclic torsion, allowing more environmental ingress and interfacial degradation.

Additionally, these findings suggest that oxygen access is correlated with the formation of conductive network discontinuities. Even small increases in local oxygen concentration can impair the $\pi - \pi$ stacking of neighbouring graphene sheets or introduce scattering centres for charge carriers, thereby deteriorating the ink's electrical conductivity. Reducing the density of oxygen-containing functional groups, whether by thermal annealing, surface functionalisation or protective coatings and

Table 9.6 The EDX mapping images for patterned GNPs ink (curve and 3 mm) for cyclic torsion

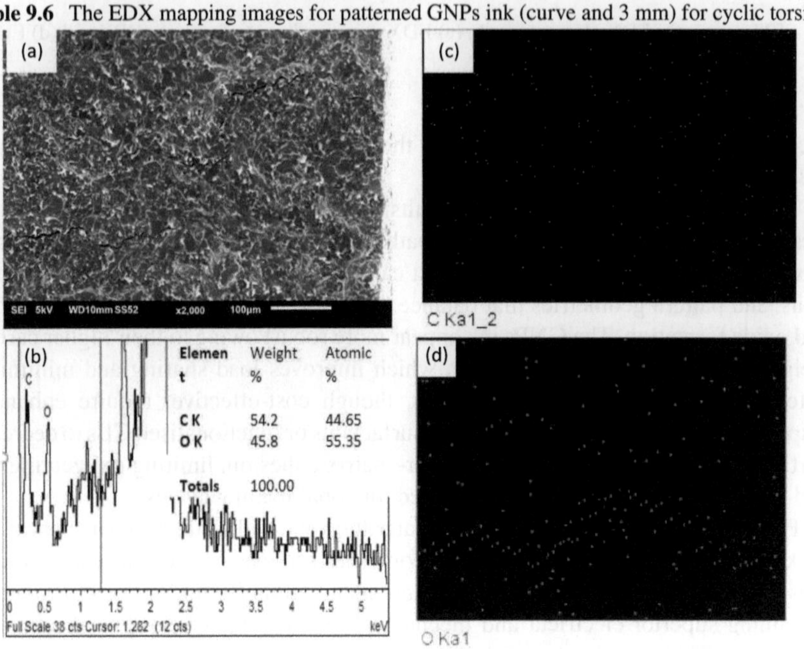

(a) SEM image, (b) EDX element graph, (c) EDX mapping image of carbon (GNPs), and (d) EDX mapping image of oxygen

Table 9.7 The EDX mapping images for patterned CB ink (curve and 3 mm) for cyclic torsion

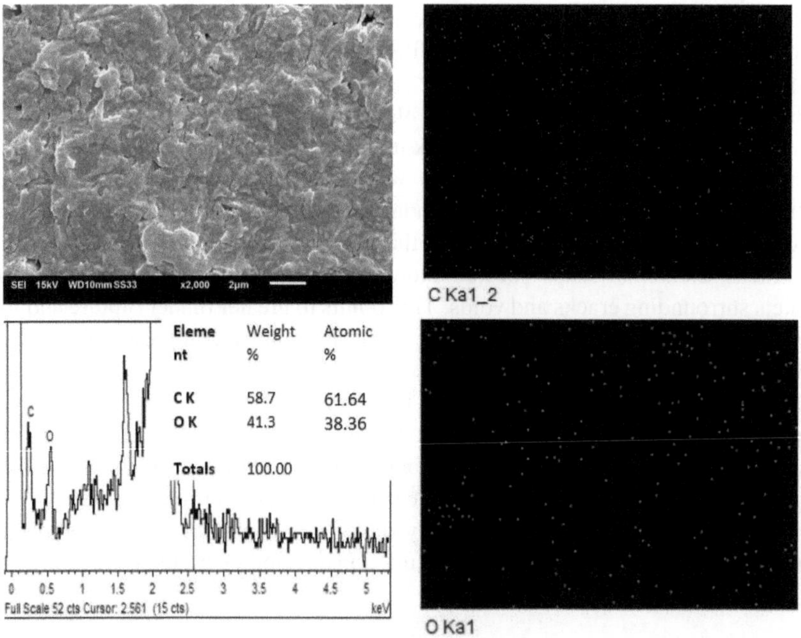

(a) SEM image, (b) EDX element graph, (c) EDX mapping image of carbon (CB), and (d) EDX mapping image of oxygen

it could mitigate this issue and enhance the long-term durability of both GNPs and CB inks under repeated torsional stress.

The EDX and elemental mapping results underscore the critical role of microstructural integrity in retaining conductive pathways under cyclic torsion. Designers of flexible and stretchable electronics must carefully choose filler types, ink formulations, and pattern geometries that balance high carbon content with minimal defect and oxide formation. The GNPs ink appear more robust owing to their higher carbon weight fraction and planar structure, which improves load sharing and minimises defect formation. Conversely, CB inks, though cost-effective, require enhanced dispersion strategies, for instance, using surfactants or functionalised CBs to decrease particle agglomeration and improve filler–matrix adhesion, limiting oxygen ingress and preserving conductivity over repeated torsional loading cycles.

Finally, the engineering takeaway is that pattern and width effects must not be overlooked. Increased width and different geometries serve as passive mitigation strategies that reduce localised stress accumulation and minimise micro-crack formation, sustaining superior electrical and mechanical integrity under extreme mechanical conditions. This combined materials and structural design approach is paramount to realising high-performance, durable conductive inks suitable for advanced flexible and stretchable electronic devices.

9.9 Summary

Following completion of the static load tests, the conductive ink traces were subsequently subjected to cyclic torsional loading. This involved applying a controlled cyclic twisting motion up to a specified angular displacement and monitoring the trace for signs of degradation. Parallel to this torsional fatigue evaluation, a comprehensive characterisation was conducted. Specifically, this involved in-depth analysis of the ink traces' electrical, mechanical, and morphological properties both pre- and post-testing. Throughout the entire process, the in-situ bulk resistance was continuously measured to assess real-time variations in electrical performance under torsional fatigue.

As consolidated in Table 9.8, the data highlight clear trends that strongly depend on ink composition, trace pattern geometry, and torsional strain amplitude. Interestingly, the curve-patterned ink trace displayed the most robust performance across all measured properties. This pattern distributed torsional stresses more uniformly, limiting crack initiation and propagation. From a physics and engineering perspective, curved geometries effectively reduce localised strain gradients and act as compliance-enhancing features, allowing the trace to accommodate more torsional deformation before microstructural damage ensues.

Basically, GNPs ink trace failed at approximately 9000 cycles across all patterns. Even though these inks retained electrical conductivity up to this cycle count, microstructural disintegration in the form of nanosheet debonding and micro-cracks was detected at and around the filler–matrix interface under repeated torsion. The relatively brittle nature of the GNPs film, which comprise stacked graphene nanosheets dispersed within a polymer binder, promotes crack nucleation at inter-flake junctions. Under cyclic torsional shear, these cracks quickly consolidate, leading to sudden electrical discontinuity beyond approximately 9000 cycles.

For CB ink trace sustained the full 10,000 cycles torsional loading without catastrophic failure, indicating a greater resilience to repeated shear deformation. This superior torsional durability can be attributed to the small spherical morphology and percolating network of ink particles that promote greater ductility under twisting

Table 9.8 Summary of electrical and mechanical characterisation of cyclic torsion load test for patterned GNPs and CB (curve pattern)

		Width, mm		
		1	2	3
GNP	Volume resistivity, Ω cm	6.19984	12.34619	13.78667
	Hardness, MPa	1.3	1.21	1.21
	Elastic modulus, GPa	0.007	0.006	0.007
CB	Volume resistivity, Ω cm	0.00789	0.00688	0.01054
	Hardness, MPa	0.5	0.6	0.72
	Elastic modulus, GPa	0.01635	0.0183	0.0223

motions. The ink clusters can redistribute torsional stresses more effectively at the nanoscale owing to their randomised 0D network and high surface roughness, which enhances particle–polymer interaction. The more isotropic packing of ink particles thus enables them to better accommodate shear strain without severing the conductive pathways.

Among all geometries tested, the curve pattern produced the most favourable outcomes for both conductive inks. This can be explained by the absence of sharp corners and abrupt directional changes that would otherwise concentrate shear stresses, as is the case with straight-line or square patterns. The smooth curvature acts to distribute torsional deformation across a larger trace length, mitigating the formation of stress concentrators and delaying the onset of plastic deformation. This, in turn, prolongs the functional lifespan of the ink traces under cyclic torsion.

From an electrical and mechanical perspective, the CB ink traces consistently demonstrated higher endurance than their GNPs counterparts across all patterns. This is reflected in the sustained lower bulk resistance and higher mechanical hardness under torsional cycling. The bulk resistance of carbon black ink traces showed $<10\%$ increase even after 10,000 cycles, suggesting that the percolating network of the ink nanoparticles is capable of reorganising under torsional shear to maintain electrical continuity. This is supported by the intrinsic resilience of sp2-hybridised carbons present in the amorphous CB structures, allowing for slip and minor reorientation without full bond rupture.

In contrast, the GNPs ink trace exhibited a substantial rise in resistance as the number of torsional cycles approached 9000 cycles, signifying a critical threshold beyond which stacked graphene nanosheets and their van der Waals interfaces cannot maintain cohesive contact. This corroborates the earlier findings that larger, planar fillers, despite their inherently higher intrinsic conductivity, are less effective under cyclic shear due to crack propagation and delamination at the filler–matrix boundaries.

On the mechanical side, nanoindentation results also indicated a retention of hardness and modulus in CB traces relative to GNPs trace under torsional cycling. The annular CB particles, being smaller and more homogeneously distributed, act as better stress-dissipation centres than plate-like GNPs filler. The GNPs sheet, under repeated twisting, tend to misalign and buckle, creating more stress risers at flake edges. This behaviour compromises both the local hardness and the modulus retention under extended cyclic torsion.

These findings underscore the engineering importance of carefully selecting filler morphology and trace pattern design for flexible and torsion-resistant electronic interconnects. The data clearly demonstrate that particle morphology matters such as nanosheet fillers offer high baseline conductivity but are inherently more fragile under shear deformation, making them susceptible to premature failure under cyclic torsion. Besides, trace geometry is critical by implementing smooth curve patterns can significantly improve torsional durability by mitigating stress concentrators, even for less ductile fillers like GNPs. This is consistent with the principles of fracture mechanics and contact mechanics, which state that sharp corners tend to accumulate high localised stresses, leading to earlier crack formation. Furthermore, structural integrity can be preserved through proper material and pattern selection, allowing both electrical and mechanical properties to be retained after extensive torsional cycling.

Finally, the comparative evaluation indicates that CB ink traces exhibit superior torsional fatigue resistance and electrical stability across all tested patterns and trace widths. This enhanced performance is fundamentally rooted in the nanoscale interaction between CB nanoparticles and the polymeric binder, their isotropic packing, and their ability to rearrange under torsional deformation. The GNPs trace, despite their excellent intrinsic conductivity and mechanical stiffness under single-cycle testing, are comparatively brittle under repeated cyclic torsion. These insights directly inform the design of next-generation flexible and wearable electronic devices, where ensuring stable and enduring electrical interconnects is paramount.

Chapter 10
Summary of Findings and Engineering Implications

10.1 Introduction

This chapter presents a concise summary of the key findings of this work. Moreover, this chapter underscores the science and engineering value of the explorations. The work not only advances scientific understanding of the structural–electrical–mechanical relationships governing ink behaviour but also offers valuable insights for future engineering designs and manufacturing processes. By demonstrating the potential of these optimised inks and trace patterns, this work lays a foundation for further innovation in the rapidly evolving area of flexible and wearable electronics. Finally, the chapter outlines carefully considered recommendations and explores promising avenues for future investigation. These suggestions highlight the need for continued refinement of ink formulations, exploration of novel conductive fillers, and enhanced cyclic durability testing to address existing performance limitations. Furthermore, the chapter outlines new opportunities and career paths that could emerge from these findings, emphasising the potential impact on both industry and academia. Together, this concluding section reinforces the relevance and aims to inspire ongoing and future work in this important and dynamic field.

10.2 Summary of Findings

This work was undertaken with the primary objective of improving the electrical, mechanical, and morphological characteristics of conductive inks based on graphene nanoplatelets (GNPs) and carbon black (CB) for flexible electronic devices. Four core objectives guided this work, covering ink formulation, pattern-level performance characterisation under static conditions, cyclic durability under torsion and tensile loads, and a holistic, comparative evaluation. Each objective yielded findings

M. A. Salim et al., *Nanotechnology in Conductive Inks*, Engineering Materials,
https://doi.org/10.1007/978-981-95-5708-0_10

that contribute valuable knowledge to the engineering of next-generation printed electronics.

For formulation and characterisation of enhanced conductive inks, the first objective involved the synthesis of optimised GNPs and CB conductive inks and their full characterisation. This was a fundamental engineering stage, as ink properties profoundly influence the subsequent stages of printing, curing, adhesion, and long-term performance. Experimental results showed substantial improvements over baseline formulations. Specifically, the enhanced GNPs ink achieved a sheet resistance of approximately 25.5×10^3 Ω/sq at a filler loading of 20 wt.%, while the improved CB ink reached 19.1×10^3 Ω/sq at 25 wt.%. These improvements underscore the effectiveness of the revised dispersions, milling parameters, and binder content.

From a materials engineering perspective, this outcome is significant. The GNPs possess a highly anisotropic 2D structure and an intrinsic high carrier mobility, which enable percolative networks at lower volume fractions if well-dispersed. Besides, CB despite its nominally lower conductivity, offers robust particle packing and dense conductive networks owing to its sub-micrometric particle size and high surface area. In both cases, carefully tuning rheology and drying processes enhances particle connectivity and reduces contact resistance across the ink film. Other than that, in engineering implications, it was improved the formulations and allow design engineers to specify thinner, lighter, and less power-hungry printed conductors without compromising electrical pathways. By increasing ink uniformity and homogeneity also minimise localised resistive hotspots that would otherwise precipitate device failure under stress or elevated temperature. Importantly, such enhancements reduce reliance on traditional metallic inks such as silver or copper, promoting more sustainable electronics with lower cost and easier scalability for roll-to-roll or inkjet printing.

In the second objective, a diverse set of ink traces which are straight-line, square, curve, and zigzag were examined to understand how trace geometry and width influence resistive and mechanical properties under static loads. This pattern-level assessment is critical in engineering practice because the electrical layout cannot always be a simple straight trace, which is functional elements like antennae, resistive heaters, or serpentine strain gauges often demand intricate paths. Findings showed that GNPs trace consistently exhibited higher resistance than their CB counterparts across all shapes, demonstrating that CB's particle arrangement yielded more efficient percolating networks at similar thicknesses. Moreover, pattern shapes introduced outstanding variations in resistance trends. Straight-line traces exhibited the most stable resistance due to uniform current paths and minimal stress concentrations at geometric transitions. Conversely, zigzag and curve shapes despite their longer path lengths and displayed more stable electrical behaviour under deformation. The more distributed geometry appeared to buffer localised strains and preserved continuous percolation routes.

In engineering perspective, the straight and square patterns, while low-resistance, may be more vulnerable to crack formation and delamination under localised stress. Curves and zigzags, by contrast, offer a favourable trade-off between increased resistance and heightened mechanical resilience which a crucial insight for design engineers optimising for both electrical performance and durability in wearable or foldable electronics. This shapes-driven resistance tuning opens new avenues for intentionally designing printed circuits with tailored impedance characteristics for signal conditioning, radio-frequency matching, or sensing.

The third objective probed the cyclic durability of the printed traces under torsional and tensile loading to simulate real-operating conditions. The key finding was that GNPs trace failed before reaching 10,000 torsion cycles across all shapes and widths, whereas CB traces withstood up to 10,000 cycles without catastrophic failure. Physically, this contrast is explained by the intrinsic microstructure and bonding between filler and binder in each ink. The GNPs, with its large flake morphology and anisotropy, is more susceptible to delamination under shear, leading to the gradual separation of conductive pathways at micro- and nano-scales. In contrast, the more isotropic, spherical morphology of CB allowed continuous particle rearrangement under cyclic shear and tensile strains. This is supported by the nanoindentation results that revealed higher hardness retention and better elastic recovery for CB under repeated deformation.

Additionally, trace width emerged as a significant factor under cycling. Narrower traces experienced higher edge stresses, encouraging crack nucleation, while wider traces distributed strain more evenly but suffered a greater absolute number of micro-cracks and debonding events owing to their larger surface areas. Interestingly, at high cycle torsion, wider CB traces showed only modest increases in volume resistivity which an evidence to CB's robust particle–matrix adhesion and resilience. The inks clearly offer longer service life in highly dynamic environments and making them suitable candidates for wearable electronics, stretchable sensors, and foldable displays where repeated flexing is expected. But, for GNPs-based ink, engineering strategies could involve incorporating secondary reinforcement phases or elastomeric binders to improve their cyclic performance. Designers of printed electronics must balance trace width against cyclic lifetime, for example, selecting slightly narrower carbon black traces can optimise between electrical resistance and fatigue life. End-use implications also include accounting for thermal effects under cycling and using protective encapsulants that further minimise crack propagation.

The final objective tied all observations together into a holistic assessment. Curve-shaped patterns consistently yielded superior durability and more stable electrical and mechanical properties across both ink types. The smooth profile of these geometries reduces localised stress raisers, which translates into longer service life and less hysteresis in electrical response under dynamic loading. Crucially, morphological analyses with Scanning Electron Microscopy (SEM) and Energy Dispersive X-ray (EDX) spectroscopy under cyclic torsion revealed further nuances. The GNPs trace displayed greater void formation and increased oxygen content at cracked interfaces, hallmarks of delamination and micro-oxidation under strain, while carbon black traces retained a more uniform microstructure. This microstructural insight is critical and engineering robust inks requires preventing filler agglomeration and binder cracking, which can be achieved by adjusted surface functionalisation of nanoparticles and optimised cure profiles.

The curve-shaped and zigzag traces can serve as standard design motifs for flexible, high-cycle-life devices like wearable antennas and electronic skin. Reducing functional group content such as oxygen and ensuring dense filler packing in GNPs ink can improve cyclic endurance. Besides, the CB ink can be further enhanced by hybridising with small quantities of metallic nanoparticles to bolster electrical conductivity without compromising its superior flexibility.

10.3 Proposed Areas for Further Investigation

This work provides a robust foundation for engineering conductive inks for flexibility and durability, especially under cyclic torsion and tensile loading. Several avenues for further exploration emerge:

Improved Ink Formulation and Processing

- Future explorations should investigate fine-tuning ink preparation procedures including ultrasonic dispersion, three-roll milling, and rheological control to achieve even greater filler homogeneity. Strategies to minimise entrapped air, control solvent evaporation rates, and promote uniform film thickness will reduce micro-void formation and improve both mechanical and electrical uniformity. Developing environmentally friendly, water-based binder systems with strong adhesion to polymer substrates could also align the ink with sustainable manufacturing principles.

Exploration of Advanced Filler Systems

- While GNPs and CB offer promising performance, hybridising these with other highly conductive fillers such as metallic nanoparticles, 2D materials, or ceramics. It would enhance specific properties like thermal conductivity or stiffness. In particular, combining GNPs with small-volume metallic particles could bridge micro-cracks as they form during cycling, mitigating resistance growth over long-term use. Such composites may also exhibit better thermal management which a critical design criterion for high-current devices and enhanced wear resistance.

Functional Pattern Design Optimisation

- From a design engineering perspective, advanced topological optimization algorithms could be applied to trace geometries that balance compliance with electrical performance. Using computational tools such as Finite Element Modelling (FEM) and multiphysics simulations would enable predictive design of patterns that distribute stresses more uniformly across the ink matrix. Curvature, trace bifurcation, and strategically placed stress-relief loops could all be explored to enhance longevity and reduce local strain concentrations.

Extended Multi-physics Characterisation

- Future work could incorporate simultaneous electrical, mechanical, and thermal measurements under cyclic loads to emulate real-device conditions. Techniques like in-situ resistance monitoring during nanoindentation or torsion, coupled with temperature mapping using infrared thermography, would further illuminate the coupled degradation processes at play. Advanced microscopy including in-situ SEM and AFM could probe crack propagation and particle rearrangement dynamically.

Field-Scale Application Testing

- Finally, real-validation on commercial substrates and under service-like operating conditions would help bridge laboratory findings to practical deployment. Application examples include stretchable interconnects in smart textiles, tactile sensing arrays for robotic skins, and foldable circuitry in next-generation display panels. Durability under combined mechanical and environmental stresses including humidity, UV exposure, and thermal cycling and it must also be evaluated to assure long-term functionality.

10.4 Summary

In summary, this work has revealed that both GNPs and CB conductive inks can be optimised to meet the stringent demands of flexible electronics. However, their performance differs markedly under cyclic mechanical loading due to differences in particle morphology, ink microstructure, and trace geometry. The enhanced inks presented here showed reduced resistances and superior microstructural integrity. The CB inks displayed especially high resilience to cyclic torsion and tensile strain owing to their dense and isotropic particle networks. Curve-shaped and zigzag patterns distributed stresses more effectively, mitigating damage accumulation and retaining electrical continuity.

From an engineering design standpoint, this investigation underscores the importance of microstructure-property relationships in selecting suitable inks for a given application. Practitioners can leverage these findings to make informed decisions when designing printed conductive traces for wearable electronics, flexible sensors, or foldable devices. Furthermore, the recommendations for future work highlight promising pathways including hybrid fillers, optimised pattern geometries, and multifunctional composite designs, that can further enhance the conductivity, mechanical compliance, and longevity of printed electronics under complex loading. By pursuing these directions, academicians and engineers can advance the frontiers of printed electronics, unlocking new commercial opportunities for lightweight, durable, and highly adaptable electronic devices.

Uncited References

1. Ab Wahid, A.K., Salim, M.A., Masripan, N.A., Photong, C., Saad, A.M., Akop, M.Z.: Finite element analysis on stretchable conductive ink materials structure under strain and stress behaviour. Int. J. Nanoelectron. Mater. **15**, 441–454 (2022)
2. Acosta, E.: Thin Films/Properties and Applications. https://www.intechopen.com/books/thin-films-properties-and-applications (2021)
3. Al-Saffar, N.L., Al-juboori, H.A., Sabri, Z.A.: Effect of the electroplating deposition conditions on the wear resistance using response surface methodology (RSM). Iraqi J. Mech. Mater. Eng. **20**(3), 239–256 (2020)
4. Amjadi, M., Pichitpajongkit, A., Lee, S., Ryu, S., Park, I.: Highly stretchable and sensitive strain sensor based on silver nanowire-elastomer nanocomposite. ACS Nano **8**, 5154–5163 (2014)
5. Anderson, S. I., Brondsted, P., Lilholt, H.: Fatigue of polymeric composites for wing blades and the establishment of stiffness-controlled fatigue diagrams. In: Proceedings of the 1996 European Union Wind Energy Conference, Goteborg, Sweden, pp. 950–953 (1996)
6. Anglin, C.: Establishing a precision stencil printing process for miniaturized electronics assembly. In: IPC APEX EXPO Technical Conference, Las Vegas, NV (2009)
7. Ansari, S., Muralidharan, M.: Electronic applications of polyurethane and its composites. In: Flexible and Stretchable Electronic Composites, pp. 87–135. Springer, Berlin (2016)
8. ASM Handbook Committee: Fatigue and Fracture, vol. 19. ASM International (1996)
9. ASTM International: Standard Test Method for Cyclic Fatigue Testing. ASTM E466-20 (2020)
10. Balguri, P.K., Samuel, D.G.H., Thumu, U.: A review on mechanical properties of epoxy nanocomposites. Mater. Today Proceed. **44**, 346–355 (2021)
11. Benight, S.J., Wang, C., Tok, J.B.H., Bao, Z.: Stretchable and self-healing polymers and devices for electronic skin. Prog. Polym. Sci. **38**, 1961–1977 (2013)
12. Biswas, S., Reiprich, J., Pezoldt, J., Hein, M., Stauden, T., Jacobs, H.O.: Stress-adaptive meander track for stretchable electronics. Flexible Print. Electron. **3**(3), 032001 (2018)
13. Bjorlin, R.W.: Electronic packaging, smart labels, the internet of things and the manufacturing paradigm. In: Flexible Packaging RSS, Flexible Packaging. https://www.flexpackmag.com/articles/88018-electronic-packaging-smart-labels-the-internet-of-things-the-manufacturing-paradigm (2016)
14. Bleu, Y., Bourquard, F., Tite, T., Loir, A.S., Maddi, C., Donnet, C., Garrelie, F.: Review of graphene growth from a solid carbon source by pulsed laser deposition (PLD). Front. Chem. **6**, 572 (2018)
15. Britannica: Polyethylene Terephthalate. Encyclopedia Britannica. https://www.britannica.com/science/polyethylene-terephthalate (2022)

M. A. Salim et al., *Nanotechnology in Conductive Inks*, Engineering Materials,
https://doi.org/10.1007/978-981-95-5708-0

16. Bulychev, S.I., Alekhin, V.P., Shorshorov, M.K., Ternovskii, A.P., Shnyrev, G.D.: Determining Young modulus from the indenter penetration diagram. Ind. Lab. **41**(9), 1409–1412 (1975)
17. Butola, R., Yuvaraj, N., Singh, R.P., Tyagi, L., Khan, F.: Evaluation of microhardness and wear properties of Al 6063 composite reinforced with yttrium oxide using stir casting process. World J. Eng. **19**, 361–367 (2021)
18. Cesano, F., Uddin, M.J., Damin, A., Scarano, D.: Multifunctional conductive paths obtained by laser processing of non-conductive carbon nanotube/polypropylene composites. Nanomaterials **11**(3), 604 (2021)
19. Chatterjee, S., Ghanta, K.C., Hens, A.: Study of multiphase flow inside straight and spiral microchannel and effect of two-phase flow on Dean's vortices. Chem. Eng. Res. Des. **165**, 398–408 (2021)
20. Chaudhuri, T.K., Ghediya, P.R.: Microwave-processed copper zinc tin sulphide (CZTS) inks for coatings in solar cells. In: Advanced Ceramic and Metallic Coating and Thin Film Materials for Energy and Environmental Applications, pp. 121–174. Springer, Berlin (2018)
21. Chen, X., Liu, S., Xiong, S., Wang, Z., Huang, W.: Influence of printing parameters on microstructure and electrical properties of silver nanoparticle conductive inkjet-printed patterns. J. Mater. Sci. Mater. Electron. **31**(5), 3569–3581 (2020)
22. Cheng, I.C., Wagner, S.: Overview of flexible electronics technology. In: Wong, W.S., Salleo, A. (eds.) Flexible Electronics: Materials and Applications, pp. 1–28. Springer, Boston, MA (2009)
23. Cherrington, R., Liang, J.: Design and manufacture of plastic components for multifunctionality. Mater. Deposit. Process. Multifunct. **14**, 19–51 (2016)
24. Choi, H.-J., Kim, M.S., Ahn, D., Yeo, S.Y., Lee, S.: Electrical percolation threshold of carbon black in a polymer matrix and its application to antistatic fibre. Sci. Rep. **9**, 6338 (2019)
25. Choi, Y., Kim, S.J., Choi, D.S.: Effects of trace width and pattern design on the microstructure and mechanical properties of conductive ink traces. J. Electron. Mater. **48**(11), 7574–7584 (2019)
26. Chortos, A., Jin, L., Lian, F., Pop, E., Linder, C., Bao, Z., Cai, W.: Microstructural origin of resistance-strain hysteresis in carbon nanotube thin film conductors. Proc. Natl. Acad. Sci. U.S.A. **115**(9), 1986–1991 (2018)
27. Corzo, D., Tostado-Blázquez, G., Baran, D.: Flexible electronics: status, challenges and opportunities. Front. Electron. **1**, 594003 (2020)
28. Cui, S., Yan, Z., Wang, X., Zhang, Y.: Influence of graphene nanoplatelet dispersion on mechanical and electrical properties of graphene/epoxy composites. Materials **14**(3), 510 (2021)
29. Dantas de Oliveira, A., Augusto Gonçalves Beatrice, C.: Polymer nanocomposites with different types of nanofiller. In: Nanocomposites—Recent Evolutions (2019)
30. Dehghanpour, H., Yilmaz, K.: The relationship between resistances measured by two-probe, Wenner probe and C1760–12 ASTM methods in electrically conductive concretes. SN Appl. Sci. **2**(10), 1664 (2020)
31. Deng, B., Hsu, P.C., Chen, G., Chandrashekar, B.N., Liao, L., Ayitimuda, Z., Peng, H.: Roll-to-roll encapsulation of metal nanowires between graphene and plastic substrate for high-performance flexible transparent electrodes. Nano Lett. **15**(6), 4206–4213 (2015)
32. Faraj, M.G., Ibrahim, K., Ali, M.K.M.: PET as a plastic substrate for flexible optoelectronic applications. Optoelectron. Adv. Mater. Rapid Commun. **8**(5), 879–882 (2011)
33. Filleter, T., Zhang, J., Persson, B.N.J.: Electrical conductivity of carbon black filled polymer composites. J. Appl. Phys. **114**(24), 243703 (2013)
34. Fizal, A.N.S., Saad, H., Salim, M.A., Khalil, N.A., Zulkifli, M., Yahaya, A.N.A.: Mechanical properties and characterization of graphene nanoparticles conductive ink at different pattern. Key Eng. Mater. **930**, 15–22 (2022)
35. Fu, Z., Hannula, M., Jauho, A., Väisänen, K.-L., Välimäki, M., Keskinen, J., Mäntysalo, M.: Cyclic bending reliability and failure mechanism of printed biodegradable flexible supercapacitor on polymer substrate. ACS Appl. Mater. Interf. **14**(35), 40145–40157 (2022)

36. Gao, F., Zhang, X., Zhao, Y., Li, Y., Li, L.: Highly stretchable and conductive composite films based on micro-structured rubber and liquid metal. J. Mater. Chem. C **7**(22), 6723–6729 (2019)
37. Gao, Y., Gao, X., Yang, J., Zhang, L., Zhang, J., Zhang, J., Jiang, T.: Hierarchical and interconnected nanostructures of Au/C nanocomposites for highly stretchable and conductive films. Chem. Eng. J. **421**, 127983 (2021)
38. Geim, A., Novoselov, K.: The rise of graphene. Nat. Mater. **6**(3), 183–191 (2007)
39. Goharshadi, E.K., Abbasi, F.: Effect of zinc oxide nanoparticles on the mechanical and electrical properties of polyurethane composites. Compos. B Eng. **63**, 209–215 (2014)
40. Grunlan, J.C.: Carbon black-filled polymer composites: property optimization with segregated microstructures. Doctoral dissertation. ProQuest Dissertations and Theses (UMI 3008690) (2001)
41. Guillonneau, G., Kermouche, G., Bec, S., Loubet, J.-L.: Extraction of mechanical properties with second harmonic detection for dynamic nanoindentation testing. Exp. Mech. **52**(7), 933–944 (2012)
42. Hentrich, A., Esquinaz, P.D.: Effects of the stacking faults on the electrical resistance of highly ordered graphite bulk samples. C **6**(3), 49 (2020)
43. Hosnie, A.N., Salim, M.A., Masripan, N.A., Saad, A.M., Dai, F., Naroh, A., Sudin, M.N.: Characterisation of mechanical-electrical properties of graphene nanoplatelets filled epoxy as conductive ink in various patterns. Int. J. Nanoelectron. Mater. **14**, 439–456 (2021)
44. Hu, N., Karube, Y., Arai, M., Watanabe, T., Yan, C., Li, Y., Fukunaga, H.: Investigation on sensitivity of a polymer/carbon nanotube composite strain sensor. Carbon **48**(3), 680–687 (2010)
45. Hutchings, I.M., Xu, Y., Sanchez, E., Ibanez, M.J., Quereda, M.F.: Optimisation of the polishing process for porcelain ceramic tiles. In: Qualicer 2006. IX World Congress on Ceramic Tile Quality, vol. 3 (2006)
46. Hwang, J., Sohn, H., Lee, S.H.: Computational characterization and control of electrical conductivity of nanowire composite network under mechanical deformation. Sci. Rep. **8**(1), 16617 (2018)
47. Jabari, E., Ahmed, F., Liravi, F., Secor, E.B., Lin, L., Toyserkani, E.: 2D printing of graphene: a review. 2D Mater. **6**(4), 042004 (2019)
48. Jasim, A.H., Joudi, W.M., Radhi, N.S., Saud, A.N.: Mechanical properties and wear characteristic of (aluminum–zinc oxide) metal matrix composite prepared using stir casting process. Mater. Sci. Forum **1002**, 175–184 (2020)
49. Jin, Y., Zhang, X., Zeng, J., Qian, L.: Enhanced electrical conductivity of polyethylene/polyaniline composite by using graphene oxide as co-filler. Polym. Compos. **39**(S1), 1804–1812 (2018)
50. Jin, L., Chortos, A., Lian, F., Pop, E., Linder, C., Bao, Z., Cai, W.: Microstructural origin of resistance-strain hysteresis in carbon nanotube thin film conductors. Proc. Natl. Acad. Sci. U.S.A. **115**(9), 1986–1991 (2018)
51. Jones, D.R.H., Ashby, M.F.: Elastic moduli. Eng. Mater. **1**, 31–47 (2019)
52. Kassim, S.A., Achour, M.E., Costa, L.C.: Prediction of the AC electrical conductivity of carbon black filled polymer composites. Mater. Res. Proceed. **1**, 114–118 (2016)
53. Keysight Technologies: Sheet Resistance/Resistivity Measurement Using a Source/Measurement Unit (SMU)—Application Note. https://www.keysight.com/my/en/assets/7018-05102/application-notes/5992-1329.pdf (2017)
54. Kim, J.Y., Yun, J.H., Lee, K.H.: Effects of cyclic tensile loading on the electrical properties of printed graphene-based conductive inks. Mater. Today Commun. **19**, 14–21 (2019)
55. Kowalewski, Z.L., Szymczak, T., Maciejewski, J.: Material effects during monotonic-cyclic loading. Int. J. Solids Struct. **51**, 740–753 (2014)
56. Kucinskis, G., Bajars, G., Kleperis, J.: Graphene in lithium ion battery cathode materials: a review. J. Power. Sour. **240**, 66–79 (2013)
57. Kumar, S., Povoden-Karadeniz, E.: Plastic deformation behavior in steels during metal forming processes: a review. Materials **14**(19), 5625 (2021)

58. Lall, P., Narangaparambil, J., Leever, B., Miller, S.: Flexure and twist test reliability assurance of flexible electronics. In: Proceedings of the ASME 2019 International Technical Conference and Exhibition on Packaging and Integration of Electronic and Photonic Microsystems InterPACK2019, October 7–9, 2019, Anaheim, CA, USA (2019)
59. Lee, H., Kim, I., Kim, M., Lee, H.: Moving beyond flexible to stretchable conductive electrodes using metal nanowires and graphenes. Nanoscale **8**(4), 1789–1822 (2016)
60. Lee, J.S., Park, W.I., Kim, S.J.: Design and analysis of impedance matched high speed PCB interconnections. IEEE Trans. Electromagn. Compat. **61**(2), 576–584 (2019)
61. Li, A., Zhang, C., Zhang, Y.F.: Thermal conductivity of graphene-polymer composites: mechanisms, properties, and applications. Polymers **9**(9), 437 (2017)
62. Liu, D., He, Y., Tang, X., Ding, H., Hu, P., Cao, P.: Size effects in the torsion of microscale copper wires: experiment and analysis. Scr. Mater. **66**(6), 406–409 (2012)
63. Liu, Z., Liu, X., Huang, S., Zhang, Z., Peng, Y., Tang, Z.: Anomalous electrical conductivity enhancement in covalent organic framework composite films. J. Mater. Chem. C **7**(45), 14011–14017 (2019)
64. Lopes, A.J., Lee, I.H., MacDonald, E., Quintana, R., Wicker, R.: Laser curing of silver-based conductive inks for in situ 3D structural electronics fabrication in stereolithography. J. Mater. Process. Technol. **214**(9), 1935–1945 (2014)
65. Md Saad, A., Salim, M.A., Al-Mola, M.H.A., Ahmad, F., Saad, H., Akop, M.Z.: Effect of the Different Printing Patterns of Graphene Nanoparticles in Conductive Ink on Electrical and Mechanical Performance. Int. J. Nanoelectron. Mater. **14**, 225–240 (2021)
66. Majadas, C.L., Peñaloga, J.M., Salvador, R.W.: Review and evaluation of nanofluids as prospect thermal energy storage material for concentrated solar power application. Resour. Eff. Technol. **4**(2020), 10–29 (2020)
67. Marsh, H., Rodríguez-Reinoso, F.: Production and reference material. Act. Carbon **14**, 454–508 (2006)
68. Maryniak, W., Uehara, T., Noras, M.: Surface resistivity and surface resistance measurements using a concentric ring probe technique. IEEE Trans. Instrum. Meas. **52**(1), 74–79 (2003)
69. Michels, J., Sena Cruz, J., Christen, R., Czaderski, C., Motavalli, M.: Mechanical performance of cold-curing epoxy adhesives after different mixing and curing procedures. Compos. B Eng. **98**, 434–443 (2016)
70. Miturska, I., Rudawska, A., Müller, M., Hromasová, M.: The influence of mixing methods of epoxy composition ingredients on selected mechanical properties of modified epoxy construction materials. Materials **14**(2), 411 (2021)
71. Mokhlis, M.: A Study on Mechanical and Electrical Properties of Hybridized Graphene-Carbon Nanotube Filled Conductive Ink (2020)
72. Monson Company: Jeffamine D-230 Technical Data Sheet [PDF]. https://monsonco.com/wp-content/uploads/2019/08/Jeffamine-D-230.-TDS.-REC-12.5.2019.pdf (2019)
73. Nazil Altay, B., Bolduc, M., Cloutier, S.G.: Sustainable advanced manufacturing of printed electronics: an environmental consideration. In: Green Energy and Environment (2020)
74. Notta-Cuvier, D., Nciri, M., Lauro, F., Chaari, F., Zouari, B., Maalej, Y.: A pragmatic approach for modelling the viscoelastic–viscoplastic behaviour of short-fibre reinforced thermoplastics coupled with anisotropic damage. Appl. Compos. Mater. **28**(2), 341–368 (2021)
75. Okada, K., Kohiki, S., Luo, S., Sekiba, D., Ishii, S., Mitome, M., Kohno, A., Tajiri, T., Shoji, F.: Correlation between resistivity and oxygen vacancy of hydrogen-doped indium tin oxide thin films. Thin Solid Films **519**(11), 3557–3561 (2011)
76. Oliver, W.C., Pharr, G.M.: An improved technique for determining hardness and elastic modulus using load and displacement sensing indentation experiments. J. Mater. Res. **7**(06), 1564–1583 (1992)
77. Papageorgiou, D.G., Kinloch, I.A., Young, R.J.: Mechanical properties of graphene and graphene-based nanocomposites. Prog. Mater. Sci. **90**, 76–127 (2017)
78. Perelaer, J., Klokkenburg, M., Hendriks, C.E., Schubert, U.S.: Microwave flash sintering of inkjet-printed silver tracks on polymer substrates. ACS Appl. Mater. Interf. **21**(47), 4830–4834 (2009)

79. Prolongo, S.G., Moriche, R., Jiménez-Suárez, A., Sánchez, M., Ureña, A.: Advantages and disadvantages of the addition of graphene nanoplatelets to epoxy resins. Eur. Polym. J. **61**, 206–214 (2014)
80. Sagu, J.S., York, N., Southee, D., Wijayantha, K.G.U.: Printed electrodes for flexible, light-weight solid-state supercapacitors—a feasibility study. Circ. World **41**(2), 80–86 (2015)
81. Spahr, M.E., Gilardi, R., Bonacchi, D.: Carbon Black for Electrically Conductive Polymer Applications. In: Rothon, R. (ed.) Fillers for Polymer Applications. Polymers and Polymeric Composites: A Reference Series, pp. 375–400. Springer, Berlin (2017)
82. Su, H., Zhang, M., Chang, Y.-H., Zhai, P., Hau, N.Y., Huang, Y.-T., Liu, C., Soh, A.K., Feng, S.-P.: Highly conductive and low-cost Ni-PET flexible substrate for efficient dye-sensitized solar cells. ACS Appl. Mater. Interf. **6**(8), 5577–5584 (2014)
83. Sun, T., Luo, W., Luo, Y., Wang, Y., Zhou, S., Liang, M., Chen, Y., Zou, H.: Self-reinforced polypropylene/graphene composite with segregated structures to achieve balanced electrical and mechanical properties. Ind. Eng. Chem. Res. **59**(24), 11206–11218 (2020)
84. Tarefder, R.A., Faisal, H.: Effect of dwell time and loading rate on the nanoindentation behavior of asphaltic materials. J. Nanomech. Micromech. **3**(2), 17–23 (2013)
85. Teodosiou, T.C., Saravanos, D.A.: Numerical investigation of mechanisms affecting the piezoresistive properties of CNT-doped polymers using multi-scale models. Compos. Sci. Technol. **70**, 1312–1320 (2010)
86. Thinky Corporation: What is a Planetary Centrifugal Mixer? AZoM. https://www.azom.com/article.aspx?ArticleID=18525 (2019)
87. Thomas, L.: How Thin is a Thin Film? AZoM. https://www.azom.com/article.aspx?ArticleID=17334 (2018)
88. Tomchenko, A.: Printed Chemical Sensors: From Screen-Printing to Microprinting (2006)
89. Tseng, W.J., Chen, C.N.: Dispersion and rheology of nickel nanoparticle inks. J. Mater. Sci. **41**(4), 1213–1219 (2006)
90. Van Hazendonk, L.S., Pinto, A.M., Arapov, K., Pillai, N., Beurskens, M.R.C., Teunissen, J.-P., Sneck, A., Smolander, M., Rentrop, C.H.A., Bouten, P.C.P., Friedrich, H.: Printed stretchable graphene conductors for wearable technology. Chem. Mater. **34**(17), 8031–8042 (2022)
91. Vena, A., Perret, E., Tedjini, S.: Implementation and measurements of chipless RFID tags. In: Chipless RFID Based on RF Encoding Particle, pp. 171–227 (2016)
92. Walter, D.: Primary particles—agglomerates—aggregates. In: Nanomaterials, pp. 9–24 (2013)
93. Wang, F., Drzal, L.T., Qin, Y., Huang, Z.: Effects of functionalized graphene nanoplatelets on the morphology and properties of epoxy resins. High Perform. Polym. **28**(5), 525–536 (2015)
94. Xu, M., Mai, Y.W.: Graphene-based conductive inks: preparation, ink-jetting printing, and applications in flexible electronics. J. Mater. Sci. **52**(6), 3019–3034 (2017)
95. Yang, L., Huang, Q.A., Li, H.: Modelling and simulation of microstrip line parameters for microstrip antennas using fractional calculus. IEEE Trans. Antennas Propag. **67**(5), 3030–3035 (2019)
96. Yang, Q., Cao, Y., He, P., Zhang, X.: Recent progress on graphene-based composites for electromagnetic interference shielding. J. Mater. Sci. Technol. **35**(6), 975–991 (2019)
97. Yang, S., Yuan, X., Zhao, H.: Graphene-based composites and their applications in energy conversion and storage. J. Mater. Chem. A **7**(34), 19536–19571 (2019)
98. Yuan, W., Liu, C., Su, Y., Gao, F., Xiong, C.: Study on electrical conductivity of graphene/polymer composite: a review. J. Mater. Sci. Technol. **49**, 122–132 (2020)
99. Yun, G., Tang, S.-Y., Lu, H., Zhang, S., Dickey, M.D., Li, W.: Hybrid-filler stretchable conductive composites: from fabrication to application. Small Sci. **1**(6), 2000080 (2021)
100. Zainal, M.Z., Buyamin, N.: Reliability performance of conductive ink subjected to hygrothermal aging. Int. J. Eng. Adv. Technol. **9**(2), 11542 (2020)
101. Zavanelli, N., Yeo, W.H.: Advances in screen printing of conductive nanomaterials for stretchable electronics. ACS Omega **6**(14), 9344–9351 (2021)
102. Zerbst, U., Bruno, G., Buffiere, J.-Y., Wegener, T., Niendorf, T., Wu, T., Zhang, X., Kashaev, N., Meneghetti, G., Hrabe, N., Madia, M., Werner, T., Hilgenberg, K., Koukolíková, M., Procházka, R., Džugan, J., Möller, B., Beretta, S., Evans, A., Wagener, R., Schnabel, K.:

Damage tolerant design of additively manufactured metallic components subjected to cyclic loading: State of the art and challenges. Prog. Mater. Sci. **121**, 100786 (2021)

103. Zulfiqar, S., Saad, A.A., Ahmad, Z., Yusof, F., Bachok, Z.: Structural analysis and material characterization of silver conductive ink for stretchable electronics. Int. J. Integr. Eng. **13**(7), 128–135 (2021)

Index

M. A. Salim et al., *Nanotechnology in Conductive Inks*, Engineering Materials,
https://doi.org/10.1007/978-981-95-5708-0

P

R

S

T

MIX
Papier aus verantwortungsvollen Quellen
Paper from responsible sources
FSC® C105338

If you have any concerns about our products,
you can contact us on
ProductSafety@springernature.com

In case Publisher is established outside the EU,
the EU authorized representative is:
Springer Nature Customer Service Center GmbH
Europaplatz 3, 69115 Heidelberg, Germany

Printed by Libri Plureos GmbH
in Hamburg, Germany